X-Ray Diffraction of Ions in Aqueous Solutions: Hydration and Complex Formation

Editor-Author

Mauro Magini
Senior Researcher
TIB-ENEA (National Council for Nuclear and Alternative Energies) Rome, Italy

Authors

Giovanni Licheri
Professor

Giorgio Paschina
Associate Professor

Giorgio Piccaluga
Professor

Gabriella Pinna
Associate Professor
Department of Chemistry
University of Cagliari
Cagliari, Italy

CRC Press
Taylor & Francis Group
Boca Raton London New York

CRC Press is an imprint of the
Taylor & Francis Group, an **informa** business

First published 1988 by CRC Press
Taylor & Francis Group
6000 Broken Sound Parkway NW, Suite 300
Boca Raton, FL 33487-2742

Reissued 2018 by CRC Press

Library of Congress Cataloging-in-Publication Data

X-ray diffraction of ions in aqueous solutions.

Includes bibliographies and index.
1. Ionic solutions. 2. X-rays—Diffraction.
3. Hydration. 4. Complex ions. I. Magini, Mauro.
II. Licheri, Giovanni. [DNLM: 1. Ions. 2. Solutions.
3. X-ray Diffraction. QD 561 X13]
QD561.X2 1988 541.3'72 87-8052
ISBN 0-8493-6945-2

A Library of Congress record exists under LC control number: 87008052

Publisher's Note
The publisher has gone to great lengths to ensure the quality of this reprint but points out that some imperfections in the original copies may be apparent.

Disclaimer
The publisher has made every effort to trace copyright holders and welcomes correspondence from those they have been unable to contact.

ISBN 13: 978-1-315-89867-4 (hbk)
ISBN 13: 978-1-351-07777-4 (ebk)

Visit the Taylor & Francis Web site at http://www.taylorandfrancis.com and the
CRC Press Web site at http://www.crcpress.com

PREFACE

The idea of writing a book about the structural investigations of ionic solutions by X-ray diffraction (XRD) wandered in the authors' minds for a long time. In fact, while research activities in this field have largely increased in the last 15 years, utilizing well-established diffractometric techniques, a **complete** review of the **up to date** results has been never attempted. More recently, the authors became aware, as it is better explained in the Foreword, of the presence of some confusion in the literature whenever results on the structure of solutions were mentioned. For these reasons, and thanks to the interest of CRC Press the authors came to the decision of examining all the material presently available, with the aim to present to the interested reader an exhaustive and reasoned selection of it. In this perspective, the authors felt that their work had to include, beside the rational classification of the results, also a comprehensive treatment of the theoretical principles of the X-ray diffraction approach as used when dealing with nonordered systems, as well as a detailed description of experimental and interpretative methods. Therefore, this book should be useful not only to scholars interested in results, but also to students approaching the field of nonordered systems. To them, the authors hope to convey the right feeling about the reliability of XRD information.

The work has been divided into 4 chapters. The first chapter is devoted to the description of the basic principles of X-ray diffraction on noncrystalline systems and to the correlation between diffracted intensities and structure. The chapter also includes a detailed description of operative procedures to follow in the treatment of the experimental data which may be useful to beginners.

The interpretation of the experimental data and their transformation into structural data, like bond lengths, coordination numbers, and geometries, is dealt with in the second chapter. Note that the mathematical formalism, as well as the data handling, exposed in the first two chapters are valid for any nonordered system other than solutions. Therefore, people working in other fields, in particular, those interested in the vitreous state of the matter, may find of utility the content of the first part of the book.

The third chapter is dedicated to a critical examination of the literature about the structure of the solutions. Selective criteria have been adopted to order all the existing material and they are extensively illustrated in the introduction of the same chapter.

The last chapter, finally, is dedicated to a critical comparison between the results obtained by XRD and the ones obtained by neutron diffraction, EXAFS, and simulation methods (molecular dynamics and Monte Carlo). We wish to underline the usefulness of this comparison, never attempted to this time, by which potentialities, limits, and complementariness with XRD of the mentioned techniques can be better understood.

The book is the result of a common work of two research teams (operating at Cagliari University and at ENEA Center of Casaccia, Rome) who had been working in strict contact for about 10 years in the field of the structure of solutions and who are now interested in the structure of the vitreous state. Therefore, the text is the result of a common work and all the authors are responsible for the various chapters.

The authors hope to fill a gap existing in the literature concerning reviews on the structure of ions in solution which can be both exhaustive and critical. The readers will judge if they have been able to reach their aim.

THE EDITOR

Mauro Magini, Ph.D., degree in 1965 from the University of Rome, Italy, in physical chemistry, is senior scientist at ENEA, the Italian Agency for Research and Development of Nuclear and Alternative Energies. Within the frame of nuclear fuel reprocessing activities, he approached first fundamental aspects of solution chemistry by physico-chemical methods (1966 to 1972). In a second period, from about 1973 to 1982, he has been engaged in systematic investigations on the structure of electrolyte solutions which had both technological and basic interests. The structural research was developed in close contact with the Sardinian group at Cagliari University and the Swedish group at the Royal Institute of Technology (KTH) of Stockholm. At KTH he has been invited for several months as visiting scientist. In recent years he has moved his attention towards the field of amorphous systems. He is actually the project leader of an ENEA program for research and development of new amorphous materials. He is author of about 60 papers, most of which concern the structure of electrolyte solutions and amorphous solids.

FOREWORD

It is now widely accepted by chemists, biochemists, and physicists that ion-water interactions are responsible for the routes of many chemical and biochemical processes; it has also become evident, from the numerous studies on ionic hydration, that any further progress in this field is strongly dependent on a satisfactory structural description of solutions. To this end, the general methods used in the study of any liquid or amorphous system can be applied.

The structure of a simple liquid, that is, a liquid formed by equivalent particles, is described by the pair correlation function, g(r), whose product by the average density ρ_0 gives the value of the local density ρ (r) around a particle taken as the origin. The structure of solutions and, in general, of any system formed by nonequivalent particles can be described in terms of "partial" pair correlation functions $g_{ij}(r)$, which allow to obtain the radial distribution of j-type particles around the i-type ones. Let ρ_{0j} be the average density of j particles; then $\rho_{0j}g_{ij}(r) = \rho_{ij}(r)$ gives the local density of j-type particles around the i-type ones.

The first experimental technique employed in the study of the structure of liquids has been X-ray diffraction (XRD). In the case of a simple liquid, the correlation function g(r) can be obtained straightforwardly from the experimental data. On the contrary, X-ray data and, in general, diffraction data from a single experiment, do not lead to the pair correlation functions $g_{ij}(r)$ required for the description of multicomponent systems; the result is instead a weighted sum G(r) of all the pair functions. Several approaches have been pursued in order to develop experimental methods apt to isolate the "partial" pair correlation functions, or at least to obtain a sum formed by a minimum number of terms. Two of them, namely neutron diffraction with isotopic substitution (NDIS) and extended X-ray absorption fine structure spectroscopy (EXAFS), have been widely used in the last decade and several aqueous solutions of strong electrolytes have been recently examined by these techniques. This fact in turn, aroused the interest of the theoreticians of ionic hydration, who need experimental "partial" distribution functions to compare with those calculated through classical, quantum mechanical and computer simulation methods.

Nevertheless, a glance to the papers appearing recently in the literature shows an apparently paradoxical situation: single diffraction experiments, mainly through the use of X-ray sources, have increased together with the progress of methods capable of directly obtaining single $g_{ij}(r)$ functions. The reader, who is not an expert in structural methods, may legitimately be disconcerted by this fact and ask questions about the usefulness of XRD experiments on complex systems. Our answer is that X-ray studies are still useful and up to date, at least for two reasons: the first, although as trivial as it may sound, is that X-ray diffraction technique is nowadays general, diffuse, and accessible to every laboratory with in-house conventional equipment; the second one, by far the most important, is that this technique, if used correctly, gives useful and successful, though not exhaustive, results, as we will try to prove in the present work.

We must admit that from the analysis of some recent review articles and books, which treat the structure of solutions in general and ionic hydration in particular, our point of view may sound overoptimistic and rather unjustified. In our opinion, this is mainly due to the general pattern of these articles, where little space is reserved to XRD studies of solutions. Therefore, an overall review, which goes beyond a simple enumeration of results and critically analyzes their reliability, should be very useful to the nonexpert scholars, who may have found it difficult to select results sometimes appearing conflicting and uncertain. Worst of all, conflicts and uncertainties have been sometimes overestimated, in order to strengthen the opinion of low reliability of the XRD technique.

We are deeply convinced that XRD alone is not sufficient to solve all the structural problems of solutions and that the concurrent use of different techniques is the correct approach; therefore we have recently undertaken parallel XRD and EXAFS studies of solutions. At the same time, we consider incorrect the attitude of those who, examining discrepancies in results obtained by XRD and by other techniques, without a critical analysis, blame XRD for its difficulty to provide the single $g_{ij}(r)$s.

Let us consider the case of $NiCl_2$ aqueous solutions; they have been widely examined and results from XRD, EXAFS, and NDIS have been presented. Illustrative radial distribution functions obtained by these techniques are given in the Figure. It is clear that the first peak, due to Ni-O distances, is definitely resolved even in the case of X-ray data, so that the first coordination sphere around Ni^{2+} can be determined with the same accuracy as from EXAFS or NDIS data. In other words, the radial function from XRD in the range 1.75 to 2.30 Å is essentially the distribution of O atoms around Ni^{2+} ions, $4\pi r^2 \rho_0 \, g_{Ni^{2+}-O}(r)$. In this case, if discrepancies are found among quantitative results obtained from different techniques, they are not to be ascribed to their different perspectives, but to some sort of experimental error present in any of the experiments. Under these circumstances, the experimental error of XRD, which is a simple and well-established technique, is not, by any means, any more probable than that in other experimental setups.

In the light of the above considerations, we deemed it useful to include in the present work: (1) a critical review of XRD results obtained for electrolyte solutions together with their reliability; (2) a comparison of XRD results with those obtained by other "homogeneous" techniques, with the aim of evaluating the limits of the various studies and of stressing the utility of reciprocal integrations. Since, as it will become clear later, XRD studies give information essentially on the short range order, this work will be mainly centered on the nearest coordination of ions by water molecules and/or small anions.

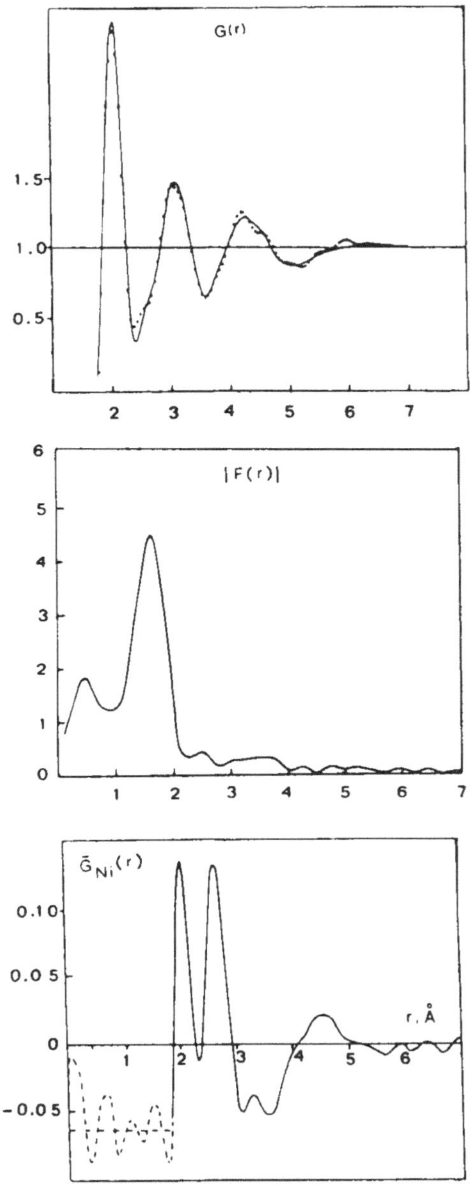

Radial curves of a 4-*M* $NiCl_2$ solution from XRD, EXAFS, and NDIS data (top to bottom). (From (top) Caminiti, R., Licheri, G., Paschina, G., Piccaluga, G., and Pinna, G., *Z. Naturforsch.*, 35a, 1361, 1980; (middle) Licheri, G., Paschina, G., Piccaluga, G., Pinna, G., and Vlaic, G., *Chem. Phys. Lett.*, 83, 384, 1981; (bottom) Neilson, G. W. and Enderby, J. E., *Proc. R. Soc. London*, A390, 353, 1983. With permissions.)

ACKNOWLEDGMENTS

The authors are particularly grateful to Prof. C. Dejak, who first stimulated their interest for XRD investigations on the structure of the solutions in a time in which only crystallography looked remunerative. One of the authors is also indebted to Profs. G. Johansson and M. Sandström who guided him in his first steps of his research activity. Many thanks are due to Prof. G. Marongiu for friendly suggestions and revision of the text. Permission by various journals, publishers, and authors to reproduce figures and diagrams is acknowledged.

G. Licheri
M. Magini
G. Paschina
G. Piccaluga
G. Pinna

TABLE OF CONTENTS

Chapter 1
X-RAY DIFFRACTION OF NONCRYSTALLINE SYSTEMS: THEORY, EXPERIMENTAL, AND DATA TREATMENT 1

I. Introduction 1

II. Distribution Functions 1
- A. One-Component Systems 2
- B. Multicomponent Systems 3

III. X-Ray Diffraction Theory for Noncrystalline Systems 6
- A. Scattering Process 6
- B. Scattering Factors 6
- C. Diffracted Intensity 8
- D. Structure Function 9
- E. One-Component System 11
- F. Partial Structure Functions 11
- G. Radial Functions 12
- H. WKM Approximation 15
- I. Neutron Diffraction 16

IV. Instrumental Apparatus and Data Collection 16

V. Data Treatment 18
- A. Preliminary Corrections 18
  - 1. Background and Air Scattering 18
    - a. Background 18
    - b. Air Scattering 20
  - 2. Absorption 21
    - a. Reflection Geometry 21
    - b. Transmission Geometry 22
  - 3. Polarization 23
- B. Evaluation of the Independent Scattering 23
  - 1. Incoherent Scattering with the Monochromator on the Diffracted Beam 25
- C. Scaling to Absolute Units: The Structure Function 26
  - 1. Scaling Methods 26
    - a. High-Angle and Integral Methods 26
    - b. Habenschuss-Spedding Method 27
  - 2. Multiple Scattering Correction 29
- D. Radial Distribution Function 29
  - 1. Fourier Transform 29
  - 2. Sources of Errors 30
    - a. Termination Errors 30
    - b. Errors in the i(s) Function 32
  - 3. Correction of Low r Spurious Peaks 33
- E. Reproducibility of XRD Results 34

Chapter 2
INTERPRETATION OF THE EXPERIMENTAL DATA 41
I. Introduction 41

II. Detectability of Partial Contributions 42

III. Analysis of the Radial Distribution 44

IV. Curve Fitting 47
A. Mathematical Modeling 47
B. Analysis of Dominant Contributions 50
C. Interpretative Models 52

References 54

Chapter 3
HYDRATION AND COMPLEX FORMATION: A CRITICAL REVIEW 57
I. Introduction 57

II. H_3O^+ Ion 59

III. Alkali Ions 60
A. Lithium 60
B. Sodium 62
C. Potassium 65
D. Cesium 65
E. Comments 65

IV. Alkaline Earth Ions 66
A. Magnesium 66
B. Calcium 70
C. Strontium 73
D. Comments 75

V. Group IIIA 77
A. Aluminium 77
B. Indium 79
C. Thallium 81
1. Hydration 81
2. Halide Complex Formation 81

VI. Group IVA (Sn,Pb) 83

VII. Rare Earth Ions 85

VIII. Group VIB 88
A. Chromium 89
1. Hydration 89
2. Chloride Complex Formation 91

IX. Group VIIB 97

A. Manganese ... 97
1. Hydration ... 97
2. Complex Formation ... 97
a. With Chloride Ions ... 97
b. With Sulfate and Nitrate Ions ... 97

X. Group VIII (Fe, Co, Ni) ... 100
A. Iron(II) ... 100
B. Iron(III) ... 102
1. Hydration ... 102
2. Complex Formation ... 102
a. With Chloride Ions ... 102
b. With Sulfate Ions ... 109
C. Cobalt ... 112
1. Hydration ... 112
2. Complex Formation ... 113
a. With Chloride Ions ... 113
b. With Bromide Ions ... 117
D. Nickel ... 119
1. Hydration ... 119
2. Complex Formation ... 120
a. With Chloride Ions ... 120
b. With Bromide Ions ... 123
c. With Sulfate and Phosphate Ions ... 125

XI. Group VIII (Rh, Pt) ... 127
A. Rhodium ... 127
B. Platinum ... 127

XII. Group IB ... 127
A. Copper ... 127
1. Hydration ... 127
2. Complex Formation ... 132
a. With Chloride Ions ... 132
b. With Bromide Ions ... 134
c. With Nitrate Ions ... 136
B. Silver ... 137
C. Gold ... 138

XIII. Group IIB ... 142
A. Zinc ... 142
1. Hydration ... 142
2. Halide Complex Formation ... 142
B. Cadmium ... 148
1. Hydration ... 148
2. Complex Formation ... 148
a. With Halide Ions ... 148
b. With Sulfate Ions ... 150
c. With Nitrate Ions ... 151
d. With Phosphate Ions ... 151

C. Mercury (Hg_2^{2+}) 152
D. Mercury (Hg^{2+}) 155
1. Hydration 155
2. Complex Formation 156

XIV. Halide Ions 160
A. Fluorine 160
B. Chlorine 160
C. Bromine 164
D. Iodine 165

XV. Oxyanions 165

XVI. General Discussion 169
A. First Coordination Shell 170
1. Bond Lengths 170
2. Coordination Numbers 170
3. Root Mean Square Deviations 171
4. Geometry 172
5. The "Bond Valence" Point of View 172
B. Second Coordination Shell 174
1. $Me\text{-}X_{2nd}$ 174
2. The $H_2O_I\text{-}H_2O_{II}$ Problem 177
C. Information on Complex Formation 179
1. Inner Complex Formation 179
2. Outer Complex Formation 181

D. Continuous Distribution of Distances 181
E. Final Comments 181

References 182

Chapter 4
STRUCTURAL INFORMATION ON AQUEOUS SOLUTIONS FROM NEUTRON DIFFRACTION, EXAFS, AND SIMULATION METHODS 189
I. Introduction 189

II. Neutron Diffraction 189
A. Single Experiments 189
1. General Considerations 189
2. Results from Single Experiments 191
3. Single Experiments Used in Connection with XRD 192
B. Neutron Diffraction with Isotopic Substitution (NDIS): General Principles 192
C. NDIS: Results and Comparison with XRD Studies 194
1. Li^+ 194
2. K^+ 195
3. Ca^{2+} 196
4. Rare Earth Ions 196
5. Ni^{2+} 197
6. Cu^{2+} 199
7. Ag^+ 201
8. Cl^- 202
9. NO_3^- 203
D. Final Comments 203

III. EXAFS Spectroscopy 210
A. X-Ray Absorption Spectroscopy 210
B. EXAFS: Basic Concepts and Relations 212
C. Data Analysis 215
D. EXAFS vs. XRD 218
E. Review of the Results 221
1. Ion-Oxygen Interactions 221
2. Ion-Halide Interactions 224
F. Final Comments 228

IV. Computer Simulation Studies 230
A. Introduction 230
B. Molecular Dynamics and Monte Carlo Technique 231
C. Interaction Potentials 232
1. Theoretical Approach 232
2. Empirical Approach 233
D. Molecular Dynamics Simulations of Aqueous Solutions 236
1. One-Ion Studies 237
2. Solution Studies 237
a. The 1:1 Salt Solutions 237
b. The 2:1 Salt Solutions 242
E. Monte Carlo Simulations 246
F. Final Comments 248

References 251

Index 259

Chapter 1

X-RAY DIFFRACTION OF NONCRYSTALLINE SYSTEMS: THEORY, EXPERIMENTAL, AND DATA TREATMENT

I. INTRODUCTION

In this chapter we deal with the atomic scale structure of noncrystalline systems., i.e., those condensed states of matter which are made up of atoms packed rather densely, but not ordered in a lattice. The distribution of the atomic positions in these materials is then largely different either from the extreme degree of topological disorder present in an ideal gas, or from the perfect spatial order present in an ideal crystal.

In an ideal gas the atoms are point masses randomly distributed and the atomic position vectors $\vec{\mathbf{r}}_i$ constitute a set of independent random variables within the whole sample volume. On the contrary, in any rather dense collection of real atoms, the interatomic forces affect the distribution of the vectors $\vec{\mathbf{r}}_i$ so that disorder is no longer merely random, and some sort of atomic ordering must appear.

This order cannot, however, be described through the formalism used for crystals, where we refer to the invariance under the translational operations of a lattice. In a perfect crystal every atomic position $\vec{\mathbf{R}}_i$ is periodically reproduced at the points:

$$\vec{\mathbf{r}}_{(m_1,m_2,m_3)} = \vec{\mathbf{R}}_i + \vec{\mathbf{L}}_{(m_1,m_2,m_3)} = \vec{\mathbf{R}}_i + m_1\vec{\mathbf{a}}_1 + m_2\vec{\mathbf{a}}_2 + m_3\vec{\mathbf{a}}_3 \quad (1)$$

where the **crystal axes** $\vec{\mathbf{a}}_1$, $\vec{\mathbf{a}}_2$, and $\vec{\mathbf{a}}_3$ are three noncoplanar vectors and m_1, m_2, and m_3 range through all integers. The **crystalline long range order** is thus described through the lattice symmetry and the atomic positions $\vec{\mathbf{R}}_i$ within the **unit cell,** spanned by the crystal axes. In some cases of disorder, an underlying lattice can still be identified and can be used in some way for a structural description. However, in topologically disordered systems, as liquids and glasses, the lattice translational invariance is definitely lost, no long range order remains, and some alternative approach is required in order to describe the main features of the atomic arrangement and to gain experimental information about it.

Our purpose in this chapter is to give a view of the theoretical and experimental bases of the X-ray diffraction (XRD) approach when used to study the atomic scale structure of noncrystalline materials.

Most of this matter can be considered standard and can be found in textbooks and review articles.[1-7] However, we present it with a detail which goes beyond the simple purpose of fixing notations, with the aim to give a useful tool to people starting research in the field.

II. DISTRIBUTION FUNCTIONS

The structure of noncrystalline materials is described through the use of **atomic distribution functions.** Let

$$dn_k(\vec{\mathbf{r}}_1, \vec{\mathbf{r}}_2, \ldots \vec{\mathbf{r}}_k) = n_k(\vec{\mathbf{r}}_1, \vec{\mathbf{r}}_2, \ldots \vec{\mathbf{r}}_k)\, dv_1 dv_2 \ldots dv_k \quad (2)$$

be the probability of finding one atom located at the point $\vec{\mathbf{r}}_1$ in the volume element dV_1

another atom at $\vec{\mathbf{r}}_2$ in the volume element dV_2, and so on, irrespective of the positions of the remaining N-k atoms. The probability densities $n_1(\vec{\mathbf{r}}_1)$, $n_2(\vec{\mathbf{r}}_1, \vec{\mathbf{r}}_2)$ . . . , constitute the hierarchy of one-body, two-body, . . . , N-body distribution functions which characterize many aspects of the statistical behavior of an assembly of N particles.[8]

We are interested only in the two-body distribution function $n_2(\vec{\mathbf{r}}_1, \vec{\mathbf{r}}_2)$, because the experimental evidence from diffraction measurements concerns only the two-body structural characteristics of a noncrystalline material.

A. One-Component System

Let us consider for simplicity, a one-component system containing N atoms in the volume V. Being N, an infinitely large number, surface effects are negligible and the sample can be considered a statistically homogeneous infinite object, where an average number N of atoms is contained in any sufficiently large volume v. We then define the **average number density** as $\rho_0 = N/v = N/V$, and the average volume available for each atom as $v_o = v/N = 1/\rho_0$. Homogeneity implies that the probability of finding an atom in each volume element dv_i is uniform and equal to dv_i/v_0, i.e.,

$$n_1(\vec{\mathbf{r}}_i) = \rho_0 = N/V \tag{3}$$

independently of the position $\vec{\mathbf{r}}_i$.

Let us now write the two-body probability density as

$$n_2(\vec{\mathbf{r}}_i, \vec{\mathbf{r}}_j) = n_1(\vec{\mathbf{r}}_i)\, n_2(\vec{\mathbf{r}}_j/\vec{\mathbf{r}}_i) \tag{4}$$

where $n_2(\vec{\mathbf{r}}_j/\vec{\mathbf{r}}_i)$ is the **conditional probability density** of finding an atom located at $\vec{\mathbf{r}}_j$ when another atom is present at $\vec{\mathbf{r}}_i$. In the case of a completely random atomic arrangement, where the atomic positions are independent of each other, this function would be $n_2(\vec{\mathbf{r}}_j/\vec{\mathbf{r}}_i) = n_1(\vec{\mathbf{r}}_j)$, which means that the occurrence of an atom at $\vec{\mathbf{r}}_i$ is irrelevant to the occurrence of an atom at $\vec{\mathbf{r}}_j$. If this is not the case, because interatomic forces produce positional correlations between atoms, the difference of $n_2(\vec{\mathbf{r}}_j/\vec{\mathbf{r}}_i)$ from $n_1(\vec{\mathbf{r}}_j)$ can be evidenced by

$$n_2(\vec{\mathbf{r}}_j/\vec{\mathbf{r}}_i) = n_1(\vec{\mathbf{r}}_j)\, g(\vec{\mathbf{r}}_i, \vec{\mathbf{r}}_j) \tag{5}$$

where the function

$$g(\vec{\mathbf{r}}_i, \vec{\mathbf{r}}_j) = n_2(\vec{\mathbf{r}}_j/\vec{\mathbf{r}}_i)/n_1(\vec{\mathbf{r}}_j) = n_2(\vec{\mathbf{r}}_i, \vec{\mathbf{r}}_j)/n_1(\vec{\mathbf{r}}_i)\, n_1(\vec{\mathbf{r}}_j) \tag{6}$$

is called the **pair correlation function.** In the case of a homogeneous sample, Equation 5 becomes

$$n_2(\vec{\mathbf{r}}_j/\vec{\mathbf{r}}_i) = \rho(\vec{\mathbf{r}}) = \rho_0 g(\vec{\mathbf{r}}) \tag{7}$$

where the dependence is only on the relative vector separation $\vec{r}$ between the point $\vec{r}_i$ and the point $\vec{r}_j$, as homogeneity implies that the atomic distribution with respect to one origin atom is statistically independent from the position of the chosen atom.

The function, $\rho(\vec{r})$, represents the average local number density of atoms at the vector distance $\vec{r}$ from an origin atom; $g(\vec{r})$ is a dimensionless function and is related to the probability of finding an atom centered at the distance $\vec{r}$ from another one, relative to the probability of a uniform distribution. In other words, $g(\vec{r})$ measures the extent of positional correlation of the atoms among each other; $g(\vec{r}) = 1$ when the surrounding structure is uncorrelated to the presence of the origin atom, and $g(\vec{r}) = 0$ when it is impossible to find an atom separated from another by the vector $\vec{r}$. For example, in the case of a perfect crystal, $g(\vec{r})$ is zero everywhere except when $\vec{r}$ is one of the lattice vectors:

$$\rho(\vec{r}) = \rho_0 g(\vec{r}) = \sum \delta(\vec{r} - \vec{l}_{m_1,m_2,m_3}) \tag{8}$$

while in any low density random arrangement of hard sphere of radius R

$$\begin{aligned} g(\vec{r}) &= 0 \quad |\vec{r}| < 2R \\ g(\vec{r}) &= 1 \quad |\vec{r}| \geqslant 2R \end{aligned} \tag{9}$$

If we introduce the additional hypothesis of randomly oriented local environments, which makes such a sample macroscopically isotropic, $g(\vec{r})$ and $\rho(\vec{r})$ become **radial distribution functions**, g(r) and ρ(r), which depend only on the modulus of $\vec{r}$. This situation is generally found in gases, liquids, and noncrystalline solids, but can also be extended to microcrystalline samples and finely ground crystalline powders, which contain an infinitely large number of anisotropic grains in all possible orientations.

Since g(r) is related to the probability of finding an atom at the scalar distance r from another one, the **radial number density** $\rho(r) = \rho_0 g(r)$ can be interpreted by saying that $4\pi r^2 \rho(r)dr$ gives the average number of atoms located in the spherical shell between r and r + dr from an origin atom.

Figure 1 shows the basic features of g(r) for typical cases. In crystals, the behavior of g(r) is given by averaging Equation 8 over all possible orientations and is represented by a succession of delta peaks, which broaden due to imperfections and thermal effects present in any real sample. In low density gases, g(r) behaves as stated in Equation 9. Noncrystalline materials lie between these two extremes: g(r) is zero within the closest distance of approach of the atoms, then reveals the existence of a local structure by exhibiting one or more peaks and oscillations which damp out to unity at long distances where the positional correlation is progressively lost.

B. Multicomponent Systems

The extension to a multicomponent system of the subject discussed in the preceding section is straightforward.

Let us consider an M-component system containing N_i atoms of i-type (i = 1, 2, . . . M)

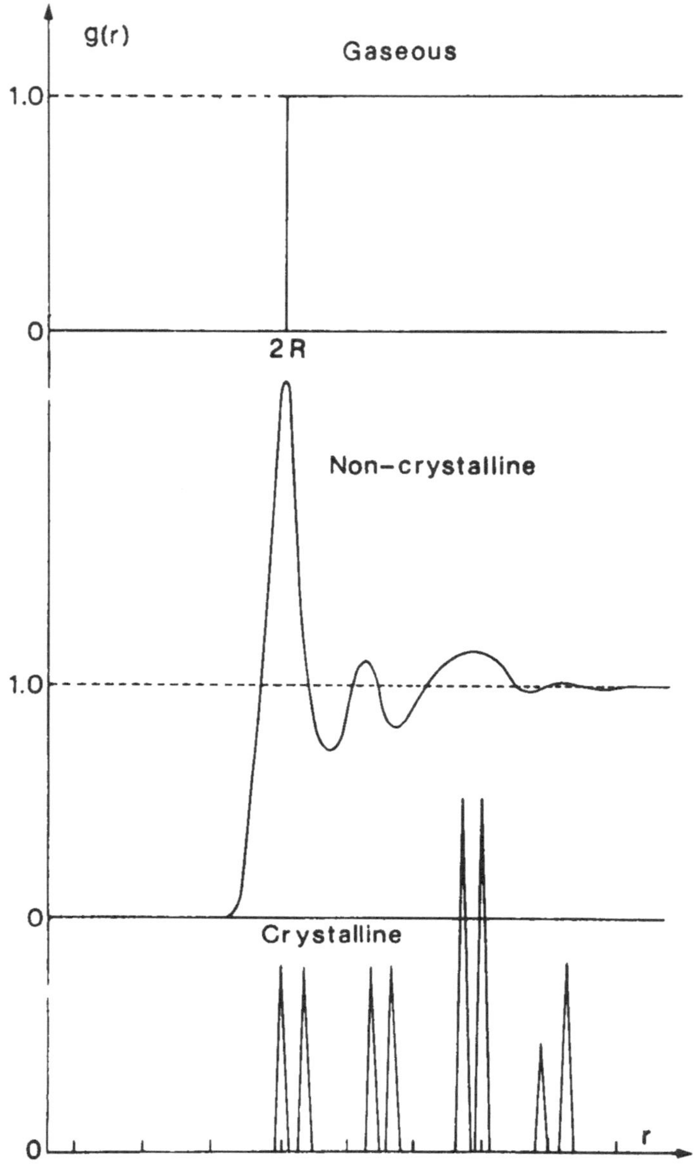

FIGURE 1. Typical correlation functions for gaseous, noncrystalline, and crystalline substances.

in the volume V. The chemical composition of such a system can be given in terms of a **composition unit** (CU), which is representative of the whole sample, i.e.,

$$(1)_{x_1}(2)_{x_2}\ldots(M)_{x_M}$$

where the **composition coefficient** x_i is the number of i-type atoms in CU.

Let us define the following quantities:

- N_a = total number of atoms in the volume V

- N_u = number of CU in the volume V
- ρ_a = N_a/V = average number density of atoms
- ρ_0 = N_u/V = average number density of CU
- ρ_{oi} = N_i/V = average number density of i-type atoms

Consequently, it is immediate to state these equations:

$$x_i = N_i/N_u$$

$$\rho_{0i} = x_i\rho_0$$

$$\Sigma N_i = N_a$$

$$\Sigma x_i = N_a/N_u$$

$$\Sigma\rho_{0i} = \rho_a$$

Note that CU can be defined in different, but equivalent ways. For example, the composition of a $CaCl_2$ aqueous solution can be described as $(CaCl_2)_x(H_2O)_{1-x}$, where χ is the salt molar fraction, or as $(CaCl_2)_{x/3}(H_2O)_{(1-x)/3}$; in the second case the composition coefficients x_i are the atomic ratios, N_i/N_a and $\Sigma x_i = 1$.

We can now make use of the above concepts about the pair distribution functions and can extend that approach in order to describe the structure of a multicomponent system in terms of **partial pair correlation functions**, $g_{ij}(r)$. In this picture, $g_{ij}(r)$ is related to the probability of finding an atom of j-type at distance r from another of i-type, and the partial radial density

$$\rho_{ij}(r) = \rho_{oj}g_{ij}(r) = x_j\rho_0 g_{ij}(r) \tag{10}$$

measures the average number of atoms of j-type per unit volume in a spherical shell between r and r + dr from an atom of i-type, so that

$$\int 4\pi r^2\rho_{ij}(r)\,dr = 4\pi\rho_0 x_j\int r^2 g_{ij}(r)\,dr = N_j \tag{11}$$

is the total number of j-type atoms contained in the sample.

Note that not all the M^2 functions g_{ij} (i,j = 1,2, . . . M) are independent from each other. It is in fact immediate to state that

$$N_i\rho_{ij} = N_j\rho_{ji} \tag{12A}$$

$$g_{ij}(r) = g_{ji}(r) \tag{12B}$$

so that the independent partial functions number is $M(M+1)/2$.

The ultimate goal of the experimental studies about the atomic scale structure in non-crystalline systems might be considered the extraction of all the $M(M+1)/2$ partial pair correlation functions.

III. X-RAY DIFFRACTION THEORY FOR NONCRYSTALLINE SYSTEMS

A. Scattering Process

The schematic picture of a diffraction experiment involves an incident beam of wave vector $\vec{\mathbf{k}}_0$ and energy E_0, which interacts locally with the sample before reemerging in free space with wave vector $\vec{\mathbf{k}}$ and energy E. The scattering process is thus generally characterized by a momentum transfer $\hbar\vec{\mathbf{s}} = \hbar\vec{\mathbf{k}} - \hbar\vec{\mathbf{k}}_0$ and an energy transfer $\hbar\omega = E - E_0$.

Since we are interested in gaining information about the **static structure,** only elastic scattering is taken into account, while inelastic effects, as well as others disturbing contributions, are either neglected or evaluated and properly subtracted. For example, it may happen that X-photons are inelastically scattered with emission or absorption of one or more phonons; the energy transfer is, however, very small compared with the incident X-ray energy, so that the effects of this phonon inelastic scattering can be ignored. On the contrary, the strong interaction of X-photons with electrons produces a Compton contribution due to inelastically scattered electrons which must be evaluated by theoretical or experimental procedures and then subtracted from experimental data (see Section V).

The primary scattering mechanism in X-ray diffraction is the interaction of photons with electrons. We can therefore assume the sample as an electron density $\rho_e(\vec{r})$ which interacts with the incident beam; in a sense, $\rho_e(\vec{r})$ behaves as a local potential towards the beam of incident photons.

If the sample is sufficiently small, it is possible to assume that (1) the absorption is negligible, so that each volume element within the sample receives the same incident intensity; and (2) both the incident and scattered photons can be approximated to a plane wave in the diffraction geometry shown in Figure 2. Under these conditions, the **scattering amplitude** in **electron units** that is expressed in terms of the amplitude scattered by a classical free electron in the same situation as the sample, is given by[2,3]

$$A_{eu}(\vec{\mathbf{s}}) = \int \rho_e(\vec{\mathbf{r}})\, e^{-i\vec{s}\cdot\vec{r}}\, dr \tag{13}$$

and depends only on the **scattering vector** $\vec{\mathbf{s}} = \vec{\mathbf{k}} - \vec{\mathbf{k}}_0$ which, for elastic scattering ($|\vec{\mathbf{k}}| = |\vec{\mathbf{k}}_0| = k$), has the magnitude

$$s = 2k\sin\vartheta = 4\pi\sin\vartheta/\lambda \tag{14}$$

where 2 ϑ is the **scattering angle** between $\vec{k}$ and $\vec{k}_0$, and λ is wavelength of X-ray.

The scattering amplitude (Equation 13) appears to be the Fourier transform of the electron density from real space into **reciprocal space** where the variable complementary to $\vec{\mathbf{r}}$ is the scattering vector $\vec{\mathbf{s}}$.

B. Scattering Factors

Since we are interested in gaining information about the atomic positions $\vec{\mathbf{r}}_n$, we have to look at the total electron density in the sample as a superposition of electron densities centered at the atomic sites, i.e.,

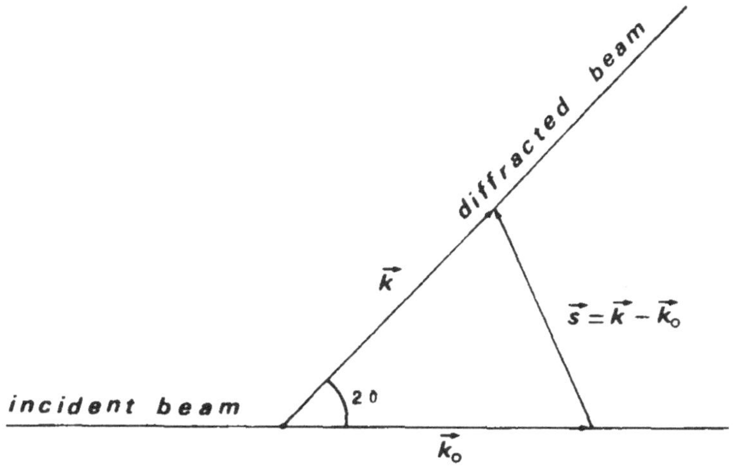

FIGURE 2. Diffraction geometry.

$$\rho_e(\vec{r}) = \sum_n \rho_n(\vec{r} - \vec{r}_n) \tag{15}$$

Thus, the scattering amplitude becomes

$$\begin{aligned} A_{eu}(\vec{s}) &= \int \sum_n \rho_n(\vec{r} - \vec{r}_n)\, e^{-i\vec{s}\cdot\vec{r}}\, dr \\ &= \sum_n e^{-i\vec{s}\cdot\vec{r}_n} \int \rho_n(\vec{r} - \vec{r}_n)\, e^{-i\vec{s}\cdot(\vec{r} - \vec{r}_n)}\, d(\vec{r} - \vec{r}_n) \\ &= \sum_n f_n(\vec{s})\, e^{-\vec{s}\cdot\vec{r}_n} \end{aligned} \tag{16}$$

where the **scattering factor** of atom n

$$f_n(\vec{s}) = \int \rho_n(\vec{r})\, e^{-i\vec{s}\cdot\vec{r}}\, dr \tag{17}$$

is the Fourier transform of the electron density of that atom, that is, the atomic scattering amplitude measured in electron units.

Scattering factors can be calculated from the electronic wave functions of the atoms and are available in tabulated form.[9] Usually, calculations are made for free atoms and ions treated as spherically symmetric so that the scattering factor depends only on the modulus of the scattering vector. This assumption is not correct for atoms with incomplete shells or involved in covalent bonding, but luckily, the scattering factors seem to be not too sensitive to distortions in electron density. However, care should be taken in the case of light atoms, because the electrons involved in chemical bonding can represent a significant fraction, and the actual electron density centered around the nucleus does not correspond to that used for determining the scattering factor.

The typical dependence of the atomic scattering factors on s is shown in Figure 3 for two illustrative cases. As it is evident from Equation 17, $f_i(s)$ is equal to the atomic number Z_i,

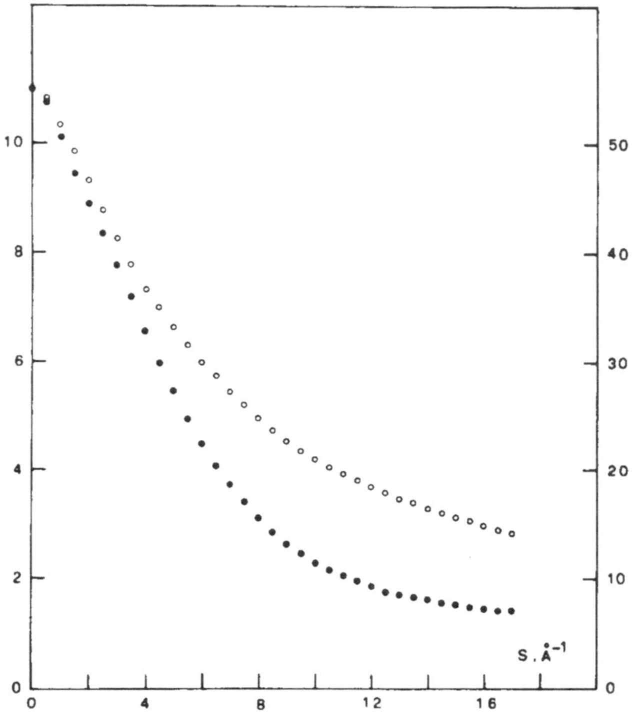

FIGURE 3. Scattering factors for Na (full dots), left y scale; Cs (open dots), right y scale.

for s = 0. At increasing s, $f_i(s)$ decreases rapidly, more rapidly in the small s region and in general, more slowly for heavier atoms. Thus, the heavier atoms contribute more heavily to X-ray scattering.

The definition (Equation 17) is valid only if the X-ray energy is much larger than the binding energy of the electrons in the innermost shell of the atom. This is because the atomic electrons have been assumed to scatter as if they were free rather than bound electrons. Thus, corrections are necessary, especially when the energy of the radiation to be scattered approaches the absorption edge. The result is that the scattering factor becomes a complex quantity:

$$f_i(s) = f^0 + \Delta f' + i\,\Delta f'' \tag{18}$$

where f^0 is the scattering factor at the high energy limit, and $\Delta f'$ and $\Delta f''$ are the real and imaginary parts of the **anomalous dispersion** term.[1] The terms $\Delta f'$ and $\Delta f''$ are small compared to f^0 and are essentially independent from the scattering angle, while they depend on the wavelength of the incident radiation and on the atomic number of the scatterer.

C. Diffracted Intensity

The experimental quantity measured in a diffraction experiment is not proportional to the amplitude, which would allow a direct accessibility to atomic position, but to its squared modulus, $A_{eu}A^*_{eu}$,

$$I^{coh}_{eu}(\vec{s}) = \left|\sum_n f_n e^{-i\vec{s}\cdot\vec{r}_n}\right|^2 = \sum_n \sum_m f_n f_m e^{-i\vec{s}\cdot\vec{r}_{nm}} \tag{19}$$

which represents the **coherent diffracted intensity** in electron units, where for simplicity, the scattering factors are supposed to be real.

Equation 19 is the most general equation for the intensity diffracted by an assembly of atoms which occupy definite sites arranged in any arbitrary way, and must be specialized according to the characteristics of the system under investigation. For example, in the case of a perfect lattice, the introduction in Equation 19 of the general formula (Equation 1) for the lattice sites, allows to state the three Laue equations:

$$\vec{\mathbf{s}} \cdot \vec{\mathbf{a}}_1 = 2\pi h$$

$$\vec{\mathbf{s}} \cdot \vec{\mathbf{a}}_2 = 2\pi k$$

$$\vec{\mathbf{s}} \cdot \vec{\mathbf{a}}_3 = 2\pi l \tag{20}$$

so that $I_{eu}^{coh}(\vec{\mathbf{s}})$ becomes a sum of sharp peaks at the points of the **reciprocal lattice**.[1-4]

The time scale of X-ray diffraction events is much shorter than that of atomic motions. However, in real liquids the interatomic vectors change quickly compared to the time of any experimental measurement, and moreover, the instantaneous configurations around equivalent atoms are not the same. The observed quantity is thus an average value of the diffracted intensity due to many different instantaneous and local configurations.

Homogeneity and ergodic hypothesis allows to describe such a situation through an ensemble average, that is, by averaging Equation 19 over the statistical distribution of interatomic vectors. Since we are interested in an isotropic system, where each vector $\vec{\mathbf{r}}_{nm}$ takes all possible orientation with equal probability, we can first average each term of the double sum (Equation 19) over all directions for a fixed modulus of the interatomic vectors, i.e.,

$$I_{eu}^{coh}(s) = \sum_n \sum_m f_n f_m (1/4\pi) \int_0^{\pi} \int_0^{2\pi} e^{-isr_{nm}\cos\alpha} \sin\alpha \, d\alpha \, d\phi \tag{21}$$

yielding

$$I_{eu}^{coh}(s) = \sum_{n=1}^{N_a} \sum_{m=1}^{N_a} f_n f_m \frac{\sin sr_{nm}}{sr_{nm}}$$

$$= \sum_n f_n^2 + \sum_n \sum_{m \neq n} f_n f_m \frac{\sin sr_{nm}}{sr_{nm}} \tag{22}$$

where now the intensity depends only on the modulus of $\vec{\mathbf{s}}$.

This important equation, called the **Debye Scattering Equation,** represents the basis for a simple interpretation of the measured intensity in terms of the radial pair distribution functions discussed above.

D. Structure Function

For a system containing M atomic species we can group together in Equation 22 the x_i atoms of i-type contained in the composition unit. If we consider that volume V contains N_u of such units, we can express the intensity, normalized to one composition unit, according to

$$I(s) = I_{eu}^{coh}(s)/N_u = \sum_{i=1}^{M} x_i f_i^2 + \sum_{i=1}^{M} x_i f_i \sum_{m=1}^{N_a} f_m \frac{\sin sr_{im}}{sr_{im}} \tag{23}$$

where now the sums are taken over the atomic species for i and over all atoms in the sample for m.

The intensity (Equation 23) appears composed of two terms; the first is a structure-independent term which represents the **coherent independent scattering,** that is, the elastic scattering that we would have if all atoms scatter independently without interference effects. The second term

$$i(s) = I(s) - \sum_i x_i f_i^2 = \sum_i x_i f_i \sum_m f_m \frac{\sin sr_{im}}{sr_{im}} \tag{24}$$

is called **structure function** (or **reduced intensity**) and represents the structurally sensitive part of the total scattered intensity.

Considering that, in the distance range r to r + dr from an atom i, there are $4\pi r^2 \rho_{ij} dr$ j-type atoms, and replacing the sum of distances by an integral, Equation 24 becomes

$$i(s) = \sum_i \sum_j x_i f_i f_j \int_V 4\pi r^2 \rho_{ij}(r) \frac{\sin sr}{sr} dr \tag{25}$$

where the sums are both taken over the atomic species and the integral is limited to distances within the volume of the sample. This integral is not well-behaved for large r where $\rho_{ij}(r)$ tends to ρ_{0j} and, for an ideal infinite sample, it diverges as $s \rightarrow 0$. For real samples, we have to consider the existence of a **forward scattering peak,** also called **zero angle scattering** or **volume scattering,** which is generally not observed because it is hidden by the primary beam. It is conventional to subtract this contribution from both sides of Equation 25 as the quantity

$$\begin{aligned} I^0 &= \sum_i \sum_j x_i f_i f_j \int_V 4\pi r^2 \rho_{0j} \frac{\sin sr}{sr} dr \\ &= (\sum_i x_i f_i)^2 \int_V 4\pi r^2 \rho_0 \frac{\sin sr}{sr} dr \end{aligned} \tag{26}$$

which corresponds to the intensity diffracted by an object with uniform density distribution and having the same shape as the sample. It can be shown that, if the sample is not too small, I^0 is significantly different from zero only in a small region around $s = 0$; only in very few cases it could be necessary to evaluate the volume scattering and actually perform the subtraction.[3] After having subtracted Equation 26 from Equation 25, and still calling i(s) the corrected structure function, we have

$$\begin{aligned} i(s) &= \sum_i \sum_j x_i f_i f_j \int_0^\infty 4\pi r^2 [\rho_{ij}(r) - \rho_{0j}] \frac{\sin sr}{sr} dr \\ &= \sum_i \sum_j x_i x_j f_i f_j \int_0^\infty 4\pi \rho_o r^2 [g_{ij}(r) - 1] \frac{\sin sr}{sr} dr \end{aligned} \tag{27}$$

where the upper limit of the integral can be considered to be infinite, because $g(r) - 1$ vanishes outside a distance very small compared to the size of the sample.

E. One-Component System

It is illustrative to consider the case of a one-component system, for which Equation 27 becomes

$$si(s)/f^2(s) = \int_0^\infty 4\pi\rho_0 r[g(r) - 1] \sin sr\, dr \tag{28}$$

This equation is a sine Fourier transform,[10] which can be directly inverted, yielding

$$\begin{aligned} 4\pi r^2\rho_0[g(r) - 1] &\equiv 4\pi r^2[\rho(r) - \rho_0] \\ &= 2r/\pi \int_0^\infty [si(s)/f^2(s)] \sin sr\, ds \end{aligned} \tag{29}$$

Thus, the pair distribution function is calculated in a direct way from an experimentally accessible quantity, apart from **termination errors** in the integral, arising when the experimental information stops at an s_{max} value where i(s) is still significantly different from zero.

F. Partial Structure Functions

In the case of a multicomponent system, the total structure function (Equation 27) can be written as a weighted sum of **partial structure functions,** i.e.,

$$i(s) = \sum_{i=1}^{M} \sum_{j=1}^{M} x_i x_j f_i f_j i_{ij}(s) \tag{30}$$

where

$$si_{ij}(s) = 4\pi\rho_0 \int_0^\infty rh_{ij}(r) \sin sr\, dr \tag{31}$$

having defined

$$h_{ij}(r) = g_{ij}(r) - 1 \tag{32}$$

Fourier inversion yields

$$\begin{aligned} 4\pi\rho_0 rh_{ij}(r) &\equiv (4\pi r/x_j)[\rho_{ij}(r) - \rho_{0j}] \\ &= (2/\pi) \int_0^\infty si_{ij}(s) \sin sr\, ds \end{aligned} \tag{33}$$

Thus, the knowledge of the partial pair correlation functions $g_{ij}(r)$ depends on the experimental accessibility of the corresponding i_{ij} in a sufficiently large range of s values.

Some authors define the partial components, called **partial interference functions** or **partial structure factors,** in the following symmetric form:[6]

$$s[S_{ij}(s) - 1] = 4\pi\rho_0 \int_0^\infty r[g_{ij}(r) - 1] \sin sr\, dr \tag{34}$$

$$4\pi\rho_0 r[g_{ij}(r) - 1] = 2/\pi \int_0^\infty s[S_{ij}(s) - 1] \sin sr\, ds \tag{35}$$

correspondingly, a total interference function is defined as

$$S(s) = I(s) - [\sum_i x_i f_i^2 - (\sum_i x_i f_i)^2] = i(s) + (\sum_i x_i f_i)^2 \tag{36}$$

Using Equation 30, it follows readily that

$$S(s) = \sum_i \sum_j x_i x_j f_i f_j S_{ij}(s) \tag{37}$$

and

$$i(s) = \sum_i \sum_j x_i x_j f_i f_j [S_{ij}(s) - 1] \tag{38}$$

Because $g_{ij} = g_{ji}$ implies $i_{ij} = i_{ji}$, the independent partial components number $M(M + 1)/2$ and the total structure function can be accordingly rewritten as

$$i(s) = \sum_{i=1}^{M} \sum_{j \geq i}^{M} W_{ij}(s)\, i_{ij}(s) = \sum_{i=1}^{M} \sum_{j \geq i}^{M} W_{ij}[S_{ij}(s) - 1] \tag{39}$$

where the weighting factors are

$$W_{ij}(s) = (2 - \delta_{ij})\, x_i x_j f_i(s) f_j(s) \tag{40}$$

It is obvious that from a single diffraction determination of i(s), it is not possible to obtain the partial structure functions separately. Equation 39, being a linear equation, the determination of $M(M + 1)/2$ unknowns implies an equal number of diffraction experiments involving independent sets of the coefficients W_{ij}, but the same atomic structure.

Since it is unreasonable to assume that changes in the composition coefficients x_i do not affect the structure, the only way to vary the weights W_{ij} independently from the partial structure functions is to vary the scattering factors f_i. In principle, this could be done in two ways: (1) by varying the nature (X-ray, neutron, electron) or wavelength of the incident radiation; and (2) by changing the nature of the atoms themselves, without changing the structuring, through **chemical isomorphous substitutions** or, in the case of neutron scattering, through **isotopic substitutions.**

In spite of the evident difficulties connected with this approach, some partial or approximate applications produced important results which will be discussed in the following chapters.

G. Radial Functions

When only a single diffraction determination of i(s) is available, it is still possible to obtain, by Fourier transform, radial distribution functions interpretable in terms of the structural characteristics of the sample.

Let us define the Fourier transform:

$$Q(r) = 2/\pi \int_0^{s_{max}} s\,i(s)\, M(s) \sin sr\, ds = \sum_i \sum_{j \geq i} (2 - \delta_{ij})\, x_i x_j Q_{ij}(r) \tag{41}$$

where

$$Q_{ij}(r) = 2/\pi \int_0^{s_{max}} s\,i_{ij}(s)\, f_i(s)\, f_j(s)\, M(s) \sin sr\, ds \tag{42}$$

Here, s_{max} is the upper limit of the experimentally available data and M(s) is a **modification function.**[10]

In principle, M(s) may be chosen in any arbitrary way, but in practice, it is mainly used to account for three aspects:

1. M(s) contains the step window which generates the cut-off in the integral, that is, M(s) = 0 out of the s range where i(s) has been actually measured. Therefore, Q(r) can be correctly considered the Fourier transform of the modified structure function i(s)M(s).
2. M(s) generally includes a *convergence factor,* usually $\exp(-bs^2)$ with $b = (\ln 10)/s^2_{max}$, in order to minimize termination effects with a smoother cut-off than the step window alone. Moreover, this factor gives a decreasing weight to large s values where the experimental accuracy decreases.
3. The third and most significant component of M(s) is a *sharpening factor,* $1/q^2(s)$, chosen to make the behavior of $f_i(s)f_j(s)/q^2(s)$ roughly independent from s.

An obvious choice of q(s) is some average of scattering factors like,

$$q(s) = \sum_i x_i f_i(s) \tag{43}$$

or

$$q(s) \equiv f_e(s) = [\sum_i x_i f_i(s)]/[\sum_i x_i f_i(0)] \tag{44}$$

which represents an average scattering factor per electron. Often the factor $1/q^2(s)$ is directly included in the definition of the structure function i(s).

The function $Q_{ij}(r)$, defined by Equation 42, is the Fourier transform of $si_{ij}(s)$ times $f_i(s)f_j(s)M(s)$. Thus, the convolution theorem for Fourier transforms allows to express $Q_{ij}(r)$ as a convolution integral involving the partial correlation functions $g_{ij}(r)$,[10,11] i.e.,

$$\begin{aligned} Q_{ij}(r) &= 4\pi\rho_0 \int_{-\infty}^{\infty} uh_{ij}(u)\, T_{ij}(r - u)\, du \\ &= 4\pi\rho_0 \int_0^{\infty} uh_{ij}(u)[T_{ij}(r - u) - T_{ij}(r + u)]\, du \end{aligned} \tag{45}$$

where $4\pi\rho_0 rh_{ij}(r)$ is the sine Fourier transform of $si_{ij}(s)$, shown in Equation 33, and $2T_{ij}(r)$ is the cosine Fourier transform of $f_i(s)f_j(s)M(s)$, i.e.,

$$T_{ij}(r) = 1/\pi \int_0^{s_{max}} f_i(s)\, f_j(s)\, M(s) \cos sr\, ds \tag{46}$$

which is called the **peak-shape function.**

Equation 46 can be inverted, yielding

$$f_i(s)\, f_j(s) = 2 \int_0^{\infty} T_{ij}(r) \cos sr\, dr \tag{47}$$

from which, by considering that $T_{ij}(r)$ is an even function and setting s = 0,

$$A_{ij}^{T} = \int_{-\infty}^{\infty} T_{ij}(r)\,dr = f_i(0)\,f_j(0)\,M(0) \tag{48}$$

which is the total area under $T_{ij}(r)$. Using this result, it is readily shown that

$$\int_0^{\infty} u[T_{ij}(r - u) - T_{ij}(r + u)]\,du = rA_{ij}^{T} \tag{49}$$

so that Equation 45 becomes

$$Q_{ij}(r) = 4\pi\rho_0 \int_0^{\infty} ug_{ij}(u)[T_{ij}(r - u) - T_{ij}(r + u)]\,du - 4\pi\rho_0 r f_i(0)\,f_i(0)\,M(0) \tag{50}$$

and

$$\begin{aligned} Q(r) &= \sum_i \sum_{j \geq i} (2 - \delta_{ij})\,x_i x_j Q_{ij}(r) \\ &= 4\pi\rho_0 \sum_i \sum_{j \geq i} (2 - \delta_{ij})\,x_i x_j \int_0^{\infty} ug_{ij}(u)[T_{ij}(r - u) - T_{ij}(r + u)]\,du \\ &- 4\pi\rho_0' \end{aligned} \tag{51}$$

where

$$\rho_0' = \rho_0[\sum_i x_i f_i(0)]^2 M(0) \tag{52}$$

It is customary to define a **total radial distribution function** D(r) and a **total correlation function** G(r) by the relations

$$D(r) = r[4\pi r\rho_0' + Q(r)] = 4\pi r^2\rho_0' + 2r/\pi \int_0^{s_{max}} si(s)\,M(s) \sin sr\,ds \tag{53}$$

and

$$\begin{aligned} G(r) &\equiv D(r)/4\pi r^2\rho_0' \equiv 1 + Q(r)/4\pi r\rho_0' \\ &= 1 + 1/2\pi^2 r\rho_0' \int_0^{s_{max}} si(s)\,M(s)\,\sin sr\,ds \end{aligned} \tag{54}$$

These functions can be expressed as

$$D(r) = \sum_i \sum_{j \geq i} (2 - \delta_{ij})\,x_i D_{ij}(r) \tag{55}$$

where

$$D_{ij}(r) = 4\pi r \int_0^{\infty} u\rho_{ij}(u)[T_{ij}(r - u) - T_{ij}(r + u)]\,du \tag{56}$$

and

$$G(r) = \sum_i \sum_{j \geq i} (2 - \delta_{ij})\, x_i x_j G_{ij}(r) \tag{57}$$

where

$$rG_{ij}(r) = 1/[\sum_i x_i f_i(0)]^2 M(0) \int_0^{\infty} u g_{ij}[T_{ij}(r - u) - T_{ij}(r + u)]\, du \tag{58}$$

The total radial functions accessible from X-ray diffraction measurements are thus weighted sums of **modified** partial functions, given by the **true** $\rho_{ij}(r)$ and $g_{ij}(r)$ convoluted with the peak-shapes $T_{ij}(r)$. Note, from Equations 45 and 46, that the modification is due to the s dependence of $f_i(s)f_j(s)M(s)$; the slower this dependence is, the sharper the peak T_{ij}. If $f_i f_j M$ = constant, T_{ij} becomes a delta peak and G_{ij} is proportional to the true g_{ij}. This explains why M(s) is mainly chosen so as to counterbalance the s behavior of the scattering factors, which is inherent to the physics of the X-ray scattering process.

H. WKM Approximation

An approximation method to obtain the total radial functions as a superposition of the true partial functions was introduced by Warren et al.[12]

The atomic scattering factor $f_i(s)$ can be written according to

$$f_i(s) = K_i f_e(s) \tag{59}$$

where $f_e(s)$ is the average scattering factor per electron defined by Equation 44, and K_i should be a slowly varying function of s, approximately equal to the atomic number, Z_i. The WKM approximation treats K_i as an effective number of electrons for each atom, given by the average value

$$K_i = 1/s_{max} \int_0^{s_{max}} f_i(s)/f_e(s)\, ds \tag{60}$$

Under this approximation, the peak-shape function defined in Equation 46 becomes

$$T_{ij}(r) = K_i K_j/\pi \int_0^{s_{max}} f_e^2(s)\, M(s) \cos sr\, ds \tag{61}$$

and by choosing $M(s) = 1/f_e^2(s)$ and neglecting cut-off effects

$$T_{ij}(r) = K_i K_j \delta(r) \tag{62}$$

We can then reconsider Equations 53 to 58, obtaining

$$D(r) \equiv 4\pi r^2 \rho_0 (\sum_i x_i K_i)^2 G(r)$$

$$= 4\pi r^2 \rho_0 (\textstyle\sum_i x_i K_i)^2 + 2r/\pi \displaystyle\int_0^{s_{max}} [si(s)/f_e^2(s)] \sin sr\, ds \tag{63}$$

$$D(r) = \sum_i \sum_{j \geq i} (2 - \delta_{ij})\, x_i K_i K_j 4\pi r^2 \rho_{ij}(r) \tag{64}$$

$$G(r) = \sum_i \sum_{j \geq i} (2 - \delta_{ij})[x_i x_j K_i K_j / (\sum_i x_i K_i)^2]\, g_{ij}(r) \tag{65}$$

I. Neutron Diffraction

It is useful to consider briefly neutron diffraction for which the condition of s independence of the scattering factors is valid without approximation.

Thermal neutrons interact primarily with nuclei through an interaction potential, which can be assumed to be a superposition of **Fermi pseudopotentials** centered at the atomic sites $\vec{\mathbf{r}}_n$;

$$V(\vec{\mathbf{r}}) = \sum_n V_n(\vec{\mathbf{r}} - \vec{\mathbf{r}}_n) = \sum_n b_n \delta(\vec{\mathbf{r}} - \vec{\mathbf{r}}_n) \tag{66}$$

The amplitude of the elastic scattering is then given by[13]

$$A(\vec{\mathbf{s}}) = \int V(\vec{\mathbf{r}})\, e^{-i\vec{s}\cdot\vec{r}}\, d\vec{\mathbf{r}} = \sum_n b_n e^{-i\vec{s}\cdot\vec{r}_n} \tag{67}$$

A comparison with Equation 13 reveals that the formalism of neutron diffraction is the same as for X-ray (and electron) diffraction, while important practical consequences come from differences in the physics of the scattering process. In neutron diffraction, the **atomic scattering factor** b_i, usually referred to as **coherent scattering length,** is characteristic of the nucleus i and of the energy of the incident neutrons, but it is independent from the scattering vector $\vec{\mathbf{s}}$. As a consequence, the weighting factors, which enter in defining the total structure function in terms of partial contributions (Equation 39), are not s dependent.

$$i(s) = \sum_i \sum_{j \geq i} (2 - \delta_{ij})\, x_i x_j b_i b_j [S_{ij}(s) - 1] \tag{68}$$

Therefore, the Fourier transform of the structure function produces directly total radial functions which are linear combinations of the **true** partial g_{ij} or ρ_{ij}; consequently, relations like Equations 64 and 65, which are approximated in X-ray diffraction, become exact in neutron diffraction.

IV. INSTRUMENTAL APPARATUS AND DATA COLLECTION

The first experimental investigations on liquid or glassy samples were carried out by an experimental apparatus which did not substantially differ from the familiar Debye-Scherrer camera. A monochromatic radiation beam was focused on the liquid sample (often contained in a cylindric capillary glass) and photographic methods were used to collect data in all the accessible reciprocal space.

The photographic method is nowadays largely obsolete and substituted by automatic diffractometric equipment, usually associated with electronic devices for counting rates measurements. A modern experiment consists in recording the diffracted intensity from the sample as a function of the scattering angle ϑ. The measurements are performed step by step at discrete intervals in the whole explorable angular interval. In one of the latest developments, a continuous radiation source of X-rays is used, and the scattering patterns are determined at fixed angles as a function of the energy; however, no application of energy-dispersive techniques to measurements on electrolyte solutions has been reported.

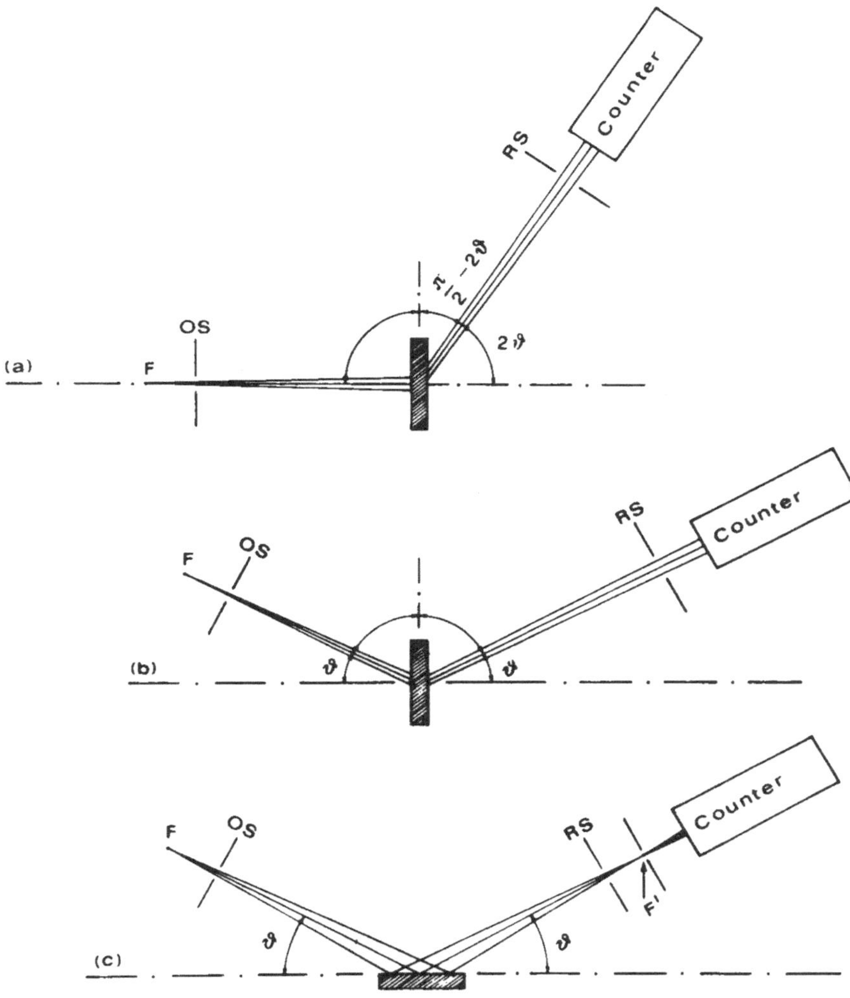

FIGURE 4. Schematic diagrams of different experimental apparatus: (a) normal-beam transmission; (b) symmetrical transmission; and (c) symmetrical reflection. OS and RS indicate opening and receiving slits, respectively, and F, the focus of the X-ray tube.

The three principal geometrical arrangements of the specimen, with respect to the X-ray source and counter, are schematized in Figure 4. Among them, the symmetrical reflection geometry (Figure 4c) is the one mostly adopted in the studies of liquid systems. A detailed description of the experimental apparatus and a particularized discussion of advantages and disadvantages of the different geometries can be found in many books (see, e.g., References 2 to 7), and we will no longer insist here on this subject. We wish just to remark that, when the focusing Bragg-Brentano reflection geometry is adopted in a ϑ-ϑ diffractometer with stationary sample, liquid samples can be handled very comfortably. This arrangement, therefore, has been purposely adopted for studying liquid systems in several commercial apparatuses, but data equally good are currently obtained also using transmission geometry.

In Figure 5, the experimental apparatus manufactured by Rich & Seifert, and used over the years by the authors, is shown in some details. The X-ray beam, coming from the X-ray tube focus (F), is collimated by an opening slit (OS) and, after diffraction by the liquid surface, passes through a receiving slit (RS), a focus slit (F′) and reflects on the mono-

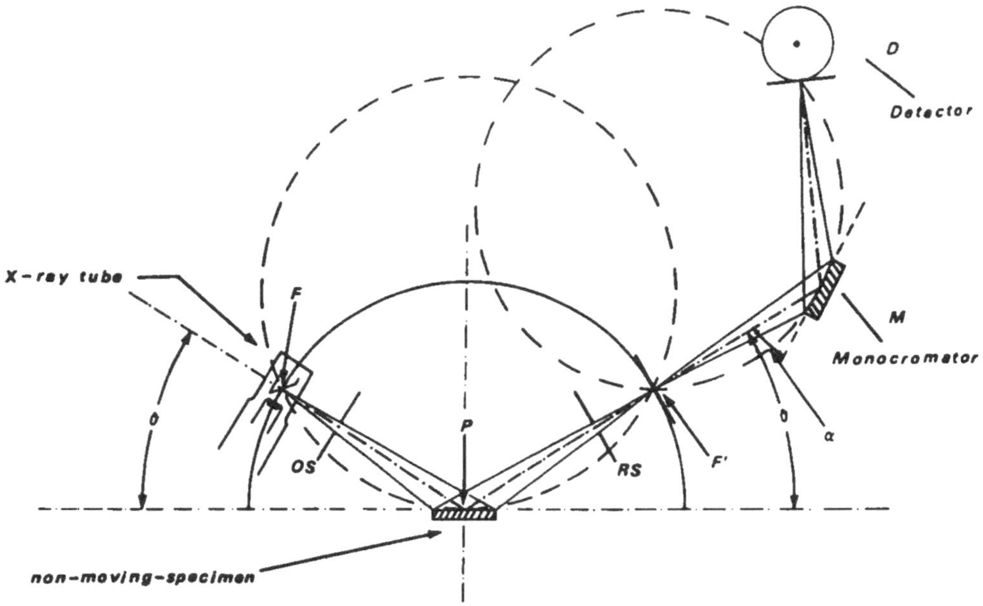

FIGURE 5. Symmetrical reflection goniometer, with horizontal nonmoving specimen, manufactured by Rich & Seifert.

chromator crystal (M). A ϑ-ϑ automatic control allows the goniometer to advance with preselected steps, so that the diffracted intensities can be recorded, point by point, on the whole angular interval (from 1 to 70 ϑ degrees in Rich & Seifert apparatus).

A limitation of symmetrical reflection technique is the excessive length of the specimen required at small Bragg angles. This fact forces the use of different combinations of slit widths, that is, narrow slits at small angles imposed by the finite sizes of the sample, and large slits at high angles in order to increase the diffracting volume and, consequently, to reduce the reading times. Usually two or three combinations of slit widths are enough to cover the entire ϑ range. Of course, overlapping regions between the various combinations of slits must be considered to scale the experimental data to a unique slit width. In order to ensure the required statistical accuracy, at least 100,000 counts at each value must be accumulated. Then, an experimental data collection, consisting of 200 to 250 values, takes approximately 1 week with modern electronic counting equipment.

Figure 6 shows a typical example of experimental data collected by the authors with a ϑ-ϑ reflection apparatus and three slit widths.

V. DATA TREATMENT

In this section, we will follow the logical flow-sheet, which, starting from the experimental data, allows to obtain the radial distribution functions discussed in Section III. We will describe the flow-sheet, mainly referring to the experimental data given in Figure 6. However, for the sake of completeness, we will also describe the data handling which is needed when a transmission geometry is adopted.

A. Preliminary Corrections

1. Background and Air Scattering

a. Background

The background can be defined as the scattering produced by every point of the apparatus

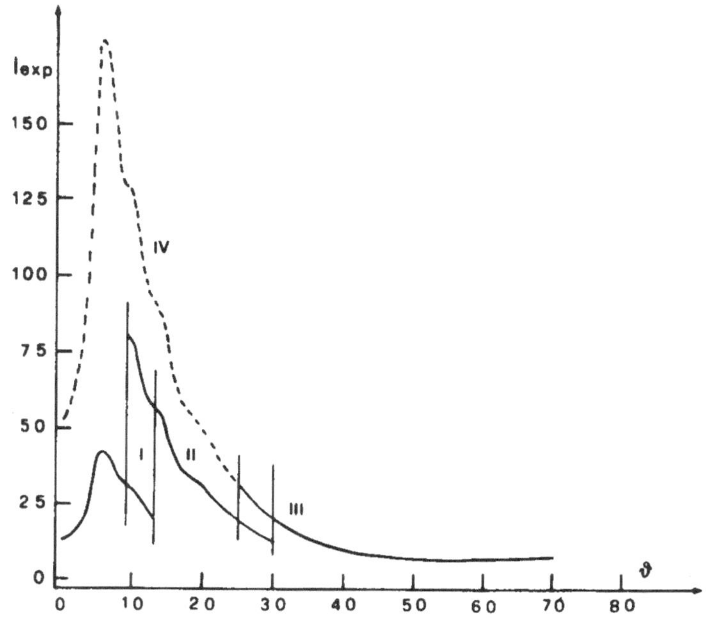

FIGURE 6. Experimental data in arbitrary units (counts per second) from a 4 *M* aqueous $NiBr_2$ solution. I, II, and III (continuous lines) refer to the experimental data obtained by using three different combinations of opening and receiving slits. The vertical lines limit the overlapping regions between different combinations of slits. The dashed line IV shows the scaling of the experimental data to a common slit width. Note that the data shown by continuous lines are already corrected for background and absorption.

(slits, sample container, filters, etc.), except the sample under examination; to this extraneous scattering, the natural non-X-ray background due to cosmic radiation and radioactivity must be added. A simple evaluation of most of the background can be performed with the apparatus set in identical conditions as when collecting data, but with a lead plate inserted in front of the receiving slit (RS in Figures 4 and 5). Usually, with well-aligned and shielded apparatus, the background is practically independent from the scattering angle and should be less than 1% of the total counts at high angles, where counting rates are at minimum. The background correction is performed by simply subtracting the background value from the data for each combination of slits used in data collection.

A special case of background can be considered the fluorescence radiation that originates when the energy of the incident X-rays is sufficient to eject electrons from inner shells of the atoms of the sample. This fluorescence background can be intense enough to obscure the diffraction effects and, if possible, should be avoided by a proper choice of the incident beam wavelength, that is, a wavelength as far removed as possible from the K or L absorption edges of the elements in the sample.[4] If this is impossible, the use of a monochromator placed on the beam diffracted by the sample is advisable. In fact, generally, fluorescence wavelengths are not able to pass through the monochromator and are eliminated.

If fluorescence cannot be removed, an empirical correction can be performed. Although an estimation of fluorescence intensity is impossible, it is accepted that fluorescence background is uniformly distributed over the entire angular range. Then, an arbitrary constant term can be subtracted from the experimental data, until the i(s) structure function (see Subsection C) is well aligned on the s axis. An example of such a correction is shown in Figure 7.[14]

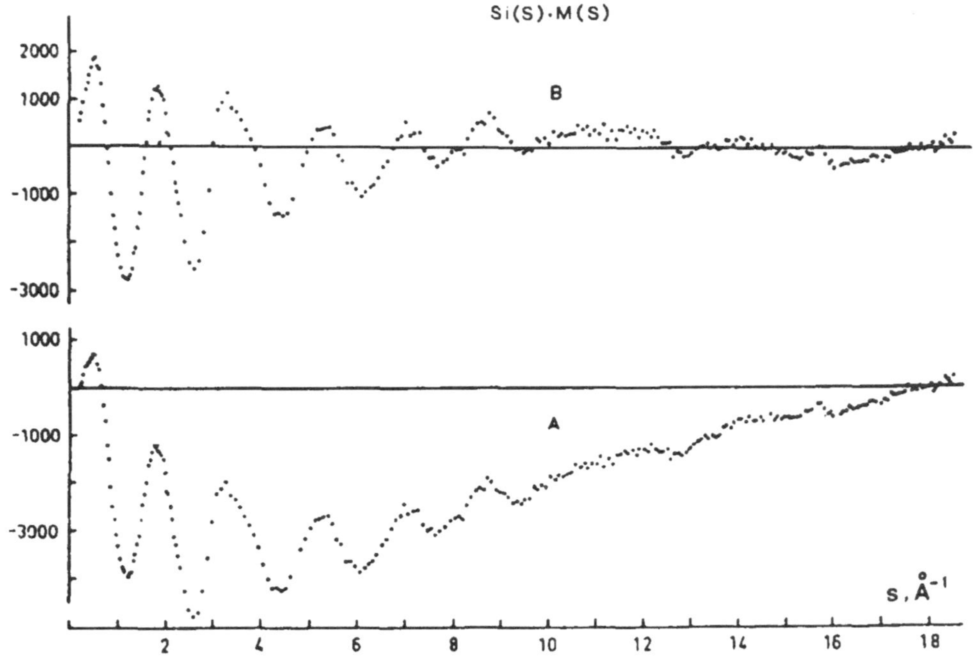

FIGURE 7. Structure functions, in the si(s)M(s) form, obtained for a uranyl(VI) chloride solution. Curve A gives the uncorrected si(s)M(s) function and curve B, the same function after fluorescence correction. (From Aberg, M., *Acta Chem. Scand.*, 24, 2901, 1970. With permission.)

b. Air Scattering

Either in reflection or in transmission geometry the detector receives the scattering from a volume of air in the vicinity of the sample. The experimental elimination of air scattering can be carried out by either sealing the liquid sample in a container and evacuating the diffraction camera, or substituting air with H or He gas. However, these procedures are not usually convenient and a geometrical correction must be applied to the experimental collected intensities.

Ergun[15] has proposed a treatment of the air scattering, which is valid for either transmission or reflection geometry. In Figure 8, the cross section of the air volume seen by the detector is indicated by the quadrilater CDEF; α and β are the angles subtended at the specimen by the X-ray tube and detector slits. The insertion of sample of thickness t reduces the volume occupied by the air and attenuates part of intensity scattered. Ergun has calculated the ratios of the air scattered intensity with and without the sample for the reflection (a_R) and for the transmission (a_T) geometry:

$$a_R = 1/2 + (1/2 - t\cos\vartheta/r\beta)\exp(-2\mu t/\sin\vartheta) \quad (69)$$

$$a_T = (1 - t\sin\vartheta/r\beta)\exp(-\mu t/\cos\vartheta) \quad (70)$$

where r is the radius of the goniometer and μ is the absorption coefficient of the sample. Therefore, the air scattering under working conditions can be obtained through a measure of the air scattering without the sample; then, Equations 69 or 70 give the contribution to be subtracted from the experimental data for each slit width. Usually this correction has some importance only at very small scattering angles.

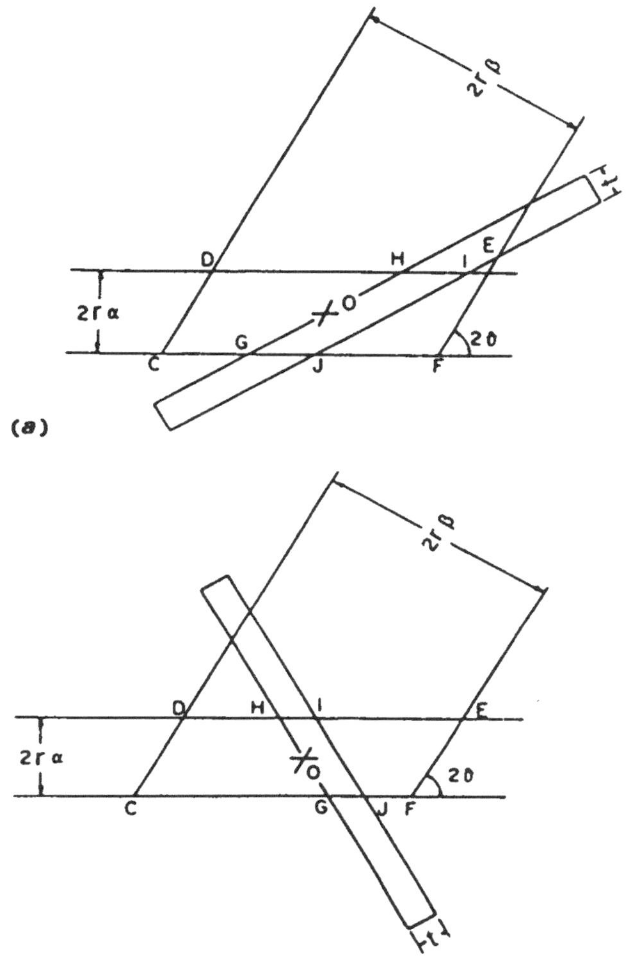

FIGURE 8. Cross section of the scattering air volume (DECF) in reflection (a) and transmission (b) geometry. α and β are the equatorial angles subtended at specimen by the tube and detector slits, respectively.

2. Absorption

In any actual diffraction experiment, the incident and diffracted X-rays are partially absorbed by the atoms of the specimen. Then, the theoretical value of intensity scattered from the sample is reduced by the absorption effect, and, if the reduction is angular dependent, a correction must be introduced. The absorption of X-rays depends critically on the diffraction geometry, so that various equations have been given for the different cases (flat sample in transmission or reflection, cylindrical sample, flat or cylindrical sample enclosed in a container, and so on). We will limit the treatment to the essential formulas for the two most common techniques.

a. Reflection Geometry

Milberg[16] has described the general expressions used for the symmetrical reflection geometry, with the only approximation that the beam irradiating the sample is parallel. By referring to Figure 9, and indicating by S and S + 2s, the cross sections of the X-ray incident and diffracted beams, respectively, it can be shown that the angular dependent part

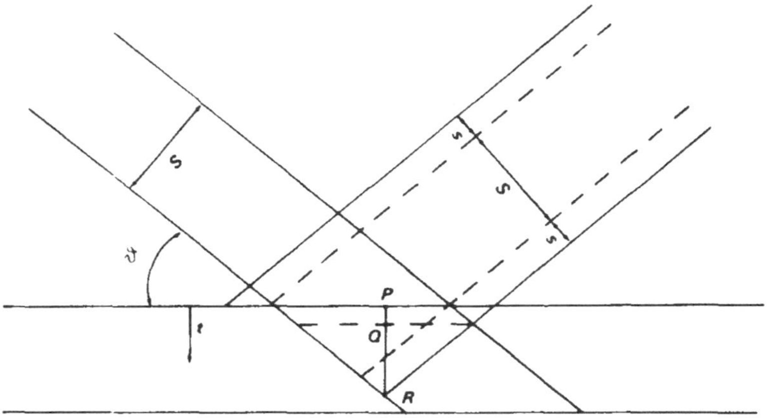

FIGURE 9. Schematic representation of different scattering volumes as a function of the sample thickness t. The thicknesses $\overline{PQ}$ and $\overline{PR}$ define the limits of validity of the equations derived by Milberg for the absorption corrections (see the text).

of the absorption coefficient is

$$ABS(\vartheta) = 1 - \exp(-2\mu t/\sin\vartheta)$$

$$0 < t < \mathbf{PQ} \tag{71}$$

$$ABS(\vartheta) = 1 - \exp(-2\mu t/\sin\vartheta)$$

$$+ [(2t\cos\vartheta - s)/S + 1/\alpha] \exp(-2\mu t/\sin\vartheta) - \exp(-\beta)/\alpha$$

$$\mathbf{PQ} < t < \mathbf{PR} \tag{72}$$

$$ABS(\vartheta) = 1 - [1 - \exp(-\alpha)] \exp(-\beta)/\alpha$$

$$t > \mathbf{PR} \tag{73}$$

where t is the thickness of the sample and $\alpha = 2\mu S/\sin(2\vartheta)$, $\beta = 2\mu s/\sin(2\vartheta)$, and μ is the linear absorption coefficient of the sample (μ is calculated from the tabulated absorption coefficients of the elements and from the composition of the sample).

When μ is high, all the expressions give an absorption correction independent of the scattering angle (that can be demonstrated proportional to $1/(2\mu)$). The angular independence of the absorption factor is one of the most important benefits of the reflection technique; in practice, this means that no correction at all is needed on the experimental data. An absorption coefficient μ of about 10 cm^{-1} is sufficient to fulfill the requirement for a constant correction, and this value is easily reached with concentrated salt aqueous solutions.

The absorption correction must be often calculated through Equation 73. The α and β depend on S and s and, as a consequence, on the slit widths used. In this case, if the scattering data have been collected with combinations of different slits, the absorption corrections must be applied separately on scattering data for each slit width before scaling to a unique slit width.

b. Transmission Geometry

For symmetrical transmission, the following correction is to be applied,[4] which is independent of the thickness of the sample:

$$ABS(\vartheta) = \sec\vartheta/\exp[-\mu t(1 - \sec\vartheta)] \quad (74)$$

The thickness t should be chosen in order to give the maximum diffracted intensity.

Finally, for transmission with the direct beam, normally incident upon the flat specimen, we have

$$ABS(\vartheta) = \{\exp[\mu t(1 - \sec 2\vartheta)] - 1\}/\{\mu t(1 - \sec 2\vartheta)\} \quad (75)$$

Whatever the geometrical arrangement used, the absorption correction is performed by dividing the diffracted intensities by the proper $ABS(\vartheta)$.

3. Polarization

The characteristic radiation from any target element is practically unpolarized. After diffraction (from the sample or from the monochromator), the radiation becomes polarized so that the scattered intensity, collected at any angle ϑ, is lower than the hypothetical intensity which would be collected in absence of polarization.

The general expression for the polarization factor is given by

$$P(\vartheta) = (1 + k\cos^2 2\vartheta)/(1 + k') \quad (76)$$

Depending on the instrumental apparatus, Equation 76 takes three different forms: (1) without monochromator, $k = k' = 1$,

$$P(\vartheta) = (1 + \cos^2 2\vartheta)/2 \quad (77)$$

(2) with the monochromator placed before the sample, $k = k' = \cos^2 2\alpha$,

$$P(\vartheta) = (1 + \cos^2 2\alpha \cos^2 2\vartheta)/(1 + \cos^2 2\alpha) \quad (78)$$

where α is the Bragg angle at the monochromator planes; and (3) with the monochromator placed after the sample, $k = \cos^2 2\alpha$, $k' = 1$,

$$P(\vartheta) = (1 + \cos^2 2\alpha \cos^2 2\vartheta)/2 \quad (79)$$

The typical shape of the polarization correction factor is shown in Figure 10. The experimental diffracted intensity must be divided by the $P(\vartheta)$ values properly chosen depending on the experimental arrangement used.

B. Evaluation of the Independent Scattering

After the above corrections have been applied to the total intensity $I_{exp}(\vartheta)$ collected by the detector, a corrected intensity is obtained according to

$$\begin{aligned} I_{corr}(\vartheta) &= [I_{exp}(\vartheta) - \text{Back.} - \text{Air Scatt.}]/[ABS(\vartheta)\,P(\vartheta)] \\ &= I_T(\vartheta)/[ABS(\vartheta)\,P(\vartheta)] \quad (80) \end{aligned}$$

where I_T represents the total intensity scattered by the sample. Both I_T and I_{corr} are expressed in arbitrary units. Conversion to the desired absolute units (electron units per composition unit) requires multiplication by a scale constant (see Subsection C), yielding

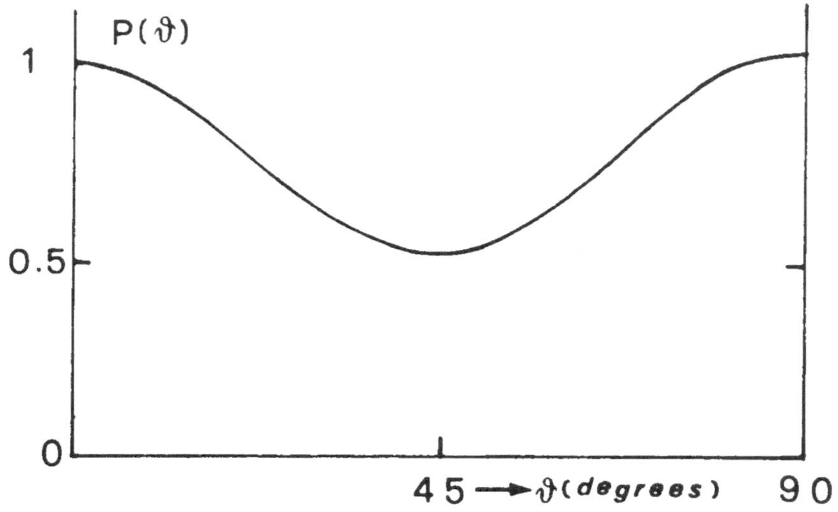

FIGURE 10. Polarization factor as a function of ϑ angle.

$$I_{eu}(\vartheta) = I_{corr}(\vartheta)\,K = I_T(\vartheta)\,\alpha(\vartheta) \tag{81}$$

where $\alpha(\vartheta) = K/[ABS(\vartheta)P(\vartheta)]$. In turn, I_{eu} is composed of different contributions:

$$\begin{aligned} I_{eu}(s) &= I(s) + I_{inc}(s) + I_{mult}(s) \\ &= i(s) + I_{coh}(s) + I_{inc}(s) + I_{mult}(s) \end{aligned} \tag{82}$$

where now the angular variable has been substituted with $s = 4\pi\sin\vartheta/\lambda$. I(s) is the elastic scattering discussed in Section III, $I_{inc}(s)$ is the **incoherent contribution** due to **Compton scattering,** and $I_{mult}(s)$ is a possible contribution from **multiple scattering** processes. Note that the intensity I(s) can be written in terms of total intensity scattered by the sample, $I_T(s)$, according to

$$I(s) = I_T(s)\,\alpha(s) - \beta(s) \tag{83}$$

which represents a quite general formulation, where the **calibration parameters** α and β are defined according to the nature of the scattered radiation and to the experimental setup. The splitting of I(s) in i(s) and $I_{coh}(s)$ has been discussed before: i(s) is the structure function which contains the structural information we are looking for, $I_{coh}(s)$ is given by $\Sigma\, x_i f_i^2(s)$. The terms I_{coh} and I_{inc} are also called **independent scattering** because they are structure independent terms, representing the scattering which should occur if all atoms scatter independently.

According to Equations 81 and 82, the independent scattering must be evaluated in order to convert intensities to absolute units and to obtain the structure function. The coherent independent scattering is calculated according to the composition coefficients and to the scattering amplitudes of the species in the sample. The incoherent contribution, or Compton scattering, is an inelastic effect due to the absorption of a small amount of the energy of the incident photon by a loosely bound or free electron in the sample. As a consequence, the scattered photon has a wavelength slightly longer than the incident one; the increase is a function of the angle 2ϑ, between the primary beam and the direction of scattering, according to[4]

$$\Delta\lambda = \lambda' - \lambda = (2h/mc)\sin^2\vartheta = 0.024(1 - \cos 2\vartheta) \quad (84)$$

where 2h/mc = 0.04852 Å (h, m, and c are the Plank's constant, the mass of the electron and the speed of light respectively).

When the monochromator is placed on the incident beam, the incoherent scattering is simply given by $\Sigma\, x_i I_{inc,i}(s)$; values of Compton scattering of the various atoms can be found in several tables. These values must be corrected for the so-called, **Breit-Dirac recoil factor:**

$$(\lambda'/\lambda)^n = [1 + (2h\lambda/mc)(\sin\vartheta/\lambda)^2]^n \quad (85)$$

According to Ergun,[15] the exponent n should be 3 or 2, depending on the device used in the measurements, that is, ionization chambers or counters.

1. Incoherent Scattering with the Monochromator on the Diffracted Beam

When the monochromator is placed on the diffracted beam, a correction must be applied to the $I_{inc}(s)$ values from the tables. Due to $\Delta\lambda$ dependence on the scattering angle, and the presence of a monochromator on the diffracted beam, a great part of the total incoherent radiation produced by the characteristic $K\alpha$ line used is cut off. Precisely, at small ϑ values, $\Delta\lambda$ is very small and the incoherent radiation reaching the detection system is just that produced by the $K\alpha$ line selected by the monochromator. On the contrary, at large ϑ values, the incoherent radiation admitted by the monochromator pass-band is produced by the continuous X-ray spectrum, which has much lower intensity than the $K\alpha$ line (see point P in Figure 11); the incoherent contribution at high angles is therefore drastically reduced.

The experimental determination of the incoherent radiation actually reaching the detecting system is difficult and would require the exact knowledge of the resolving power of the monochromator and of the spectrum of the X-ray tube in the working conditions. However, the quality of the results is not seriously affected by small uncertainties in knowledge of the true Compton radiation, and a semiempirical method proposed by Levy et al.[17] has given satisfactory $I_{inc}(s)$ functions in many cases. The method consists in the experimental determination of the incoherent radiation at high angles and in connecting, by an S-shaped curve, the determined values with the theoretical incoherent values at low angles. The correct S-shape is fixed, searching for the alignment of the structure function on the s axis (see Subsection C). Dividing the incoherent radiation values obtained in the experimental determination by the theoretical values, the so-called **monochromator discrimination function,** Mon(s), is obtained. Mon(s) can then be used to calculate actual incoherent intensities in all the practical cases as $\Sigma x_i I_{inc,i}(s)\,\mathrm{Mon}(s)$. This product is then the correct expression of $I_{inc}(s)$ to be used in Equations 82 and 83.

The determination of the incoherent radiation at high angle is performed by a filter measurement technique, properly choosing the filter according to the radiation employed; for instance, Zirconium filter must be used with $MoK\alpha$ radiation. In fact, a measurement of the diffracted intensity at a given (high) angle, with a Zr filter placed between the X-ray source and the sample, gives only the coherent scattering (the source for Compton scattering being eliminated by the filter, see point P in Figure 11). A second measurement, taken at the same angle, with the Zr filter placed between the sample and the monochromator, gives the sum of coherent and incoherent radiation passing through the monochromator pass-band. Then:

$$I^1_{Zr}/I^2_{Zr} = I_{coh}(s)/[I_{coh}(s) + I_{inc}(s)\,\mathrm{Mon}(s)] \quad (86)$$

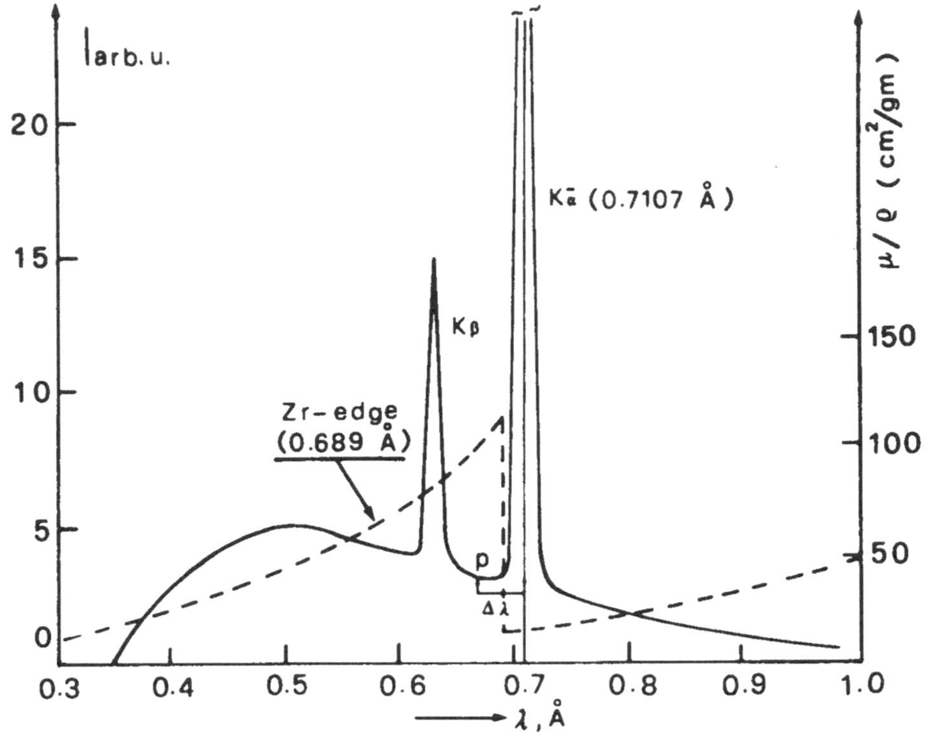

FIGURE 11. Characteristic X-ray spectrum of a molybdenum target. The dashed line indicates the position of the zirconium K-edge.

Expressing $I_{coh}(s)$ and $I_{inc}(s)$ by the tabulated coherent and incoherent scattering factors, Mon(s) can be determined. Typical Mon(s) functions are shown in Figure 12.

Total elimination of the incoherent radiation can be obtained using an experimental arrangement proposed by Warren and Mozzi.[18] The fluorescence radiation of a suitable element induced by the radiation scattered from the sample is measured. Since only coherent scattered radiation from the sample can excite the fluorescence of the element chosen, the final recorded data are completely free from Compton scattering. This method has been rarely used because of its experimental difficulties, which are not accompanied by decisive benefits.

C. Scaling to Absolute Units: The Structure Function

1. Scaling Methods

To scale the intensity values expressed in arbitrary units to absolute units in Equation 81, the scaling constant K must be determined. The absence of multiple scattering will be initially assumed.

a. High-Angle and Integral Methods

In the high-angle method, the experimental intensities in arbitrary units are matched to the independent scattering in electron units per composition unit. In fact, in this angular region, oscillations of the structure-dependent i(s) functions are vanishing, therefore, from Equations 81 and 82, for each point i in this region, a factor K_i can be evaluated as:

$$K_i = [I_{coh}(s) + I_{inc}(s)]/I_{corr}(s) \qquad (87)$$

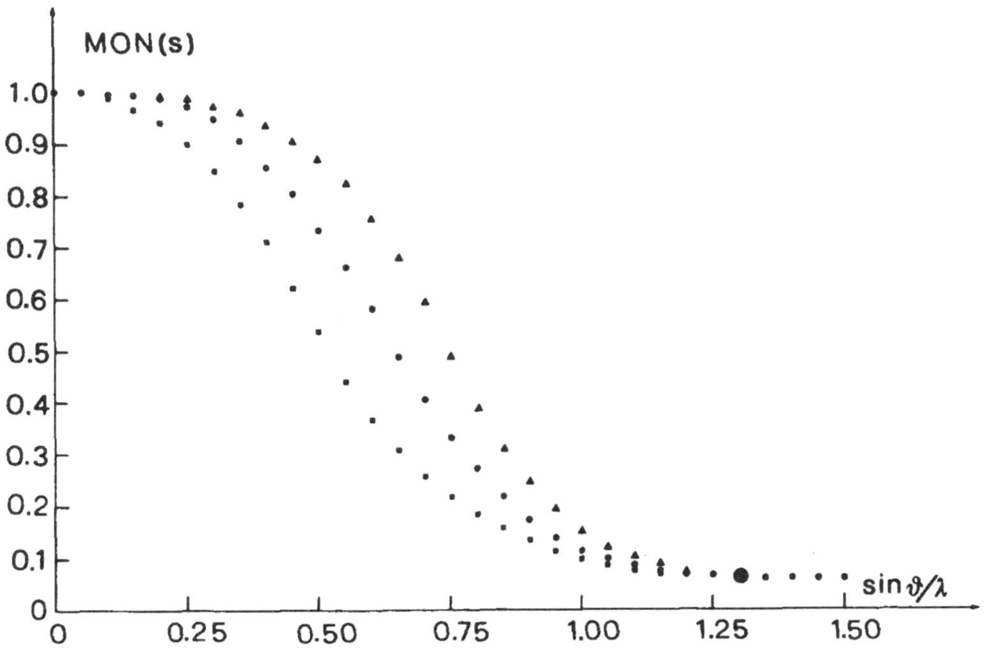

FIGURE 12. Typical monochromator discrimination functions, Mon(s).

and the scaling constant K is given by the average value over the n points included in the chosen scaling region:

$$K = (\Sigma K_i)/n \tag{88}$$

In the integral method, proposed by Norman[19] and Krogh-Moe,[20] a relationship between density, value of the radial function at r = 0, and diffracted intensity is exploited to calculate the constant K, using the complete experimental scattering curve, according to:

$$K = \frac{\int_{s_{min}}^{s_{max}} s^2M(s)[I_{coh}(s) + I_{inc}(s)]\, ds - 2\pi^2\rho_0'}{\int_{s_{min}}^{s_{max}} s^2M(s)\, I_{corr}(s)\, ds} \tag{89}$$

where M(s) and ρ'_0 are the quantities defined in Section III. When a monochromator on the diffracted beam is used, $I_{inc}(s)$ in the Equations 87 and 89 is given by $\Sigma x_i I_{inc,i}(s)Mon(s)$, and a correct choice of Mon(s), especially at high angles, is required for a good normalization.

It is a common practice to use both scaling methods for comparative purposes. When a good monochromator function has been introduced, the two methods give coincident scaling factors. Once the scaling factor has been obtained, the structure function i(s) can be calculated from Equations 81 and 82. Figure 13 shows the scaling procedure for the experimental data of Figure 6; the i(s) function from the same data is given in Figure 14.

b. The Habenschuss-Spedding Method

Recently, Habenschuss and Spedding[21] have proposed a scaling method which does not require the knowledge of the monochromator discrimination function Mon(s). In their notation the corrected intensity $I_{corr}(s)$ can be written as:

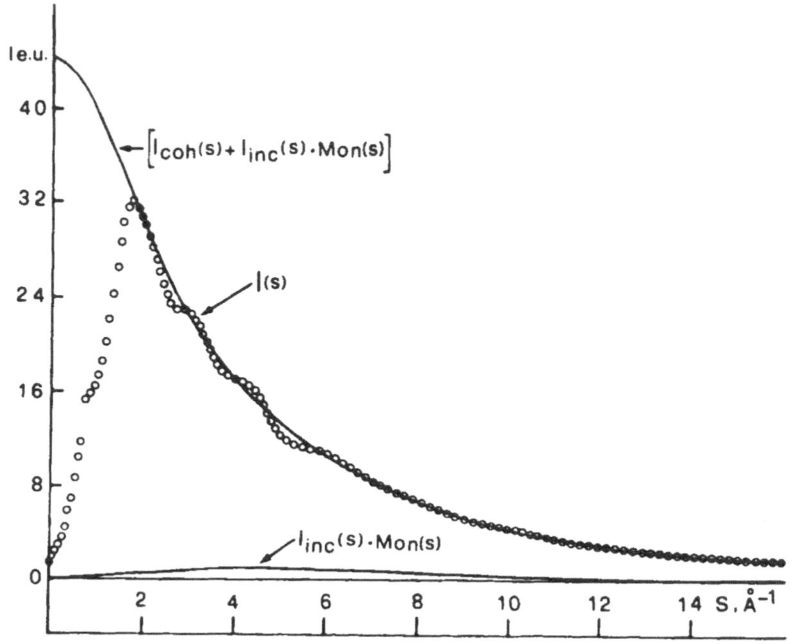

FIGURE 13. The 4 *M* $NiBr_2$ experimental data, given in Figure 6, are scaled to absolute unit by comparison with the total independent scattering $I_{coh}(s) + I_{inc}(s)Mon(s)$.

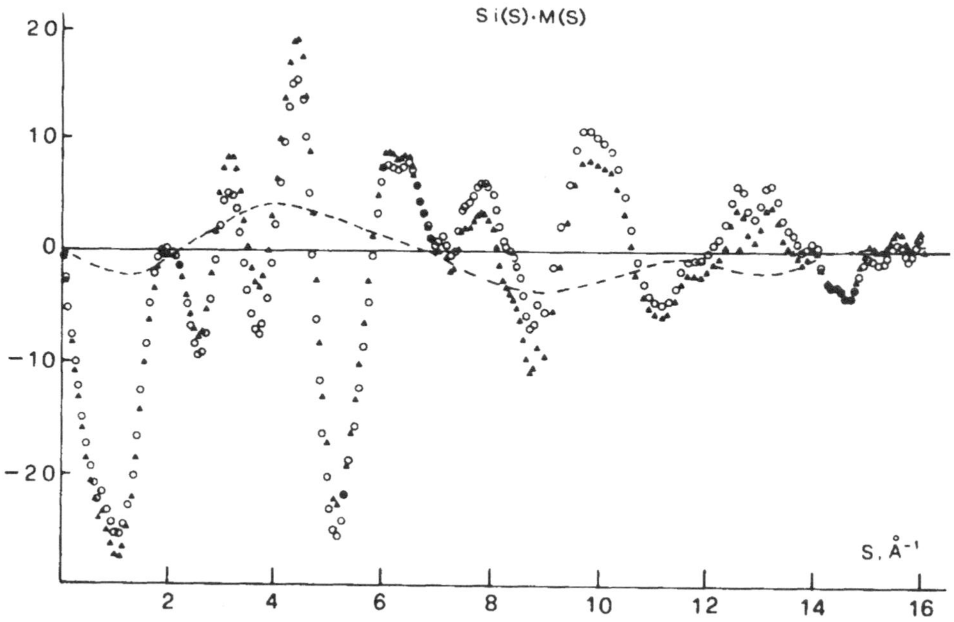

FIGURE 14. 4 *M* $NiBr_2$ structure functions, in the si(s)M(s) form, $M(s) = \exp(-0.005s^2)[\Sigma\chi_if_i(0)]^2/[\Sigma\chi_if_i(s)]^2$, before (open dots) and after (triangles) the correction procedure for systematic errors (see the text). The dashed line represents the correction term; note its long period angle dependence.

$$I_{corr}(s) \simeq [I_{coh}(s) + I_{inc}(s)][A + \Sigma B_j exp(-C_j s^{D_j})] \quad (90)$$

where $I_{inc} = \Sigma x_i I_{inc,i}$, and the sign $\simeq$ means that only when the oscillations of the structure function i(s) vanish, $I_{corr}(s)$ approaches the value of the right side. In this formulation, the monochromator function is absorbed into the empirical smoothed term E(s):

$$E(s) = A + \Sigma B_j exp(-C_j s^{D_j}) \quad (91)$$

where A, B_j, C_j, and D_j, are empirical parameters.

These parameters can be obtained by usual least squares procedures. Equation 90 can be rewritten as

$$F(s) = I_{corr}/[I_{coh}(s) + I_{inc}(s)] \simeq E(S) \quad (92)$$

and the smooth E(s) function is fitted to the experimental F(s) function to obtain the parameters A, B_j, C_j, and D_j, by which, in turn, the scaling constant is obtained. In fact, when s tends toward zero, E(s) tends to $A + \Sigma B_j$, which is the inverse of the scaling constant, $K = (A + \Sigma B_j)^{-1}$. We have therefore:

$$KI_{corr}(s) = K[I_{coh}(s) + I_{inc}(s)]\ E(s) + i(s) \quad (93)$$

where the i(s) function introduced in Equation 93 restores the equal mark in Equation 90. It finally follows that

$$i(s) = K\{I_{corr}(s) - [I_{coh}(s) + I_{inc}(s)]\ E(s)\} \quad (94)$$

If the method is correctly applied, the value of K comes out very similar to the ones obtained by the high angle and integral methods.

2. Multiple Scattering Correction

X-ray photons which reach the detector coming from the sample can also be due to a rescattering of photons previously scattered within the sample itself. Then, the experimental intensity may contain contributions from multiple scattering. Unless samples with a very low absorption coefficient are examined (e.g., pure water), the correction for this effect is small and is normally neglected. In any case, it is sufficient to consider only the doubly scattered intensity.

Even when double scattering correction is to be applied, initially the data treatment proceeds as if $I_{mult}(s)$ in Equation 82 in zero, obtaining an approximate $I_{eu}(s)$ function. Warren and Mozzi have proposed an expression of the ratio of the second-order diffracted intensity to the first-order intensity, I_2/I_1.[22,23] This ratio is multiplied by the approximate $I_{eu}(s)$ and a preliminary value of I_2 is obtained; I_2 is then converted to arbitrary units through the scaling constant and a first correction is applied to the experimental values. The new data can be rescaled and the whole procedure iterated until no changes occur in the correction term.

D. Radial Distribution Function

1. Fourier Transform

The structure function i(s) is now Fourier transformed to give the radial distribution function

$$D(r) = 4\pi r^2 \rho_0' + (2r/\pi) \int_{s_{min}}^{s_{max}} si(s)\ M(s) \sin(rs)\ ds \quad (95)$$

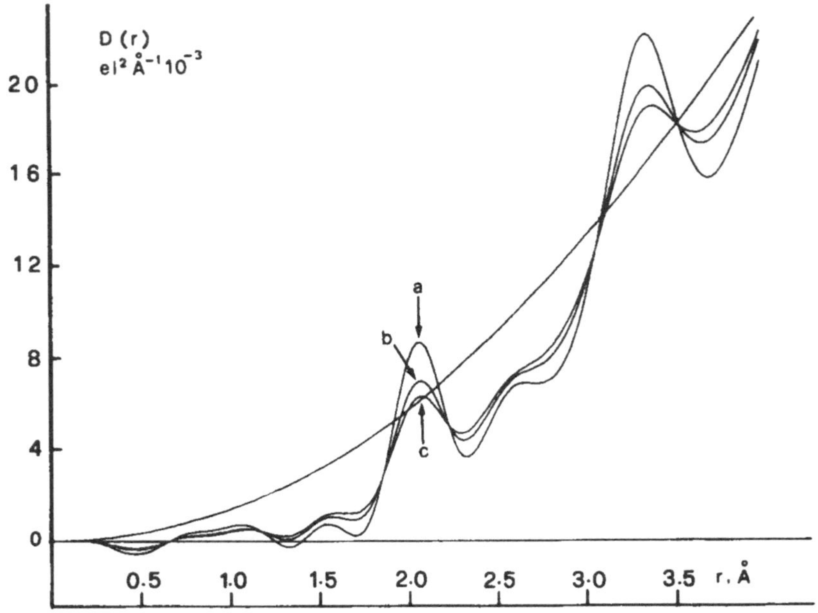

FIGURE 15. 4 *M* $NiBr_2$ radial distribution functions, D(r), obtained by Fourier transform of the si(s) function modified with three different M(s):

a) $M(s) = \exp(-0.005s^2)[\Sigma\chi_i f_i(0)]^2/[\Sigma\chi_i f_i(s)]^2$
b) $M(s) = \exp(-0.005s^2)[\Sigma\chi_i f_i^2(0)]/[\Sigma\chi_i f_i^2(s)]$
c) $M(s) = \exp(-0.005s^2)f_{Br}^2(0)/f_{Br}^2(s)$

In Section III, Equation 95 has been obtained and the quantities appearing in it (ρ'_0, M(s)) defined.

The numerical evaluation of the integral can be done with standard quadrature formulas. Care must be taken in adopting an integration step Δs small relative to the period in sin(rs) at the largest r values considered in the analysis; generally, $\Delta s = 0.05$ Å^{-1} is a suitable choice.

Since D(r) depends on the choice of the modification function M(s) and of the stoichiometric unit, full comparison of different results can be hindered by different choices in this regard. As examples, Figure 15 shows three D(r) functions obtained by Fourier transforming the i(s) function of Figure 14 with different formulas for the sharpening factor in the M(s) function, while Figure 16 shows the effect of differences in the convergence factor on the Fourier transform of the same i(s).

2. Sources of Errors

The D(r)'s coming from the processing of experimental data can be affected by errors, partly arising from inaccuracies in the function i(s) to be transformed (and hence from data collection and treatment), partly due to **termination errors,** because of the presence in the integral of Equation 95 of finite limits, s_{min} and s_{max}, in place of the ideal 0 and ∞ limits. The finite range of experimental data is unavoidable and is often the most serious trouble in diffractometric work.

a. Termination Errors

The experimentally accessible data start at $\vartheta = 1.0$ to 2.0 that is at $s_{min} = 0.3$ to 0.5 Å^{-1} for the typical MoKα. Thus, the so-called small angle scattering is excluded, and this

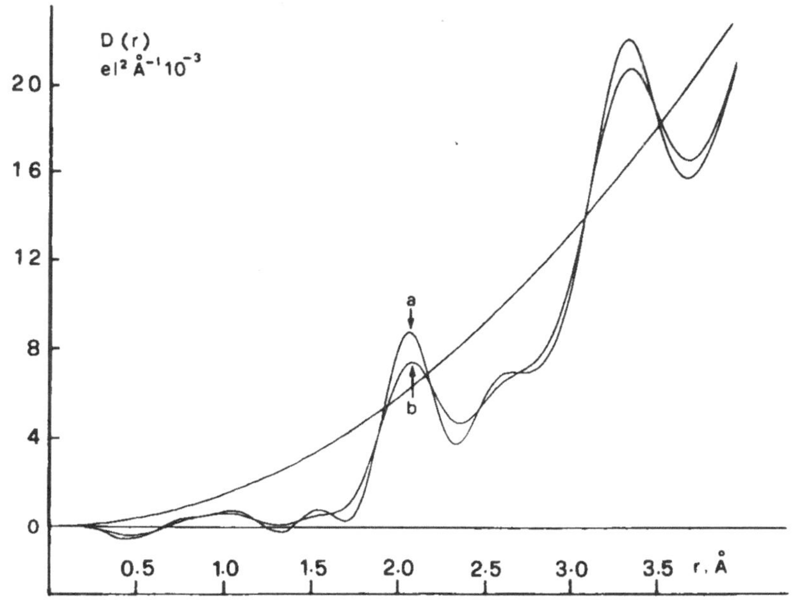

FIGURE 16. 4 *M* $NiBr_2$ radial distribution functions, D(r), obtained by Fourier transform of the si(s)M(s) function using different k values in the convergence factor of the modification function M(s): a) k = 0.005; b) k = 0.01.

produces long-period oscillations in the radial functions.[4] Furthermore, according to Cargill III,[24] the neglect of small angle scattering gives rise to radial functions which appears to correspond to a material of greater average atomic density than that of the sample being studied. These effects are not dramatic and, in addition, can be minimized through extrapolations of the data to s = 0. These extrapolations are often suggested by the trend of I_{exp} at the smallest angles examined or can be made according to a relationship between the zero angle limit of coherent scattering and isothermal compressibility of the system.[2] In any case, since multiplication by s forces the function to be integrated in Equation 95 to be null at s = 0, the uncertainty in $i(s)_{extrap}$ is almost negligible.

The truncation of the experimental data at $s = s_{max}$ is much more serious; because of this effect, spurious ripples appear on both sides of each major peak in D(r) and, moreover, the shape of real peaks, which become broadened, lowering s_{max}, is affected. Quantitative analyses of the details of radial functions can be compromised and, worst of all, a false peak could be mistaken for a true one. Therefore, different attempts for artificially extending the i(s) data beyond the upper experimental limits were performed, which did not result in any generally valid method. Extension of the measurements to very high s values is the only way to suppress termination ripples.

It is important to notice that the condition for termination error to be important is that modulations in the structure functions are still appreciable at s_{max}. This means that the same range of experimental data can be completely satisfactory in some cases and inadequate in other situations; examples of both eventualities for electrolyte aqueous solutions were given by Triolo et al.[25] As a consequence, tests of Fourier transform evaluation with variable s_{max} are ordinarily performed in order to identify spurious termination ripples. When truncation error is likely to be present, a damping factor may be introduced in the modification function M(s), which minimizes spurious ripples without excessive loss of resolution.

Figure 17 shows the effect of different termination values on the Fourier transform of the i(s) given in Figure 14. A useful expedient to depress spurious modulations is the choice of the s_{max} value near a null value of the structure function to be Fourier transformed.

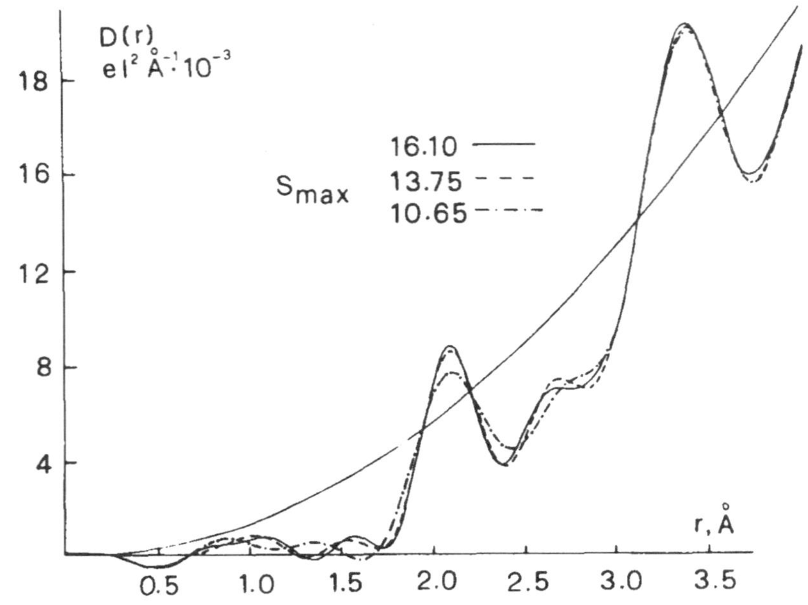

FIGURE 17. 4 *M* $NiBr_2$ radial distribution functions, D(r), obtained by Fourier transform of the si(s)M(s) function of Figure 14 (open dots) with three different s_{max} values.

b. Errors in the i(s) Function

Errors in the i(s) function can arise from inaccuracies in intensity counting and from approximations in data treatment.

When a counter is used to measure counting rates, the statistical errors, due to random fluctuations of X-ray tube emission, depend upon the total number of quanta counted. Precisely, if N events have been observed, the standard deviation from the true average value N is given by [4] $\sigma_N = N^{1/2}$, and the relative standard deviation is $\sigma_N^{rel} = N^{1/2}/N$. Then, the percentual uncertainty of the measured intensity is 1% when 10,000 counts are accumulated, and becomes definitely smaller than 1% at the counting rates usually adopted (e.g., ~ 0.3% accumulating 100,000 counts). At the low background level noticed in well-alligned diffractometers, this is practically all the statistical error. Thus, in modern equipments, for mere propagation of statistical error, i(s) functions [and D(r) functions] do not suffer by serious uncertainties.

In the step-by-step counting method, problems may arise from the long-period variations in the X-ray tube emission, or from the occurrence of some drift in the efficiency of electronic counting equipment. Long-period deformations can be then produced in i(s). Similar consequences can be produced by the following error sources:

- Misalignement of the diffractometer.
- Inadequacy of the formulas adopted for the preliminary corrections.
- Neglect of some correction, such as air scattering or multiple scattering.
- Imprecisions in the tabulated scattering factors.
- Uncertainty in the monochromator function Mon(s).
- Approximations in scaling constant evaluation.

Low frequency perturbations in i(s), contrary to termination errors, produce spurious oscillations in the low r region of the radial functions which sometimes can be sharp and of large amplitude. In Figure 18, typical forms of the errors produced on the radial function

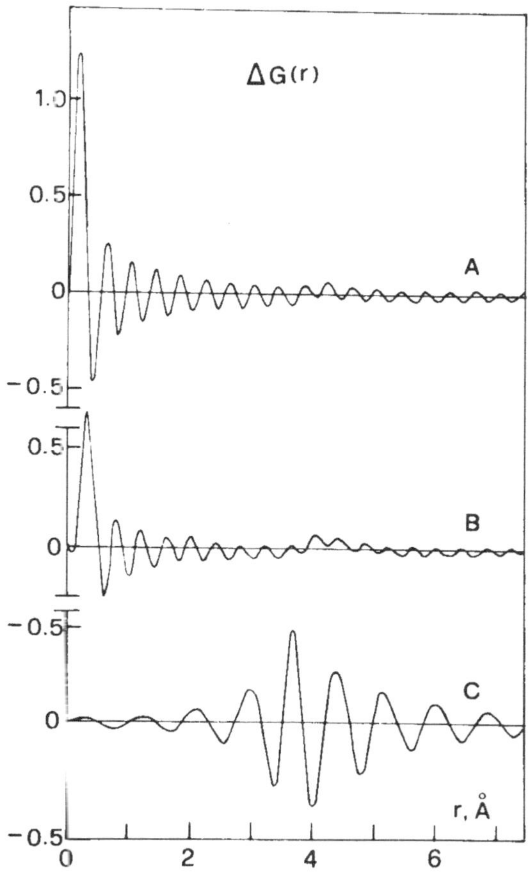

FIGURE 18. Typical spurious ripples in the radial function resulting from errors in the structure function: A) error in the scale factor; B) long period error, e.g., from incorrect alignment of the apparatus; C) termination errors. (From Kaplow, R., Strong, S. L., and Averbach, B. L., *Phys. Rev. A*, 138, 1336, 1965. With permission.)

by some of the above listed sources are reported and compared with the termination error. It is clear that the truncation can give rise to spurious ripples in the region of the real maxima, causing serious interpretative difficulties, while the other errors produce spurious ripples which, in general, do not superimpose on real peaks.

3. Correction of Low r Spurious Peaks

The previous discussion suggests that a valuable criterion for verifying the accuracy of a diffractometric experiment on liquid or amorphous substances is the absence of oscillations of radial curve in the low r region. When peaks from short intramolecular distances are present (e.g., N – O distances in nitrate ions), the accuracy of the experiment can be judged also from the occurrence of the peak at the correct known distance and with the amplitude corresponding to the known coordination.

When the presence of spurious peaks near r = 0 indicates some long period deformation of the structure function, a correction can be applied by a back-transforming (reinversion) procedure that imposes the D(r) to be correct at low r values. The correction term is defined as:

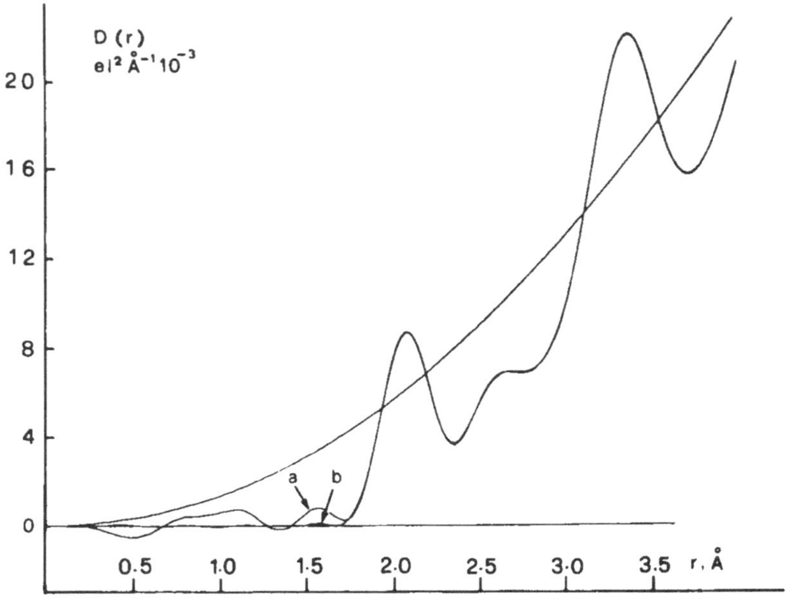

FIGURE 19. 4 *M* $NiBr_2$ radial distribution functions, D(r), obtained by using the structure functions of Figure 14: (a) before and (b) after the correction for systematic errors.

$$\Delta i(s) = M(s)^{-1} \int_0^R [D(r)_w - D(r)_{exp}] \sin(sr)/(sr)\, dr \tag{96}$$

$D(r)_w$ represents the distribution function "wanted" between 0 and R, which will be a null function as far as no real peak is present in this interval. Figure 14 shows the $\Delta i(s)$ term and its effect on the structure function; Figure 19 shows the D(r) function before and after the correction procedure.

It must be emphasized that the correction procedure described above has no effect on the physically meaningful peaks of the distribution function as far as the distortions present in the noncorrected i(s) are minimized by a correct diffractometer alignment and proper data corrections. The most positive result of a reinversion procedure consists in a good alignment on the s axis of the structure function and this, in turn, will allow a least square refinements of the i(s) function itself, as we will describe in the next chapter.

One of the most troublesome causes of i(s) distortion is the imprecise knowledge of the Mon(s) function in the region in which experimental I_{inc} data at high angle must be connected to theoretical incoherent intensities at low angle. Figure 12 shows a set of different Mon(s) functions, all passing through the measured high-angle value. It is easy to recognize the strong effect produced by different monochromator discrimination functions on the i(s) functions shown in Figure 20A. However, if we look at Figure 20B, we see that the Fourier transforms of so much different i(s)s produce D(r) curves that, apart from the spurious ripples region, practically coincide.

In conclusion, on one hand the physical peaks of the distribution curves are affected very little by the uncertainty due to the monochromator function or other minor sources of i(s) distortion; on the other hand these distortions can be removed without perturbation of the structural details of D(r).

E. Reproducibility of XRD Results

In view of the critical analysis of XRD results on aqueous solutions we intend to carry

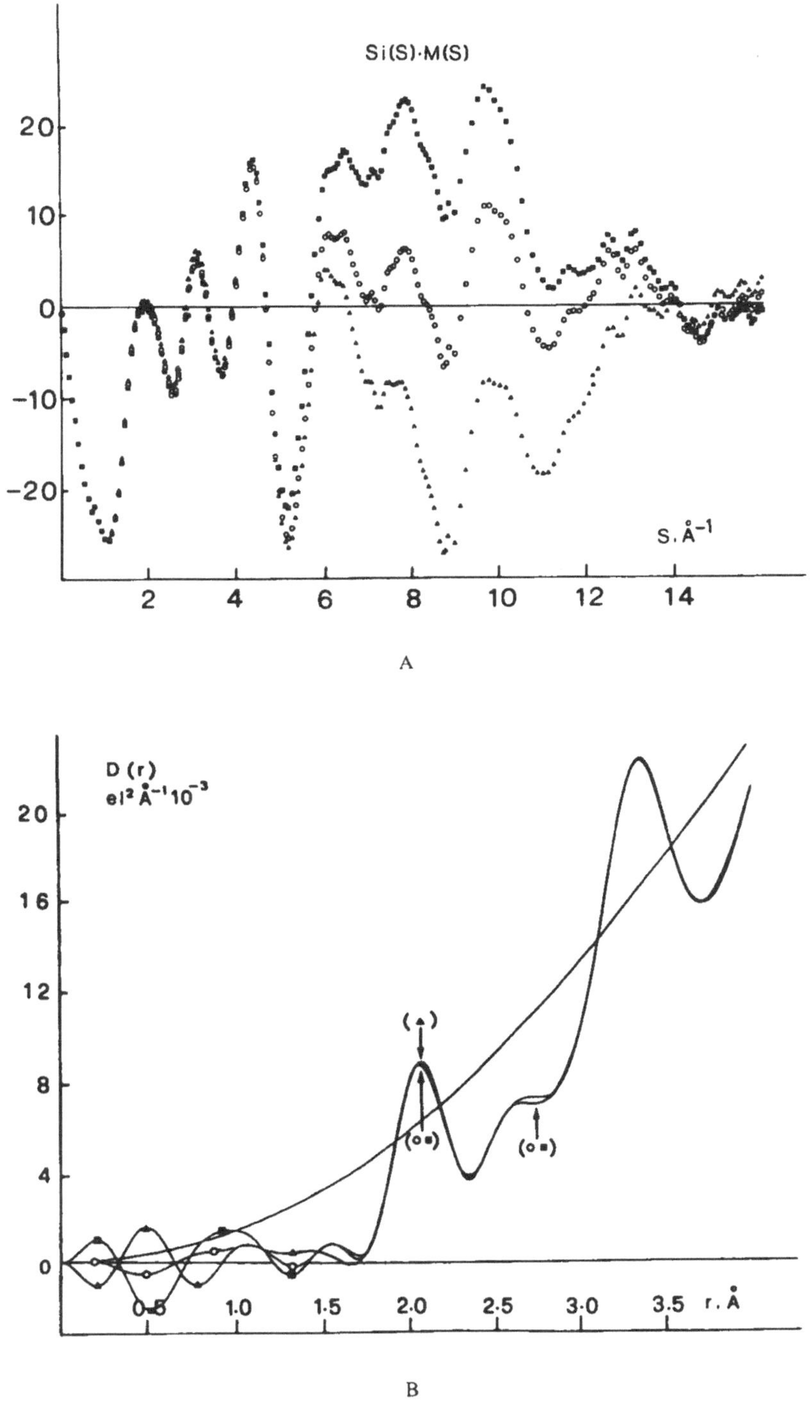

FIGURE 20. 4 *M* $NiBr_2$: (A) The structure functions, si(s)M(s), corresponding to the three Mon(s) functions shown in Figure 12. 4 *M* $NiBr_2$: (B) The radial distribution curves, D(r), obtained from the structure functions given in A.

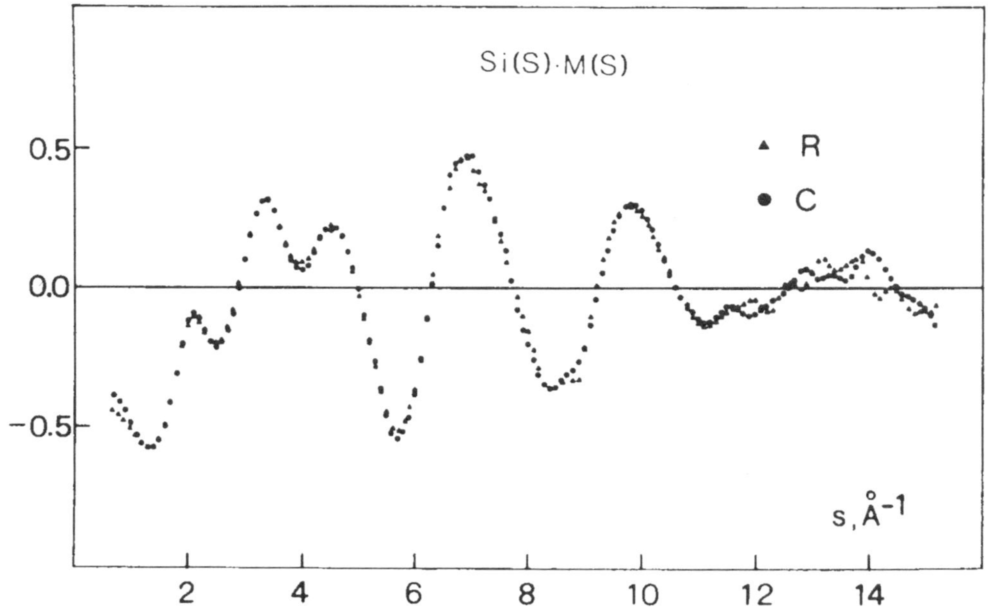

FIGURE 21. Experimental structure functions, si(s)M(s), obtained for a ≃ 3 *M* $Cu(NO_3)_2$ solution in two different reflection diffractometers: C (Cagliari data), R (Rome data). (From Licheri, G., Musinu, A., Paschina, G., Piccaluga, G., Pinna, G., and Sedda, A. F., *J. Chem. Phys.*, 80, 5308, 1984. With permission.)

out in the following chapters, it is essential to know the reproducibility of experimental results. We are actually interested in the reproducibility of the functions used in the structural interpretation, i.e., radial functions and, principally, structure functions; they are not the experimental data, but follow from them through the above described procedure and may suffer from errors due to improper data treatment.

If we look at all possible sources of error, we see that the greatest difference among the radial functions of a given system can be produced by the different range of experimental data explored. The other sources are less critical. Working with modern apparatus, the statistical error can be kept within reasonable limits, even if this requires long times of data collection. The other errors potentially produce distortion in the structure functions which do not affect dangerously the real peaks of D(r)s and can be easily removed. We should then expect that under the same termination of the Fourier integral, the conditions exist to obtain similar results for the same sample through different experiments, as far as real peaks of the radial curves or structure functions, after correction for low r spurious peaks, are compared.

Unfortunately, the possibilities of such a comparison are scarce, since the systems examined by more than one worker are few. Furthermore, the comparisons are hampered by the fact that different treatments can be equally correct, but, notwithstanding, can lead to different i(s) or D(r) functions, as a consequence of different choices in data processing (for example, in the definition of the composition unit, in the use of the modification functions, and so on).

Some of the literature examples of diffraction data, collected for identical (or very similar) solutions and processed using identical definitions and units, are reported below. The first case (Figure 21) refers to two i(s)s obtained for a 3 *M* solution of $Cu(NO_3)_2$, starting from data collected in two different diffractometers, using reflection geometry and treated exactly in the same way. Two other examples (Figures 22 and 23) regard one solution of $MgCl_2$ and one of $CaCl_2$, both 1 *M*, separately measured in a reflection diffractometer and in a

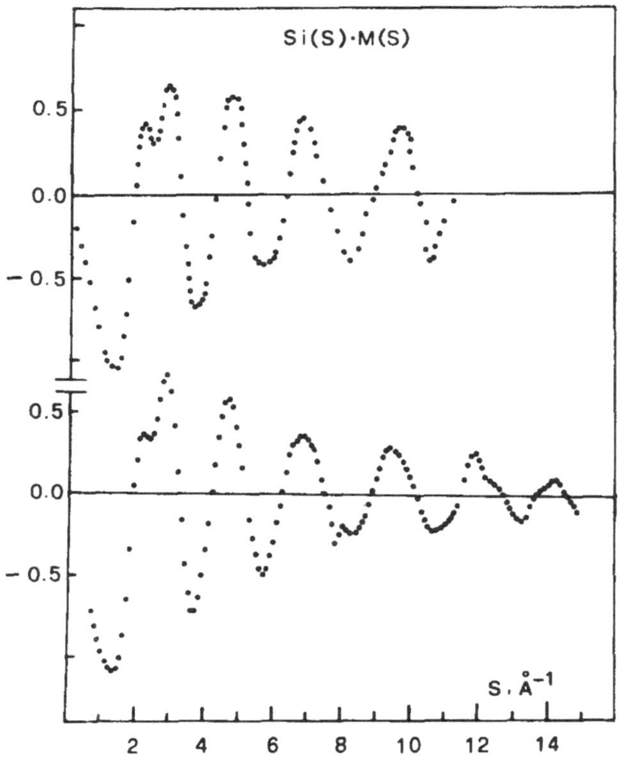

FIGURE 22. Experimental structure functions, si(s)M(s), obtained for two ≃ 1 *M* $MgCl_2$ solutions in a transmission diffractometer (top, Budapest data), and in a reflection diffractometer (bottom, Cagliari data). (Top data from Pálinkás, G., Radnai, T., Dietz, W., Szász, Gy, I., and Heinzinger, K., *Z. Naturforsch.*, 37a, 1049, 1982. With permission. Bottom data from Caminiti, R., Licheri, G., Piccaluga, G., and Pinna, G., *J. Appl. Cryst.*, 12, 34, 1979. With permission.)

transmission one. The agreement between the pairs of si(s)s appears good; this is particularly encouraging in the last two cases, since the structure functions result from apparatus using different geometries and experimental devices (for example, the monochromator is placed in the direct beam in one case, and in the diffracted beam in the other), and, as a consequence, they are obtained through treatments in which the various corrections had to be performed in different ways. Further examples will be given in Chapter 3.

These examples match our hope that modern measurement apparatus allow to gain reproducible experimental data. Obviously, the reproducibility will be judged more or less good according to the delicacy of the structure analysis to be performed. A more detailed discussion of this point is postponed to the specific cases discussed in Chapter 3. Here we may anticipate that, in the light of the previous conclusions, the disagreements in XRD studies results, we will talk about in Chapter 3, for the most part originate from differences in the interpretation of the experimental data (due to the difficulty of deciphering single scattering experiments on complex polyatomic systems), rather than from differences in the data.

Finally, when differences in the experimental data appear, criteria of evaluation of the experiment quality should be helpful. Unfortunately, the only general criterion seems to be the minimization or absence of spurious ripples at low r, as previously discussed. However, this criterion does not work on the differences affecting the real peaks, about which, prac-

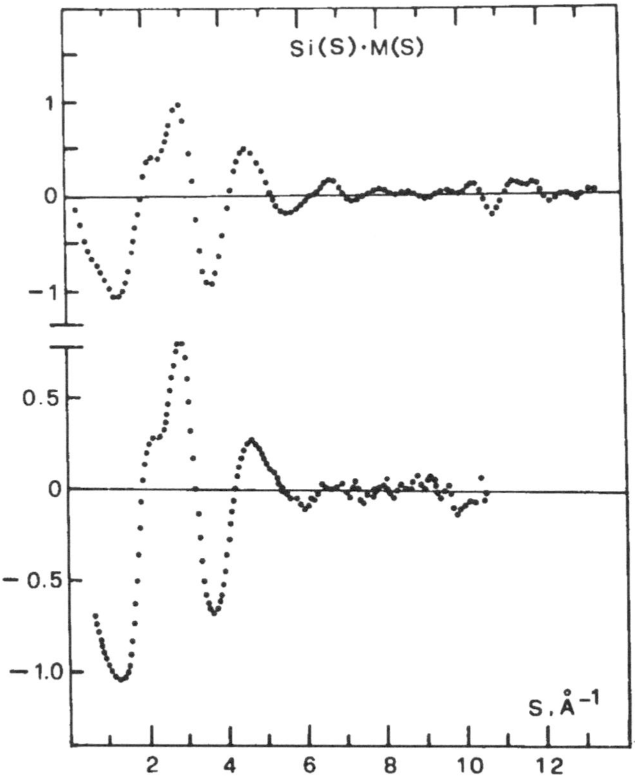

FIGURE 23. Experimental structure functions, si(s)M(s), obtained for two ≃ 1 *M* $CaCl_2$ solutions in a transmission diffractometer (top, Budapest data), and in a reflection diffractometer (bottom, Cagliari data). (Top data from Probst, M. M., Radnai, T., Heinzinger, K., Bopp, P., and Rode, B. M., *J. Phys. Chem.*, 89, 753, 1985. With permission. Bottom data from Licheri, G., Piccaluga, G., and Pinna, G., *J. Chem. Phys.*, 64, 2437, 1976. With permission.)

tically no *a priori* judgment is possible. In this case, some judgment about the correctness of the different studies will eventually come from consistency of the structural results with the overall chemical physical knowledge of the system investigated.

REFERENCES

1. **James, R. W.,** *The Optical Principles of the Diffraction of X-Rays,* Bell and Sons, London, 1962.
2. **Guinier, A.,** *X-Ray Diffraction,* Freeman, Cooper & Co., San Francisco, 1963.
3. **Warren, B. E.,** *X-Ray Diffraction,* Addison-Wesley, Reading, Mass., 1969.
4. **Klug, H. P. and Alexander, L. E.,** *X-Ray Diffraction Procedures For Polycrystalline and Amorphous Materials,* John Wiley & Sons, New York, 1974.
5. **Pings, C. J.,** Structure of simple liquids by X-ray diffraction, in *Physics of Simple Liquids,* Temperley, H. N. Y., Rowlinson, J. S., and Rushbrooke, G. S., Eds., North-Holland, Amsterdam, 1968, chap. 10.
6. **Wagner, C. N. J.,** Direct methods for the determination of atomic-scale structure of amorphous solids (X-ray, electron, and neutron scattering), in *Atomic Scale Structure of Amorphous Solids,* Cargill, G. S., III and Chaudhari, P., Eds., North-Holland, Amsterdam, 1979.
7. **Waseda, Y.,** *The Structure of Non-Crystalline Materials,* McGraw-Hill, New York, 1980.

8. **Hill, F. L.,** *Statistical Mechanics,* McGraw-Hill, New York, 1956.
9. **Kaspar, John S. and Lonsdale, K., Eds.,** *International Tables of X-ray Crystallography,* Vol. 3, Kynoch, Birmingham, 1962.
10. **Sneddon, I. N.,** *Fourier Transform,* McGraw-Hill, New York, 1951.
11. **Waser, J. and Schomaker, V.,** The Fourier inversion of diffraction data, *Rev. Mod. Phys.,* 25, 671, 1953.
12. **Warren, B. E., Krutter, H., and Morningstar, O.,** Fourier analysis of X-ray patterns of vitreous SiO_2 and B_2O_3, *J. Am. Ceramic Soc.,* 19, 202, 1936.
13. **Bacon, G.,** *Neutron Diffraction,* Clarendon, Oxford, 1975.
14. **Aberg, M.,** On the structure of the predominant Hydrolysis products of Uranil (VI) in solution, *Acta Chem. Scand.,* 24, 2901, 1970.
15. **Ergun, S.,** *Chemistry and Physics of Carbon,* Vol. 3, Welker, P. L., Jr., Ed., Marcel Dekker, New York, 1968, 211.
16. **Milberg, M. E.,** Transparency factor for weakly absorbing samples, *J. Appl. Phys.,* 29, 64, 1958.
17. **Levy, H. A., Danford, M. D., and Narten, A. H.,** Data collection and evaluation with an X-ray diffractometer designed for the study of liquid structures, ORNL 3960, July 1966.
18. **Warren, B. E. and Mozzi, R. L.,** Corrections for intensity measurements from glass samples, *J. Appl. Cryst.,* 3, 59, 1970.
19. **Norman, N.,** The Fourier transform method for normalizing intensities, *Acta Cryst.,* 10, 370, 1957.
20. **Krogh-Moe, J.,** A method for converting experimental X-ray intensities to an absolute scale, *Acta Cryst.,* 9, 951, 1956.
21. **Habenschuss, A. and Spedding, F. H.,** The coordination (hydration) of rare earth ions in aqueous chloride solutions from X-ray diffraction. I. $TbCl_3$, $DyCl_3$, $ErCl_3$, $TmCl_3$ and $LuCl_3$, *J. Chem. Phys.,* 70, 2797, 1979.
22. **Warren, B. E. and Mozzi, R. L.,** Multiple scattering of X-rays by amorphous samples, *Acta Cryst.,* 21, 459, 1966.
23. **Warren, B. E.,** *X-ray Diffraction,* Addison-Wesley, Reading, Mass., 1969, chap. 10.
24. **Cargill, G. S., III,** Influence of neglected small-angle scattering in radial distribution function analysis, *Acta Cryst.,* 4, 277, 1971.
25. **Triolo, R. and Ruffo, I.,** Termination errors in Fourier analysis of diffraction data of aqueous electrolyte solutions, *J. Chem. Soc. Faraday Trans. I,* 73, 186, 1977.
26. **Licheri, G., Musinu, A., Paschina, G., Piccaluga, G., Pinna, G., and Sedda, A. F.,** Coordination of Cu(II) in $Cu(NO_3)_2$ aqueous solutions, *J. Chem. Phys.,* 80, 5308, 1984.
27. **Pálinkás, G., Radnai, T., Dietz, W., Szász, Gy, I., and Heinzinger, K.,** Hydration shell structure in an $MgCl_2$ solution from X-ray and MD studies, *Z. Naturforsch.,* 37a, 1049, 1982.
28. **Caminiti, R., Licheri, G., Piccaluga, G., and Pinna, G.,** X-ray diffraction study of $MgCl_2$ aqueous solutions, *J. Appl. Cryst.,* 12, 34, 1979.
29. **Probst, M.M., Radnai, T., Heinzinger, K., Bopp, P., and Rode, B. M.,** Molecular dynamics and X-ray investigation of an aqueous $CaCl_2$ solution, *J. Phys. Chem.,* 89, 753, 1985.
30. **Licheri, G., Piccaluga, G., and Pinna, G.,** X-ray diffraction study of the average solute species in $CaCl_2$ aqueous solutions, *J. Chem. Phys.,* 64, 2437, 1976.
31. **Kaplow, R., Strong, S. L., and Averbach, B. L.,** Radial density functions for liquid mercury and lead, *Phys. Rev. A,* 138, 1336, 1965.

Chapter 2

INTERPRETATION OF THE EXPERIMENTAL DATA

I. INTRODUCTION

The principles of X-ray diffraction theory and the details of data treatment for noncrystalline materials are extensively discussed in Chapter 1; the measured diffraction pattern is corrected for various instrumental effects in order to obtain the structure function i(s) and, by Fourier transform, the total radial distribution function D(r) or the correlation function $G(r) = D(r)/4\pi r^2\rho_0'$. We recall here the main equations used in the following discussion and refer to Chapter 1 for the definition of the quantities appearing in them.

$$D(r) = 4\pi r^2\rho_0' + (2r/\pi)\int_0^{s_{max}} si(s)\, M(s)\sin(sr)\, ds \tag{1}$$

$$i(s) = \sum_{i=1}^{M}\sum_{j\geq i}^{M} W_{ij}(s)\, i_{ij}(s) \tag{2}$$

$$W_{ij}(s) = (2 - \delta_{ij})\, x_i x_j f_i(s)\, f_j(s) \tag{3}$$

$$i_{ij}(s) = (4\pi/sx_j)\int_0^{\infty} r[\rho_{ij}(r) - \rho_{oj}]\sin(sr)\, dr \tag{4}$$

$$D(r) = \sum_{i=1}^{M}\sum_{j\geq i}^{M} (2 - \delta_{ij})\, x_i D_{ij}(r) \tag{5}$$

$$D_{ij}(r) = 4\pi r\int_0^{\infty} u\rho_{ij}(u)\,[T_{ij}(r - u) - T_{ij}(r + u)]\, du \tag{6}$$

$$T_{ij}(r) = (1/\pi)\int_0^{s_{max}} f_i(s)\, f_j(s)\, M(s)\cos(rs)\, ds \tag{7}$$

Let us observe that both i(s) and D(r) are weighted sums of partial contributions, which determine the structuring between pairs of atomic species in a multicomponent system.

The analysis of experimental data can be performed either in s- or r-space by using i(s) or D(r), respectively; both contain the same structural information, although visualized in different forms. Strictly speaking, the choice of some **modification function** M(s) could affect the structural information in going from i(s) to D(r), but this effect is not to be mistaken with others which depend on inherent experimental limitations and consequently show themselves both in i(s) and D(r). For example, the **spurious details,** which affect the radial function because of **termination errors** in Equation 1, represent only the counterpart in r-space of the lack of experimental information out of the measured s range. Thus, it is merely a practical question to analyze diffraction data in r- or s-space.

The analysis in real space is more practical if we look for direct information about bond distances, because this information, which is spread through the whole s range, becomes concentrated around each distance by the effect of Fourier transform. On the contrary, working in s-space is more convenient when we are testing models against diffraction data, since, once $\rho_{ij}(r)$ has been modeled, it is simpler to obtain i_{ij} via the Fourier transform

(Equation 4) than D_{ij} via the convolution integral (Equation 6). In general, it is useful to examine both i(s) and D(r) representations, since structural aspects which may not appear in one function can produce significant effects in the other.

It is important to realize that i(s) and D(r) are only one-dimensional shadows of a three-dimensional arrangement of atoms and statistical averages of local structures over both time and volume of the sample. It means that, even in the most favorable case of a one-component system, measured without errors in the entire s range where i(s) is not zero, the information content is not sufficient to uniquely define the structure. Three-dimensional descriptions, involving, for example, coordination geometries, can only be inferred by interpreting features of the radial curves and/or can be assumed as models and showed to be consistent with experimental data.

For an M-component system, things become more and more complicated as M increases. In fact, the independent components of i(s) and D(r) increase as M(M + 1)/2, and, obviously, they cannot be disentangled using only one diffraction experiment. However, in spite of these limitations, the situation is not hopeless, and a good deal of experimental information about the structural behavior of ionic aqueous solutions comes from X-ray diffraction studies, as will be seen in the next chapter. In this chapter, we deal with some general aspects concerning the approaches which have been or could be used to infer structural information from the experimental functions which contain it.

II. DETECTABILITY OF PARTIAL CONTRIBUTIONS

The different M(M + 1)/2 terms present in Equation 2 enter with different weights in determining the total structure function. Each term contains two parts: the partial structure function i_{ij} which describes the structuring of the atomic pair i-j, and the weighting factor W_{ij} which depends on the concentrations and the nature of the atoms involved. It is apparent that i(s) is dominated by the prevalent and heavier species, so that the structural information about low concentration or light species is nearly invisible. This is why structural details involving hydrogen are not detectable in aqueous solutions, and the nearly spherical water molecule can be treated as a **quasi atom** described by its own scattering factor.[1] An ionic solution of an MX_n salt behaves towards X-ray diffraction essentially as a three-component system involving cation, anion, and water. This obviously produces a great simplification in interpreting the structure, but prevents from gaining direct information about the orientation of water molecules. Three species involve six partial structure functions; Figure 1 shows the s behavior of the normalized weighting factors $W_{ij}/(\Sigma W_{ij})$ for a typical concentrated solution. It is apparent that the weights of ion-ion contributions are significantly smaller than those coming from water-water and ion-water pairs. As a consequence, structural aspects involving ion-ion interactions become clearly detectable only when a strong structural component is present, as in the case of a substantial amount of cation-anion inner sphere complex formation. Contributions coming from less defined, longer range correlations usually do not emerge as specific, analyzable details.

The best way of extracting structural information would be to independently determine the different partial structure functions. As discussed in Chapter 1, this depends on the possibility of changing the weighting factors W_{ij} in Equation 2 without changing the set of i_{ij}. For this purpose, X-ray diffraction may exploit the anomalous scattering phenomenon, where the anomalous dispersion corrections are used as a way of changing the scattering factors.[2]

Recalling that an atomic scattering factor with dispersion corrections takes the form

$$f_i^o = f_i + \Delta f_i' + i\Delta f_i'' \tag{8}$$

the weighting factors in Equation 3 become

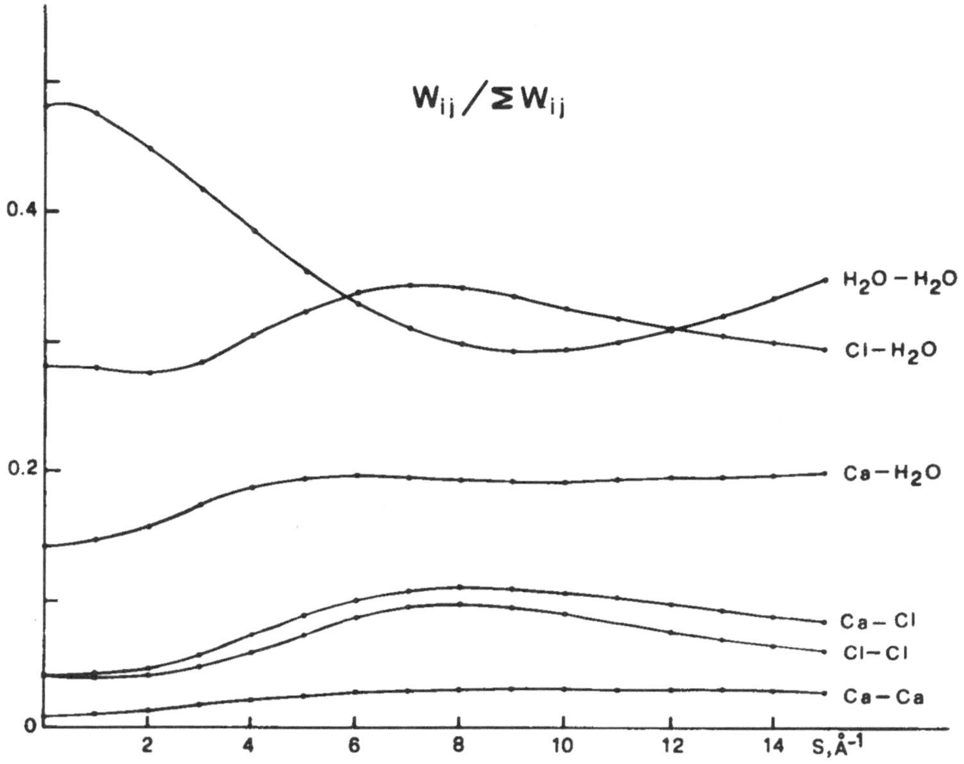

FIGURE 1. Weighting factors for 4 *M* $CaCl_2$ aqueous solution.

$$W_{ij} = (2 - \delta_{ij})\, x_i x_j [(f_i^0 + \Delta f_i')(f_j^0 + \Delta f_j') + \Delta f_i'' \Delta f_j''] \tag{9}$$

Since $\Delta f_i'$ and $\Delta f_i''$ change drastically when the energy of the incident radiation is near the absorption edges of the i species, significant changes in W_{ij} can be produced by appropriately choosing the wavelengths of the incident radiation to be used in different diffraction measurements.

Tunable X-rays are now available from synchrotron radiation sources, which allow the choice of the more appropriate wavelengths and the optimization of the variations in $\Delta f'$ and $\Delta f''$.[3] So far this technique has been employed only for binary systems, where only three experiments are needed, and the extension to a three-component system, involving six partial structure functions, does not seem at present an easily tractable problem. The basic difficulty comes from the small values of $\Delta f'$ and $\Delta f''$ compared to f^0, which prevents from obtaining appreciable differences W_{ij} so as to allow a unique solution of the six simultaneous equations. The system of linear equations we have to solve, can be rewritten according to

$$T^k = \sum_{l=1}^{M(M+1)/2} W^{kl} X^l$$

$$k = 1,\ldots,M(M+1)/2 \tag{10}$$

or in matrix notation

$$[T^k] = [W^{kl}][X^l] \tag{11}$$

where T^k is the k-th measured total structure function, and X^l, is the l-th unknown partial structure function. A unique solution exists, and it is formally obtained by

$$[X^l] = [W^{kl}]^{-1}[T^k] \qquad (12)$$

if the determinant of $[W^{kl}]$ is different from zero. When the determinant is small compared to the coefficients, an appropriate solution can not be found through Equation 12, because of the experimental uncertainties. The problem belongs to the class of **ill-posed** inverse problems, which do not display the property of stability with respect to perturbations on input data. In other words, small errors in the measured T^k may produce large distorsions in the solution.

A less ambitious, but more practicable approach aims at determining the overall structuring around a species by changing only its scattering factor. Let us consider two independent determinations, i(s) and i′(s), of the structure function for an aqueous solution considered as a three species system; cation M, anion X, and solvent S. The radiation wavelengths are chosen in order to produce an appreciable difference only in the scattering factor of one ionic species, say the cation. By taking the difference

$$\begin{aligned} \Delta i_M(s) &= i(s) - i'(s) \\ &= (W_{MM} - W'_{MM})\, i_{MM} + (W_{MX} - W'_{MX})\, i_{MX} + (W_{MS} - W'_{MS})\, i_{MS} \qquad (13) \end{aligned}$$

some advantages are evident. Δi_M allows the study of the structuring around the cation without the masking effect due to water-water correlations, which have been shown to dominate the total structure function. This approach could shift to lower values the concentration limit currently accessible to X-ray diffraction studies. Moreoever, at relatively small concentrations, the terms M-M and M-X in Equation 13 are small, and often negligible, compared to cation-water correlations. In such cases, Equation 13 gives a direct description of the cationic hydration.

This approach is analogous to the **first order difference method** introduced by Enderby et al.[4] for neutron diffraction, where the changes in scattering factors (coherent scattering lengths) are obtained by **isotopic substitutions.** This method will be discussed in Chapter 4.

Equation 13 and the above discussion are still approximately valid when i and i′ refer to solutions where the change in scattering factors is obtained by an **isomorphous substitution.** In this case, we consider two solutions identical in all respects except for the chemical nature of one ionic species, i.e., we assume that the substitution does not affect the partial structure functions i_{ij}. Apart from the difficulty of choosing solutions which can be considered **isomorphous** a priori, this approach has the advantage of a straightforward applicability with standard diffraction measurements. Some applications will be reported in Chapter 3.

III. ANALYSIS OF THE RADIAL DISTRIBUTION

The data analysis in real space is particularly useful when looking for a first-level picture of the structure. In fact, a radial function exhibits structural information in a more direct way than its counterpart in s-space, and it contains many features which have an immediate physical interpretation. As Figure 2 illustrates, G(r) must be essentially zero for distances below the minimum hard sphere diameter of the atoms involved, and must become unitary with damped oscillations at the distance which can be empirically assumed as the **range of order** of the sample. The intermediate region is descriptive of the interatomic local order. It can be interpreted as a succession of peaks which become blurred and mixed with one

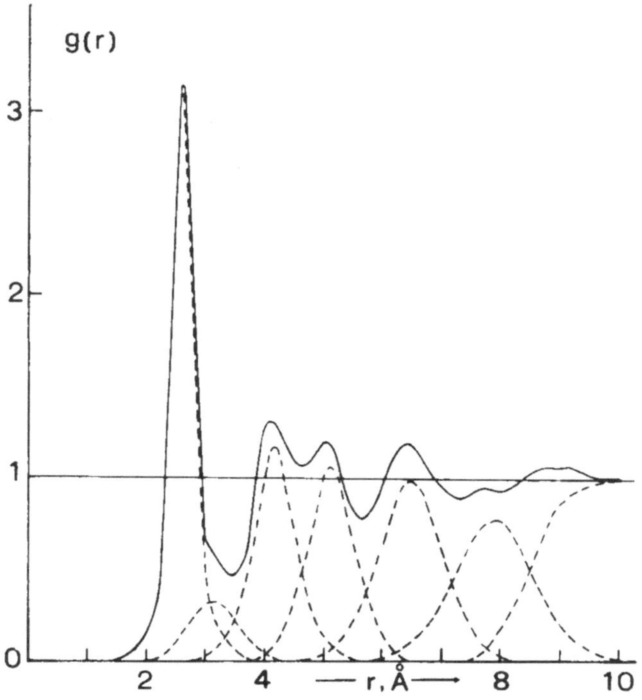

FIGURE 2. Typical behavior of the correlation function G(r).

another as r increases, until they merge into a continuum where the positional correlation is lost.

According to this picture, each partial radial distribution function $\rho_{ij}(r)$ is interpreted as a superposition of **coordination shells** centered at characteristic distances and involving specific numbers of atoms around the central one, i.e.,

$$4\pi r^2 \rho_{ij}(r) = \sum_k N_{ijk} P_{ijk}(r) \tag{14}$$

where N_{ijk} is a **coordination number,** defined as the mean number of j-type atoms at the mean distance r_{ijk} from an i-type atom, and $P_{ijk}(r)dr$ is the probability of finding an interatomic distance between R and r + dr for the atom pair i-j in the k-th shell.

The first step of data interpretation is an inspection of the radial distribution function in order to relate its features to the different shells, by the use of some *a priori* knowledge about the r regions where specific distances are expected. For example, useful indications about ion-water and ion-ion distances come from distance values found in hydrated crystals, or simply from the ionic radii of the species involved.

Once a well resolved peak has been ascribed to one coordination shell involving the atomic species i-j, the peak position R provides a direct measure of the coordination distance, and the area under it determines the coordination number. In fact, in the distance range where the peak is confined, the total D(r) can be written according to (omitting the indication of the specific shell)

$$\begin{aligned} D(r) &= (2 - \delta_{ij})\, x_i D_{ij}(r) \\ &= (2 - \delta_{ij})\, x_i N_{ij} r \int_0^{\infty} P_{ij}(u; R)/u[T_{ij}(r - u) - T_{ij}(r + u)]\, du \end{aligned} \tag{15}$$

and the area A^P under the peak becomes

$$A^P = \int_{peak} D(r)\, dr$$

$$= (2 - \delta_{ij})\, x_i N_{ij} \int_0^\infty P_{ij}/u\, du \int_0^\infty r[T_{ij}(r - u) - T_{ij}(r + u)]\, dr$$

$$= (2 - \delta_{ij})\, x_i N_{ij} A_{ij}^T = (2 - \delta_{ij})\, x_i N_{ij} f_i(0)\, f_j(0)\, M(0) \quad (16)$$

where A_{ij}^T is the area under the **peak-shape function** $T_{ij}(r)$, as it has been discussed in Chapter 1. Therefore, the area A^P gives a direct measure of the mean coordination numbers N_{ij} and N_{ji} which, in the absence of dispersion corrections and because usually $M(0) = 1$, are given by

$$N_{ij} = A^P/[(2 - \delta_{ij})\, x_i Z_i Z_j] = (x_j/x_i)\, N_{ji} \quad (17)$$

Note that the WKM approximation,[5] which introduces the **effective number of electrons** K_i as described in Chapter 1, gives an incorrect coordination number, namely

$$N_{ij}(WKM) = A^P/[(2 - \delta_{ij})\, x_i K_i K_j] \quad (18)$$

as can be easily calculated by recalling that D_{ij} in Equation 15 is substituted by $K_i K_j 4\pi r^2 \rho_{ij}$. If a peak is not well resolved, the direct evaluation of the coordination number becomes ambiguous because it is difficult to determine the area A^P univocally. Only the first shell peak, involving cation-water distances, has sometimes a resolution sufficient for a direct evaluation of the coordination number.

Important qualitative information can be gained by observing the influence on radial curves of variations in some significant factor as concentration, temperature, or composition. A few examples will illustrate this point.

In Figure 3, the radial function G(r) for three LiCl aqueous solution of different concentrations are compared with that of pure water.[6] The progressive rising of a peak ascribable to $Li^+ - H_2O$ nearest neighbor contacts, provides direct evidence of the restructuring of water molecules in the hydration shell around the cation. Moreover, the shift of the second broad peak towards high r values, reveals that this peak is doubly composed and that the longer distance component becomes prevalent with increasing concentration. This suggests the existence of anion hydration as well.

The structural behavior of the near environment around Cr^{3+} ions in the presence of water and Cl^- ions, is shown in Figure 4.[7] The three radial curves refer to three solutions where the Cr^{3+} concentrations and the Cl^-/Cr^{3+} ratios are different. The lowering of the peak at about 2 Å, diagnostic of $Cr^{3+} - H_2O$ distances, and the appearing of the hump at about 2.3 Å where $Cr^{3+} - Cl^-$ distances are expected to occur, are indicative of an increasing amount of **inner sphere complex formation.**

When comparing different cases, care should be taken in choosing the radial curves which can better reveal the effects which are supposed to occur. For example, the G(r) curves in Figure 3 correspond to the composition units $(LiCl)_x\, (H_2O)_{1-x}$; consequently the weight of the $Li^+ - H_2O$ nearest neighbor shell increases with Li^+ concentration. On the contrary, the curves in Figure 4 are calculated for composition units for which the composition coefficient of the Cr^{3+} ion is equal to one. In such a case, the peaks describing the same structure around Cr^{3+} ions should have the same shape in the three curves. If this does not happen, as in Figure 4, the conclusion is that structural changes have occurred.

Weak structural effects can be revealed by subtracting radial curves, one from the other,

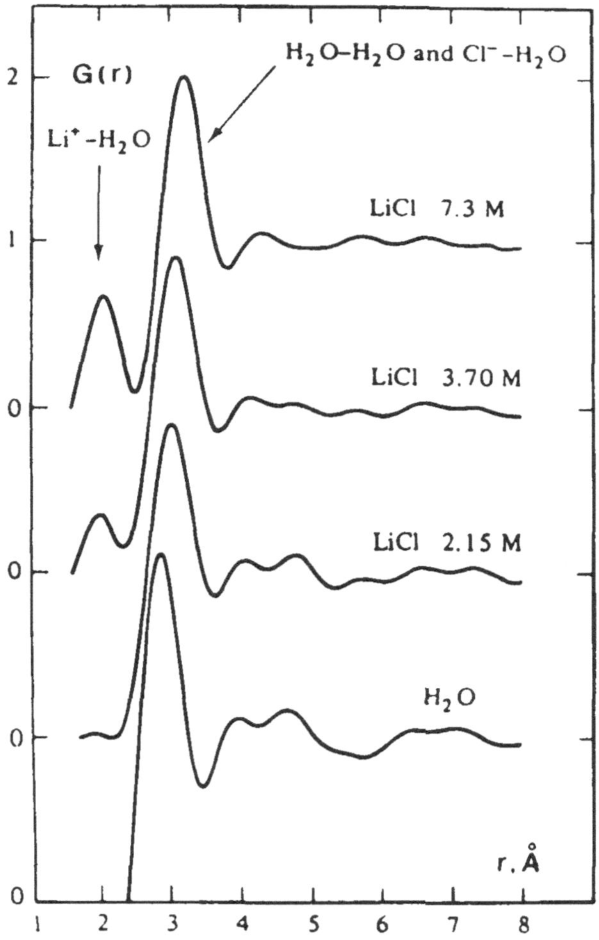

FIGURE 3. Correlation functions for water and LiCl aqueous solutions. (From Licheri, G., Piccaluga, G., and Pinna G., *J. Appl. Cryst.*, 6, 392, 1973. With permission.)

once the proper normalization has been chosen so as to make the difference meaningful. The illustrative example given in Figure 5 refers to the existence of two distances, $Cu-O_{eq}$ (equatorial) and $Cu-O_{ax}$ (axial), due to the Jahn-Teller effect in a $CuSO_4$ aqueous solution.[8] There is no definite indication in the radial distribution curve of a distance of about 2.4 Å, where the $Cu-O_{ax}$ distance should fall. However, a comparison with a $ZnSO_4$ solution, which has similar composition coefficients and scattering factors, shows a dissimilarity in the shape of the first peak which is only ascribable to different structural behaviors in the short distance range. The difference $\Delta D = D(CuSO_4) - D(ZnSO_4)$ puts into evidence two distinct contributions at the distances corresponding to $Cu-O_{eq}$ and $Cu-O_{ax}$, while the negative peak corresponds to $Zn-H_2O$ distance.

IV. CURVE FITTING

A. Mathematical Modeling

If we aim at obtaining a more complete and quantitative description of the structure, we have to resort to modeling techniques, that is, make assumptions about the structure and

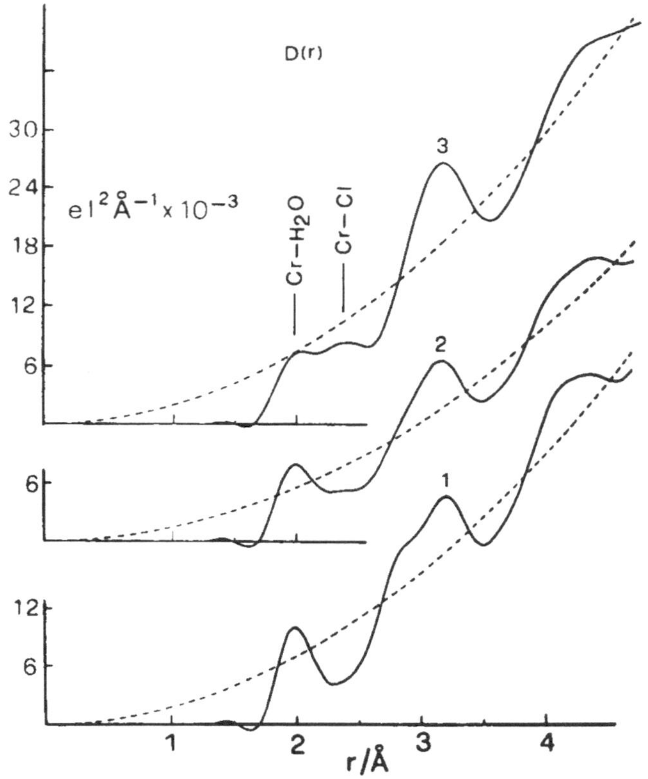

FIGURE 4. Radial distribution functions for three Cr(III) chloride solutions, where $[Cr^{3+}]$ = 2.0, 2.8, and 1.9 *M* and $[Cl^-]/[Cr^{3+}]$ = 3.0, 3.0, and 4.9, respectively, for the curves 1, 2, and 3. (From Magini, M., *J. Chem. Phys.*, 73, 2499, 1980. With permission.)

test them against diffraction data. The first step consists in assuming a specific form for the probability distribution of interatomic distances, and deriving analytical parameterized expressions for the functions to be used in a least squares fitting procedure. The simplest and most common choice is to assume harmonic motion between pairs of atoms to account for thermal disorder, and gaussian distribution to describe static disorder. In this case, the partial structure function i_{ij} is written according to[9-11]

$$si_{ij}(s) = \sum_k (N_{ijk}/x_j r_{ijk}) \exp(-\sigma_{ijk}^2 s^2/2) \sin(sr_{ijk}) \tag{19}$$

which is the corresponding, in s-space, of Equation 14 where the distribution of distances assumes the form[10]

$$P(r; r', \sigma) = (r/r')\{\exp[-(r - r')^2/2\sigma^2] - \exp[-(r + r')^2]\}/(2\pi\sigma^2)^{1/2} \tag{20}$$

The expression of the total structure function i(s) which must be compared with that obtained from the measurements, can be written as

$$si(s) = \sum_{i=1}^{M} \sum_j x_i f_i f_j (N_{ij}/r_{ij}) \exp(-\sigma_{ij}\, s^2/2) \sin(sr_{ij}) \tag{21}$$

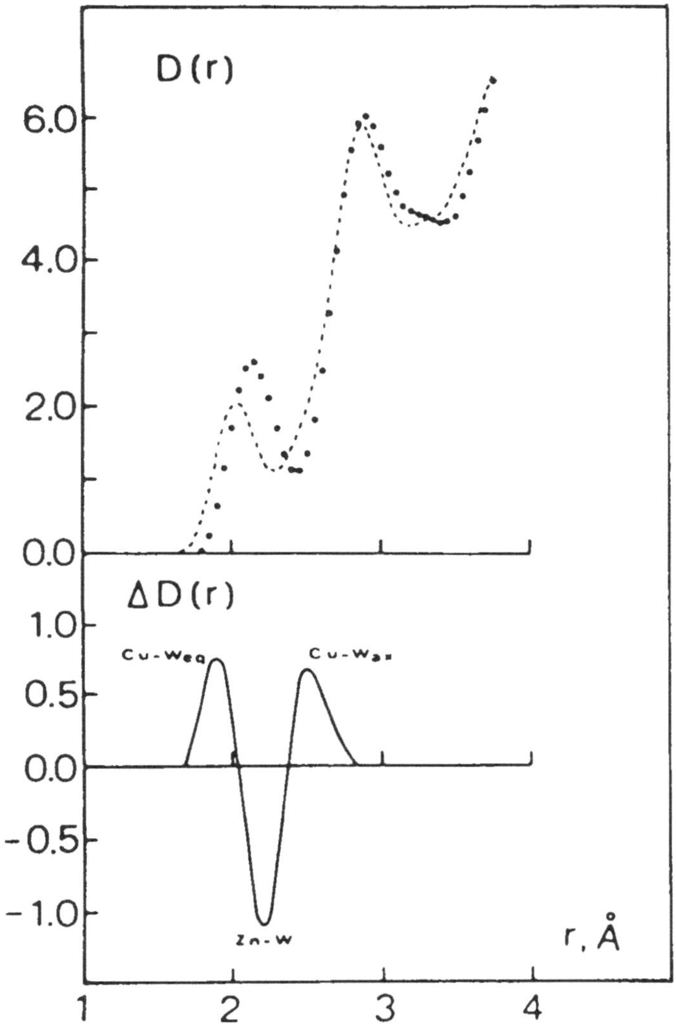

FIGURE 5. Radial distribution functions for $ZnSO_4$ (dots) and $CuSO_4$ (broken line) aqueous solutions, and difference between them (solid line). (From Musinu, A., Paschina, G., Piccaluga, G., and Magini, M., *Inorg. Chem.*, 22, 1184, 1983. With permission.)

where, for simplicity, only two sums appear; the first is over the composition unit, while the j-sum is over the coordination shells without specifying the atomic species (the j-th shell contributes a term involving N_{ij} atoms at a mean distance r_{ij} with r.m.s. deviation σ_{ij}). Equation 21 can also be written according to

$$si(s) = \sum_{i=1}^{M} \sum_{j} x_i f_i f_j (1/r_{ij}) \exp(-\sigma_{ij}^2 s^2/2) \sin(sr_{ij}) \quad (22)$$

where now the j-sum is taken over all atoms in the sample, because the grouping of atoms in coordination shells is not considered. It is worth observing that, when building a model, the i-sum in Equations 21 and 22 should be considered over structurally different atoms rather than chemically different ones. In fact, it is possible that chemically equivalent atoms are structurally nonequivalent, so that different spectra of distances should be modeled around each of them.

A model for noncrystalline systems must describe the loss of positional correlation beyond a characteristic distance which measures the **range of local order** in the system. This property can be crudely embodied in the above picture by considering around each structurally different species a finite region containing a **discrete** structuring of atoms followed by a **continuum** of distances where there is no correlation between the atomic positions.[12] For such a picture, the structure function can be written according to:

$$si(s) = \sum_{i=1}^{M} \sum_{j=i}^{n_i} x_i f_i f_j N_{ij} \exp(-\sigma_{ij}^2 s^2/2) \sin(sr_{ij})/r_{ij}$$

$$+ 4\pi\rho_0 \sum_{i}^{M} \sum_{j}^{M} x_i x_j f_i f_j \exp(-\sigma_{ij}^2 s^2)[sr_{oij}\cos(sr_{oij}) - \sin(sr_{oij})]/s^2 \quad (23)$$

The first term describes the discrete structure and arises from Equation 21 after cut off of the sum over j, in order to consider only a number n_i of **discrete shells** around each i-species. The second term is added to account for the continuum, that is, the regions where $\rho_{ij}(r) = \rho_{0j}$. The radii r_{0ij} define the boundaries of the two regions for each i-j pair and the terms involving σ_{0ij} make these boundaries not too sharp.

This semiempirical mathematical description of the structuring in noncrystalline systems, can be used straightforwardly for extracting structural parameters from diffraction data, as it will be discussed in the following sections. However, it does not describe some other structural properties of the sample, as the asymptotic behavior of the radial distribution and the limit of the structure function at $s = 0$.

B. Analysis of Dominant Contributions

The above picture is particularly suitable for discussing how the different structural details contribute to the scattering. The function si(s) is given by a superposition of sinusoidal waves coming from successive shells of atoms located at characteristic distances from another atom. Each wave is damped by the behavior of the scattering factors and by the disorder term $\exp(-\sigma_{ij}^2 s^2/2)$, and further weighted by structure dependent (N_{ij}/r_{ij}) and composition dependent (x_i) terms. It is easy to realize that the factor $1/r_{ij}$ lessens the contributions coming from longer distances more heavily, and that the disorder term damps the contributions with larger σ_{ij} more quickly. Since disorder usually increases with the distance, the scattering at high s is essentially determined by few sharp features at low r. On the contrary, the information about long range and broad features is concentrated in the low s region.

It is thus possible to obtain the structural parameters N_{ij}, r_{ij}, and σ_{ij} for the dominant nearest neighbor interactions through a least squares fitting in the region of high s values (**high angle analysis,** HAA). This procedure has been extensively used to describe the ion-water and water-water interactions within the first hydration shells, and the short cation-anion distances when inner sphere complexes are present. The example in Figure 6 shows the comparison between measured and calculated si(s) values for a $Zn(ClO_4)_2$ aqueous solution.[13] The calculated function was obtained with the contributions of the $Zn^{2+}-H_2O$ close hydration, and of $Cl-O$ and $O-O$ pairs within the ClO_4^- ion. It can be seen that the good agreement between calculated and experimental functions extends over a large s region, the difference at low s values being due to contributions from longer distances.

The same results as in high angle analysis should be obtained, in principle, by analyzing the peaks in r-space which determine the behavior of i(s) at high s. Actually, working in r-space is somewhat more complicated because we have to handle gaussian peaks broadened by the convolution with appropriate peak-shape functions T_{ij}, i.e.,

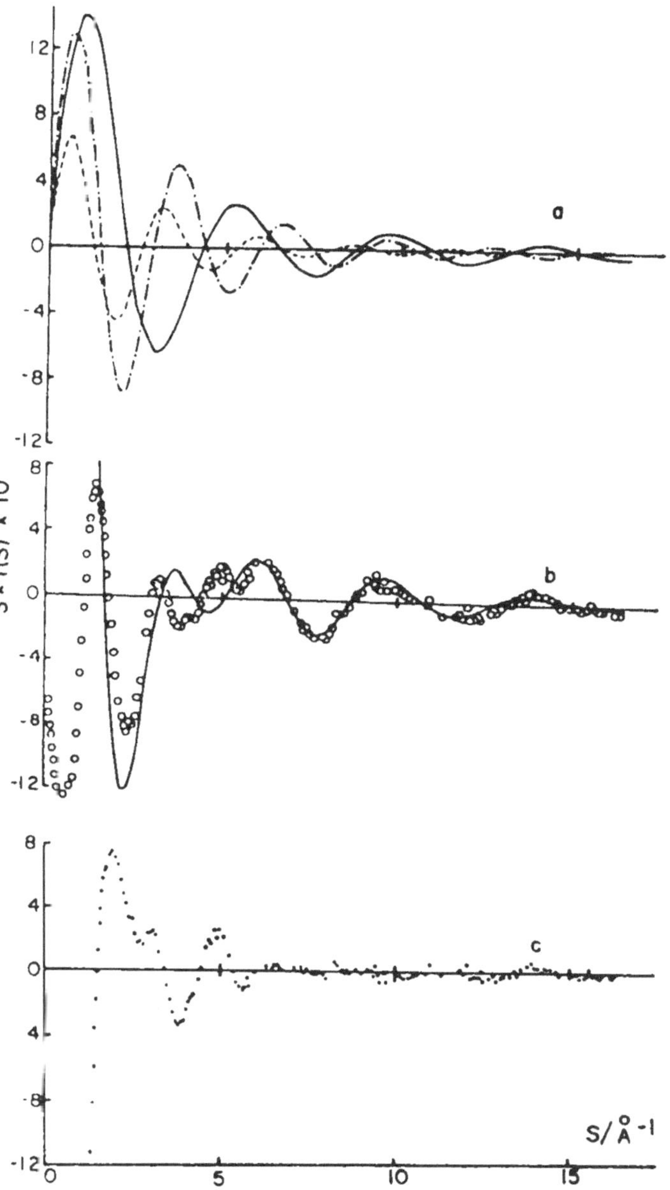

FIGURE 6. Observed and calculated si(s) values for a $Zn(ClO_4)_2$ solution: (a) contributions of the Cl-O (——), O-O (····) and Zn-O (-·-·-) atom pairs; (b) comparison between observed si(s) (circles) and calculated si(s) (solid line) for the atom pairs considered; and (c) difference between observed and calculated si(s) values. (From Ohtaki, H., Yamaguchi, T., and Maeda, M., *Bull. Chem. Soc. Jpn.*, 49, 701, 1976. With permission.)

$$D_{ij}(r) = N_{ij}r \int P_{ij}(u)/u \, [T_{ij}(r - u) - T_{ij}(r + u)] \, du \tag{24}$$

On the contrary, in the s-space we can fit a more tractable expression to the actual raw data. However, the **peaks-shape analysis** (PSA) often visualizes the goodness of the fitting better than the high angle analysis. In Figure 7, the short range peaks, corresponding to Zn^{2+} –

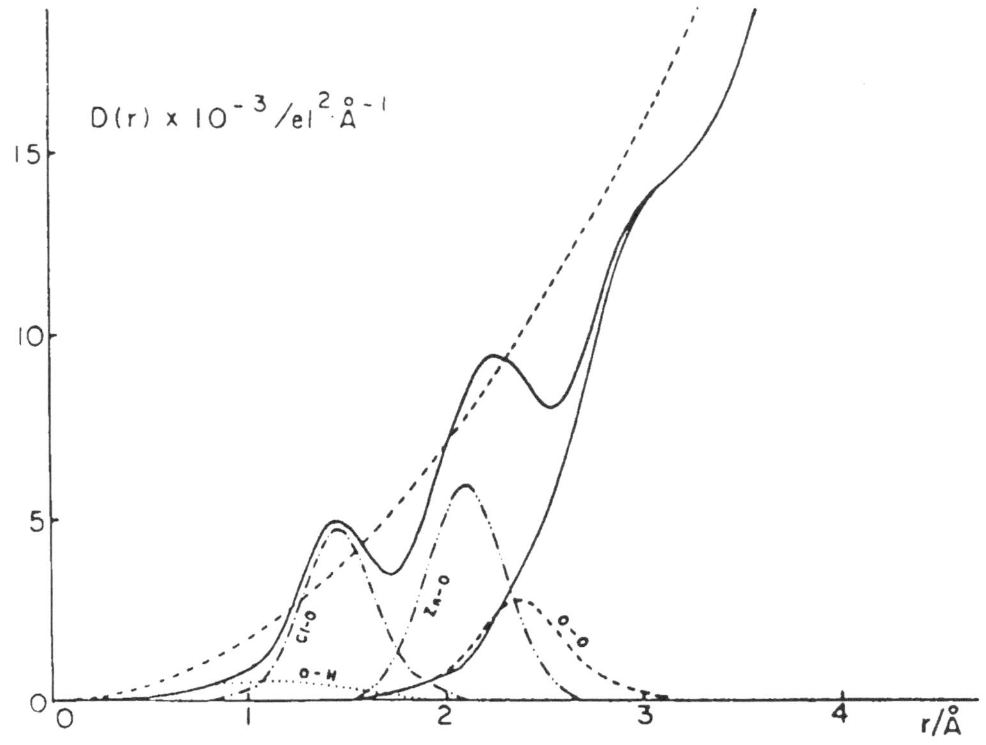

FIGURE 7. Peak shape analysis for the same solution as in Figure 6; the smooth curve (fine solid line) gives the residual distribution function obtained after subtraction of the peak shapes of the O-H, Cl-O, O-O, and Zn-O atom pairs from the D(r) curve. (From Ohtaki, H., Yamaguchi, T., and Maeda, M., *Bull. Chem. Soc. Jpn.*, 49, 701, 1976. With permission.)

H_2O, Cl – O, and O – O contributions shown in Figure 6, are compared with the total D(r).[13] The difference is only a smooth background, with no indication of residual pair correlations, meaning that the model peaks describe the short range structuring properly. In general, it is useful to compare model and experimental curves both in r- and s-space, as it may happen that differences which appear meaningless in one case, produce significant misfits in the other.

C. Interpretative Models

When the structure function has to be modeled in the whole s range, all the relevant partial contributions are to be accounted for, and the whole distance spectrum around each atomic species must be considered, including the long distance behavior which significantly influence the i(s) at low s. Therefore, we have to refer to a geometrical picture of the whole structure, which will determine how the distance distributions around the different atoms are connected one to the other. This means that when a structural model has to be described with the mathematical formalism discussed above, some building rules must be observed, and certain conditions must be met in order to produce a realistic, not contradictory, and structurally significant description of the structure.[14,15] In short, structurally equivalent atoms must have the same distance spectra, and distance spectra around different atoms must be geometrically compatible. Besides, the model must be consistent with the average bulk density of the sample, that is, when modeling the discrete structuring of j-type atoms within a sphere of radius r_{0ij} around an i atom, the number of j atoms in the sphere must be approximately $(4/3)\pi r_{0ij}^3 \rho_{0j}$.

The most general way of strictly fulfilling these requirements is to assume a **quasi-lattice model,** i.e., an atomic arrangement with the topology of a space lattice, but where the local geometries are perturbed by random displacements of atomic positions, so that the progressive loss of positional correlation at increasing distances from any origin atom is accounted for. This picture was used, for example, by the Oak Ridge group for the interpretation of the diffraction patterns of liquid water in a wide range of temperatures.[14,15] The model was a three-dimensional hydrogen-bonded network, based on a slightly expanded ice-I lattice with random vacancies in the network and random occupancy of cavities. This was the first study in which a structural model for a molecular liquid was carefully analyzed in order to be tested against diffraction data.

The quasi-lattice approach becomes hardly applicable when the complexity of the system increases and when it becomes difficult to find a realistic model lattice which one may refer to. Moreover, some ''in principle'' arguments can be found against it. Strictly speaking, a quasi-lattice model is in fact, descriptive of **hot crystals** rather than liquids or glasses,[16] where topological disorder is always present and cannot be referred to an underlying lattice. In any case, it seems that this description is not useful for analyzing the structure of ionic solutions. In fact, the radial curves obtained from diffraction data appear increasingly dominated, at increasing concentration, by ion-nearest neighbor interactions and show a swift loss of positional correlation for distances beyond these direct contacts. Each ion appears to give rise to a local restructuring of water molecules, and the solution could be thought as a collection of **quasi-molecules,** made up of ions and their nearest neighbors, which behave as almost independent units with only weak correlation with the environment. This picture, used first by Narten et al.[17] in a study of LiCl aqueous solutions, is the core of several interpretative models widely used to extract from diffraction data quantitative information about the structural parameters characterizing the local order around the ions.

The simplest model, which goes under the name of **first neighbor model 1** (FNM1), was mainly used in studying the **close hydration;** it assumes that:

1. Only the nearest neighbor shell around each ion has a discrete structure described by a definite coordination geometry.
2. No positional correlation exists between different coordinated ions.
3. The water molecules not involved in hydration (free water) are assumed to maintain the structure of pure water.

Point 1. implies that only a few contributions enter in the first term of Equation 23, i.e., those involving ion-water distances in the hydration shells. The independent parameters are limited in number, as the interatomic distances within the shell are connected one to another by the chosen coordination geometry. For example, in the case of a regular polyhedron, only one distance is to be introduced as independent parameters. Point 2. implies that uniform distributions, described by the second part of Equation 23, must be assumed for interaction distances beyond the first shell. Point 3. implies that the free water can be accounted for by properly weighting the measured i(s) of pure water in the total structure function. In this way, the data analysis is independent from a detailed knowledge of the water structure. Neglecting point 3. produces a modified version of the first neighbors model (FNM2) discussed by Pálinkás and Kálmán.[18] In FNM2 the free water is described by introducing new independent parameters describing nearest neighbor water-water interactions. Thus, FNM2 can describe possible variations in the solvent structure.

These models can also be extended to describe inner sphere complex formation, by considering independent units made up of a metal ion surrounded by a mixed shell containing water molecules and counterions at different distances. The anion can be a halide ion X, thus producing $M-X$ distances somewhat longer than $M-OH_2$ distances. It can also be an oxyanion $X'O_n$, thus giving distances undistinguishable from $M-OH_2$ ones. In this second case, however, typical $M-X'$ distances should appear which contribute significantly to the scattering. Obviously, new independent parameters should be introduced, in order to describe this more complicated situation, that is, mean numbers of $M-X$ (or $M-X'$) contacts and relative mean distances and r.m.s. deviations. The description of possible structurally different aggregates around individual ions of the same species, e.g., cations surrounded by mixed shells of water and anions with different compositions and geometries, may also be introduced in the model, but at the cost of further complications. As a matter of fact, when complex formation takes place, different complexes are simultaneously present in solution, unless suitable and particular chemical conditions favor the predominancy of a specific complex. However, X-ray diffraction data contain only an average information and cannot distinguish among possible different distributions. Only if the distribution of the existing species in solution is known a priori, e.g., from thermodynamic constants, the analysis of diffraction data can play its role of inferring some information about the structure of the complexes.

In some cases, the first neighbor model is not sufficient to describe the structure function. For example, systems involving ions of high charge density as trivalent or small divalent cations, exhibit a well defined and rather tightly bound first coordination shell, which favors the extension of the order range beyond the nearest neighbors distances. Usually, a second shell of water molecules around the cation has to be considered in order to describe the structure function. It is worth noting that X-ray diffraction studies of these systems produced the first direct experimental evidence and detailed structural determination of a second hydration shell in ionic aqueous solutions.[19-21] We refer to this structural description as two shell model (TSM), which represents a modification of the first neighbor models, where the introduction of a second cationic shell neither removes the independence of the structural aggregates around each ion, nor the ways of treating the free water; so, when we introduce the measured scattering of pure water, we have the TSM1, and when we model the free water, we have the TSM2. Obviously, this modification involves the introduction of new parameters such as cation-external water distance $r_{M-H_2O_{II}}$, nearest neighbor-external water distance, $r_{H_2O_I-H_2O_{II}}$, and number of water molecules in the second shell, $N_{M-H_2O_{II}}$.

The second shell can be treated in different ways. For example, one can consider a rather rigid situation, where all the molecules which compose the two shells are involved in rigid aggregates with fixed-bond angles between hydrogen-bonded water molecules. Alternatively, one can suppose a more loose situation, where only second shell molecules directly bonded to the same molecule in the first shell have a significant correlation with each other. Different intermediate situations can obviously be used, depending on the system studied and on chemical and physical effects which have to be translated in terms of structural descriptions.

It must be stressed that the models described above are interpretative models rather than complete pictures of the overall structural behavior of ionic solutions. They are useful tools for gaining quantitative information about the local ordering around the ions, when these structural features dominate the overall scattering. Some sort of middle-range, water-mediate structural effect due to ion-ion interactions is probably present in any ionic solution, usually hardly detected by X-ray diffraction measurements, because of little weight in comparison to the dominant terms.

REFERENCES

1. **Narten, A. H. and Levy, H. A.,** Liquid water: molecular correlation function from X-ray diffraction, *J. Chem. Phys.*, 55, 2263, 1971.

2. **Krogh-Moe, J.,** A method for the resolution of composite radial pair distribution functions, *Acta Chem. Scand.*, 20, 2890, 1966.
3. **Fuoss, P. H. and Bienenstock, A.,** X-ray anomalous scattering factors measurements and applications, in *Inner-Shell and X-ray Physics of Atoms and Solids*, Fabian, D. J., Klinpoppen, A., and Watson, L. M., Eds., Plenum Press, New York, 1981.
4. **Soper, A. K., Neilson, G. W., Enderby, J. E., and Howe, R. A.,** A neutron diffraction study of hydration effects in aqueous solutions, *J. Phys. C*, 10, 1793, 1977.
5. **Warren, B. E., Krutter, H., and Morningstar, O.,** Fourier analysis of X-ray patterns of vitreous SiO_2 and B_2O_3, *J. Am. Ceram. Soc.*, 19, 202, 1936.
6. **Licheri, G., Piccaluga, G., and Pinna, G.,** X-ray diffraction studies of alkali halide solutions, *J. Appl. Cryst.*, 6, 392, 1973.
7. **Magini, M.,** X-ray diffraction study of concentrated chromium(III) chloride solutions. I. Complex formation analysis in equilibrium conditions, *J. Chem. Phys.*, 73, 2499, 1980.
8. **Musinu, A., Paschina, G., Piccaluga, G., and Magini, M.,** Coordination of copper(II) in aqueous $CuSO_4$ solution, *Inorg. Chem.*, 22, 1184, 1983.
9. **James, R. W.,** Uber den Einfluss der Temperatur auf die Streuung der Roentgenstrahlen durch Gasmolekule, *Phys. Z.*, 33, 737, 1932.
10. **Karle, J. and Karle, I. L.,** Internal motion and molecular structure studies by electron diffraction. II. Interpretation and method, *J. Chem. Phys.*, 18, 957, 1950.
11. **Prins, J. A. and Petersen, H.,** Theoretical diffraction patterns corresponding to some sample types of molecular arrangement in liquids, *Physica*, 3, 147, 1936.
12. **Levy, H. A., Danford, M. D., and Narten, A. H.,** Data collection and evaluation with an X-ray diffractometer designed for the study of liquid structure, Report ORNL-3960, Oak Ridge National Laboratory, Tennessee, 1966.
13. **Ohtaki, H., Yamaguchi, T., and Maeda, M.,** X-ray diffraction studies of the structures of hydrated divalent transition-metal ions in aqueous solution, *Bull. Chem. Soc. Jpn*, 49, 701, 1976.
14. **Narten, A. H., Danford, M. D., and Levy, H. A.,** X-ray diffraction study of liquid water in the temperature range 4-200 C, *Discuss. Faraday Soc.*, 43, 97, 1967.
15. **Narten, A. H. and Levy, H. A.,** Observed diffraction pattern and proposed models of liquid water, *Science*, 165, 447, 1969.
16. **Ziman, J. M.,** *Models of Disorder*, Cambridge University Press, Cambridge, 1979.
17. **Narten, A. H., Vaslow, F., and Levy, H. A.,** Diffraction pattern and structure of aqueous lithium chloride solutions, *J. Chem. Phys.*, 58, 5017, 1973.
18. **Pálinkás, G. and Kálmán, E.,** Diffraction investigations of aqueous electrolyte solutions, in *Diffraction Studies on Non-Crystalline Substances*, Hargittai, I. and Orville-Thomas, W. J., Eds., Elsevier, Amsterdam, 1981.
19. **Bol, W., Gerrits, G. J. A., and van Panthaleon van Eck, C. L.,** The hydration of divalent cations in aqueous solution. An X-ray investigation with isomorphous replacement, *J. Appl. Cryst.*, 3, 486, 1970.
20. **Caminiti, R., Licheri, G., Piccaluga, G., and Pinna, G.,** Solute structuring in aqueous $[Cr(H_2O)_6]Cl_3$, *J. Chem. Phys.*, 65, 3134, 1976.
21. **Caminiti, R., Licheri, G., Piccaluga, G., and Pinna, G.,** Hydration water-external water interactions around Cr^{3+} ions, *J. Chem. Phys.*, 69, 1, 1978.

Chapter 3

HYDRATION AND COMPLEX FORMATION: A CRITICAL REVIEW

I. INTRODUCTION

Several books and reviews on hydration phenomena underline how confused the situation is in literature with regard to proposed values of hydration numbers.[1-3] In our opinion such emphasis is sometimes too severe, as the confusion arises from the unreasonable attempt of comparing quantities which are actually different. In this situation, the usefulness of studying ion-water (and ion-ion) interaction is apparent from a structural point of view, investigating by *direct techniques* the ordering of water molecules around the ions and the time evolution of such order.

X-ray diffraction (XRD) technique has been used extensively for studying the static aspect of order phenomena and the most significant results will be reviewed here. If we look at how XRD results have been sometimes presented in literature, we get the idea that the confusion typical of many studies on hydration even contaminates XRD investigations, as inconsistent results on the same or similar systems are often reported (see for example the Chapter by Enderby and Neilson in the well known "Water: A Comprehensive Treatise", Vol. 6).[4] It comes to mind that this is the price these studies have to pay to the "inherent difficulty of a single diffraction experiment to provide meaningful results on complex systems".[4] As already mentioned at the beginning of this book, this is at least partially not true, as the confusion often originates from having presented with equal dignity results which are reliable and others that are not (at least to an expert of XRD).

In this chapter, most of the diffractometric studies which have come out up to 1985 are quoted, but only the most reliable results will be priviledged. We are aware that this choice is not without risk, as it may rely in many cases on nonobjective criteria; in fact, as there is no way to evaluate diffraction experimental data, when two experiments carried out by similar procedures are different in results, it is not easy to establish *a priori* which one is less reliable. However, we have to run this risk, if we wish all the diffractometric work, made in a span of at least 30 years, not to become meaningless.

It is important to explain the criteria followed in the selection and presentation of the results. When the studies present particularly serious experimental and interpretative inadequacies, they are completely omitted from the discussion. These are data collected with poor counting statistics and/or in too limited a range. Some papers use unreliable approximations (see Chapter 2) for the evaluation of coordination numbers. There are papers in which the quantitative evaluation of ionic coordination goes through assumptions which cannot be checked and are not, actually. Most of them are the earliest papers in the field. On the other side, there are papers which for quality of experimental data, accuracy of treatment, consistency of the results as a whole, represent the actual highest level of XRD studies. All the investigations, which results have been reported in the tables of this chapter, belong to this group. In between, there are many special cases which are explicitly discussed in the text. For example, in this large group we find studies which, though of good quality from the experimental point of view, present results partially questionable because the interpretative analysis has been carried out beyond the limits of the diffractometric technique (see problems about complex formation of transition metal ions). We are well aware that this selection may have not been painless, neither do we expect that we have not been wrong in rating some of the papers. To our justification we shall add that sometimes the difficulties in classification arise from situations which are beyond our responsibility. Too often, in fact, investigations on single cases have been presented, instead of systematic studies on different solutions or different ions, and the full evaluation of consistency of the results or quality of the experiments is impossible. Therefore, we hope not to be excessively blamed if we have underevaluated some papers, or given credit to some others which the time will reveal inaccurate.

The selection of the material has been done on the ground of quality and content as well. The state of ions in solution, in fact, has been investigated by XRD in very different situations. Though the investigative effort has been largely directed towards concentrated strong electrolyte solutions, more recent studies deal with complex formation with inorganic ions in aqueous solution, with complex formation with organic ligands, with interactions ion-macromolecules, with species obtained through extreme hydrolysis of aqueous solutions up to the formation of polynuclear complexes, and with superconcentrated solutions or melted salts. Not in all cases are the characteristics of XRD particularly apt to solve the structural problems confronted. In fact, as better described in the previous chapters, XRD investigates with accuracy short-range order phenomena in noncrystalline systems, but it meets with increasing difficulties in decoding longer range order phenomena. The technique is therefore useful in cases in which few ordered atomic shells are set around the central species, while it is less suitable when heavy aggregates, big molecules, and extended structures have to be taken into account. Then, we have chosen to consider hydration phenomena and complex formation with inorganic ions, even if examples of different applications will be mentioned eventually. Nevertheless, it is our opinion that the topics covered here will be stimulating, as, among the chapters of physical chemistry of the solutions, the one that regards hydration phenomena and complex formation is surely among the most interesting for chemists and research workers of many different fields.

The material will be presented ion by ion, within their respective group in the Periodic Table. Nontransition metals will be discussed first, followed by transition metals; the most significant results about the anions will be considered next. Finally, a discussion of the entire picture of the results will be presented and some points worth mentioning will be examined. The attempt to point out the information directly rising from XRD is implied in the choice of the material. However, as in any other experimental technique, not all the details of ion-water interactions can be deduced from XRD data alone. While the most useful complementary techniques will be discussed in Chapter 4, already in the following paragraphs (when needed) will be additional studies mentioned, which have been of support to XRD investigations and have given full meaning to their analyses. The presentation per single ion implies that the results obtained for cations and anions in the same solution will be mentioned and discussed separately, with the exception of the cases in which complex formation takes place (only a hint as to how the anion has been treated will be enclosed in the tables reporting the parameters of cationic coordination). This inconvenience appeared a minor one among the many inherent to any attempt of classification, in view of the very characteristics of the most common models and procedures used in interpreting the experimental data, which actually imply separate treatment for different ionic types. We may anticipate (but we will see it better in the Discussion) that the results support such a choice, as the features of ionic aggregates have proved to be essentially independent from the other species present in the system.

Before closing this section, we recall the symbols and abbreviations used in the tables of results:

- M molar concentration
- m molal concentration
- $r_{i\text{-}j}$ distance in Å between atoms i and j
- $\sigma_{i\text{-}j}$ root mean square deviation in Å of $r_{i\text{-}j}$
- $N_{i\text{-}j}$ average number of j atoms around one i atom

- f parameter value kept fixed in the fitting procedures
- FNM1 first neighbor model 1
- FNM2 first neighbor model 2
- TSM1 two-shell model 1
- TSM2 two-shell model 2
- PSA peak-shape analysis
- HAA high-angle analysis
- CF complex formation

For the description of models and interpretative methods the reader is referred to Chapter 2.

II. H_3O^+ ION

Protons coming from dissociation of acids in water are present as oxonium ions H_3O^+. These can be further hydrated and the aim of diffractometric investigations on concentrated aqueous solutions of acids is to understand the structure of hydrated oxonium ions. Of the several studies on these systems,[5-12] only a few provide reliable information. In fact, variations of the available data range yield, in this case, dramatic modifications in the radial distribution functions,[13] so that only high-resolution investigations are meaningful.

Lee and Kaplow[7] first observed that adding HCl to pure water caused significant modifications in the radial function: the H_2O-H_2O peak at 2.8 Å broadens, then splits in one peak, centered at about 3.20 Å, and in another at a distance value which becomes smaller (from 2.75 to 2.56 Å) with increasing acid content. The peak at 3.20 Å matches the value expected for a well-defined $O-Cl$ interaction distance, while the peak at 2.56 to 2.75 Å was attributed to strongly bonded oxygen pairs, as suggested by the structures of crystalline di- and trihydrated HCl.

The evolution of the radial functions with HCl concentration was confirmed in an extensive study by Triolo and Narten[10] who examined five HCl solutions of compositions ranging from $HCl{\cdot}96H_2O$ to $HCl{\cdot}4H_2O$ (Figure 1). The experimental data showed to be consistent in the whole concentration range with the assumed regular tetrahedral configuration, $H_3O^+(H_2O)_4$, with $H_3O^+-H_2O$ distance of 2.52 Å; the increasing number of these units causes the peak to shift to shorter distances. Neutron diffraction data were used by Triolo and Narten to ascertain the orientational correlations in the groups $D_3O^+(D_2O)_4$.

The tetrahedral model was rejected by Lee et al.[11] in an investigation of a $CoCl_2$ solution in concentrated hydrochloric acid, with Cl^-/Co^{2+} molar ratio of about 21. Assuming the cobalt as completely engaged in tetrahedral tetrachlorocobaltate ions, the best fit to the structure function was obtained by means of a trigonal pyramidal configuration, $H_3O^+(H_2O)_3 \ldots H_2O$. In this model, the oxonium ion has an almost planar structure and binds three water molecules by short hydrogen bonds (r_{O-O} = 2.45 Å), while another water molecule sets above the ion, at the apex of the pyramid, at a distance of 2.90 Å from the oxygen atom of the oxonium ion. As support of this model, Lee et al. mention the absence of tetrahedral configurations around H_3O^+ in crystalline structures and the results of some theoretical calculations by the molecular orbital method.

Thus, there is general acceptance of the existence of very short H-bonds (2.5 Å) in acid solutions, connected with the presence of protons in water; the type of coordination of the oxonium ion is, however, not known.

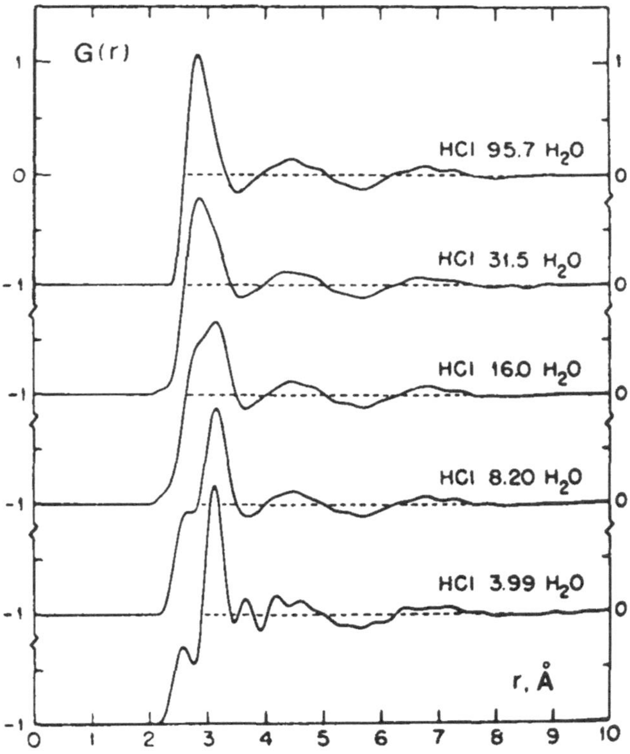

FIGURE 1. Correlation functions for HCl solutions. (From Triolo, R. and Narten, A. H., *J. Chem. Phys.*, 63, 3624, 1975. With permission.)

III. ALKALI IONS

A. Lithium

Light atoms, like lithium, are not good scatterers of X-rays. Nevertheless, XRD studies of lithium salt solutions are plentiful.[14-22] In contrast, the information obtained is scanty.

Actually, the low scattering power of Li^+ ions has been often exploited to study the structure of the counterion in more favorable conditions. Likewise, the addition of LiCl to solutions of transition metal chlorides has been used to study chloro-complexation phenomena with high Cl^-/Me^{+n} molar ratios (see solutions of $NiCl_2$, $CoCl_2$, etc.).

The difficulties coming from the low scattering factor of Li^+ cannot obviously be removed, so that not even the most recent studies have provided conclusive descriptions of Li^+ hydration. However, looking at the succession of the results, the evidence of improvement in measurements clearly stands out. In fact, while in the radial functions presented in the early investigations, $Li^+ - H_2O$ peaks were almost unobservable, masked, or strongly deformed by spurious peaks, in more recent studies, peaks appear at distances consistent with the crystal radius of the cation.[18-22] For instance, in the G(r)s of LiCl solutions (reported in Figure 3 Chapter 2), a peak at 2.1 Å, which increases with increasing concentration, is quite evident and is ascribed to $Li^+ - H_2O$ distances. Then, some attempts of investigating the coordination of Li^+ ion by XRD have been performed, sometimes supporting XRD with other structural methods, for example, neutron diffraction[19] or molecular dynamics.[22] These latter studies will be described in the next chapter in greater detail. Restricting our attention only to XRD, the uncertainty of the results still appears in the best and most complete study on lithium salt solutions, carried out by Narten et al.,[19] who examined seven LiCl solutions

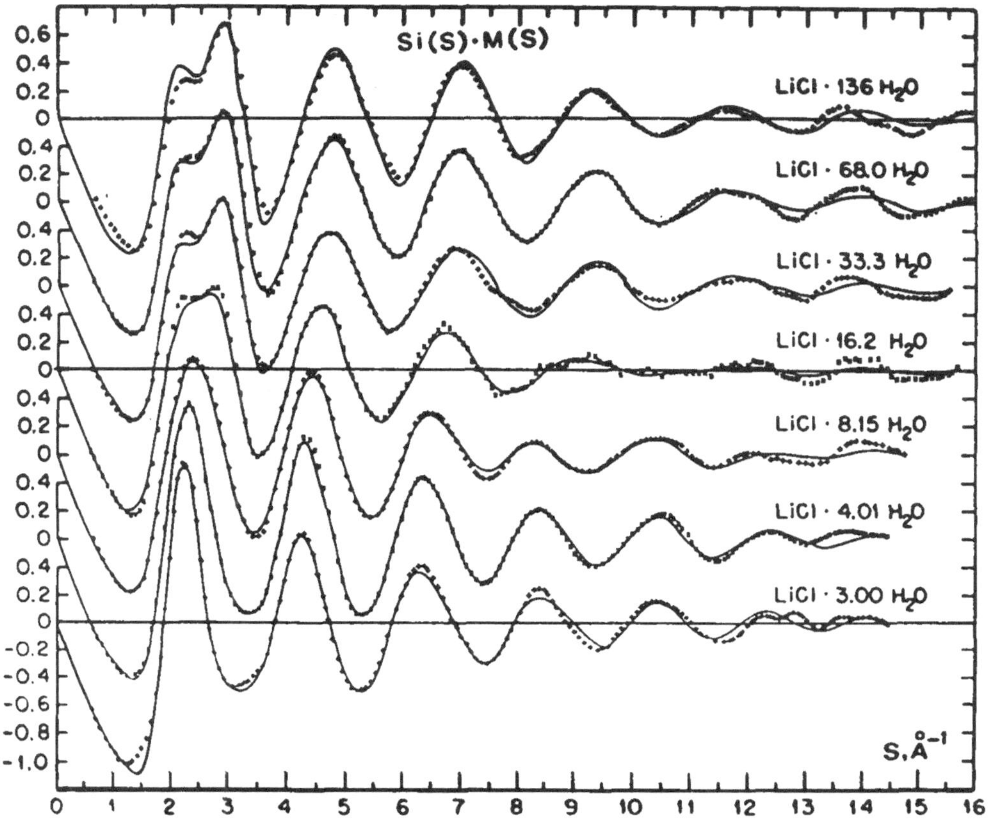

FIGURE 2. Structure functions for LiCl solutions. (From Narten, A. H., Vaslow, F., and Levy, H. A., *J. Chem. Phys.*, 58, 5017, 1973. With permission.)

in the composition range from $LiCl{\cdot}136H_2O$ to $LiCl{\cdot}3H_2O$. Thanks to exceptionally good experimental conditions set up at Oak Ridge Laboratories in the renowned investigations on liquid water,[23] and applying an equally exceptional counting statistic (600,000 counts were accumulated at each point in the high-angle region!!!), Narten et al. obtained hardly perfectible structure functions (Figure 2). In spite of this, in the correlation functions (Figure 3), spurious peaks superimpose to the $Li^+ - H_2O$ peak at ~2.0 Å, which sometimes splits, but never appears well isolated. Under these circumstances, an analysis of Li^+ coordination, exclusively based on XRD results, is problematic. However, in the study under discussion, the same solutions were also examined by neutron diffraction. Both X-ray and neutron diffraction data turned out consistent with intensity data calculated through FNM1 using the same units (tetrahedral $Li^+[H_2O]_4$ and octahedral $Cl^-[H_2O]_6$) and the same ion-water coordination parameters (distances involving D atoms were introduced only in the calculation of neutron structure functions).

The structural parameters obtained by XRD data analysis by Narten et al. are given in Table 1, together with the values proposed in other investigations on LiCl solutions. The spreading of $Li^+ - H_2O$ distance values clearly confirms the uncertainty in XRD determinations of Li^+ coordination.

Obviously, the uncertainty cannot but increase when Li^+ ions are studied in more complex solutions or with heavier counterions. This is apparent by looking at an investigation of Radnai et al.[22] on a solution of LiI. The weighted partial structure functions, derived from the proposed model, clearly show that the contribution from the $Li^+ - H_2O$ pair is almost

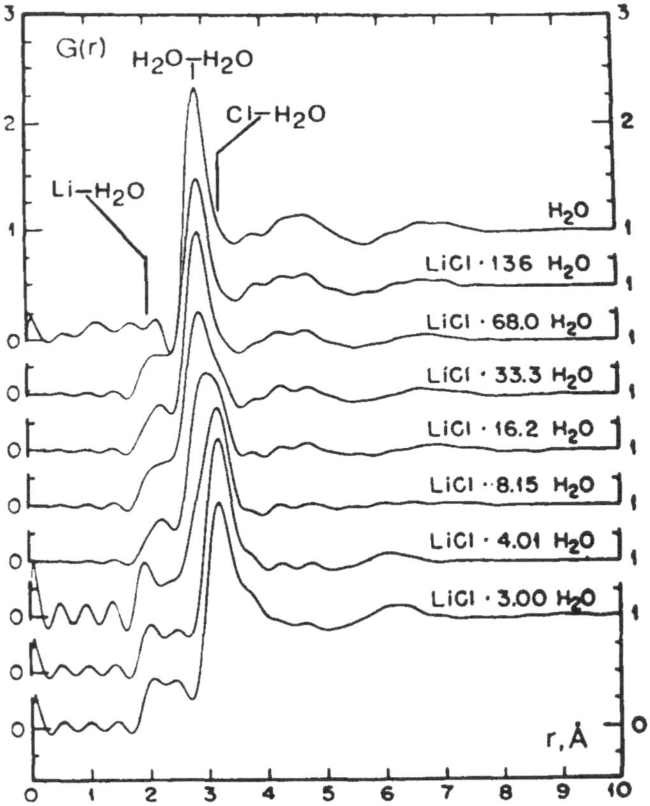

FIGURE 3. Correlation functions for LiCl solutions. (From Narten, A. H., Vaslow, F., and Levy, H. A., *J. Chem. Phys.*, 58, 5017, 1973. With permission.)

insignificant in comparison with H_2O-H_2O and $I^- - H_2O$ contributions. On this account, the analysis about Li^+ coordination in LiI solution has been, in our opinion, carried too far.

B. Sodium

The Na^+ ion has ten electrons and is detectable by X-ray diffraction; ions with equal or similar number of scatterers (for instance, Mg^{2+}) have been well characterized in solution. However, the radius of Na^+ ion is greater than that of Li^+ and, consequently, the average $Na^+ - H_2O$ distance approaches that of H_2O-H_2O pair and hardly becomes separable from it. The positive effects of the first fact are then counterbalanced by the lesser resolution, so that the information from XRD studies on Na^+ hydration is not more accurate than that concerning lithium.

Several quantitative determinations of ionic coordination have been performed in solutions of Na halides.[15,17,21,25-29] Apart from a study of Bosi et al.,[29] who present a radial curve of the system $NaI.16H_2O$ in total disagreement with those published for sodium halide solutions (including another NaI solution), the most recent investigations indicate concordant values of average $Na^+ - H_2O$ distances (2.40 Å in a 7 *M* NaI solution,[27] 2.42 Å in 2 *M* and 4 *M* NaCl solutions,[21] and 2.41 Å in a 5 *M* NaCl solution,[28] respectively). In contrast, there is no univocal indication about the coordination numbers; some studies suggest tetracoordination,[21,27] others hexacoordination,[28] and, in any case, alternative values are not definitely ruled out. The coordination number of 6, with $Na^+ - H_2O$ distance near 2.45 Å, has been

Table 1
COORDINATION OF Li^+ ION

Sample no.	Sample	Ref.	Li^+ conc. m	r_{Li^+-H2O}	σ_{Li^+-H2O}	N_{Li^+-H2O}	Geometry	Method	Anion
1	LiCl 136.0 H_2O	19	0.4						
2	LiCl 68.0 H_2O	19	0.8						
3	LiCl 33.3 H_2O	19	1.7	1.95	0.25	4	Tetrahedral	FNM1	$Cl(H_2O)_6^-$
4	LiCl 16.2 H_2O	19	3.4						
5	LiCl 8.15 H_2O	19	6.8						
6	LiCl 4.0 H_2O	19	13.9	2.22	0.31	4	Tetrahedral	FNM1	$Cl(H_2O)_6^-$
7	LiCl 3.0 H_2O	19	18.5	2.25	0.25	4	Tetrahedral	FNM1	$Cl(H_2O)_6^-$
8	LiCl 27.8 H_2O	21	2.0	2.17	0.51	4	Tetrahedral	FNM2	$Cl(H_2O)_6^-$
9	LiCl 13.9 H_2O	21	4.0	2.08	0.25	4	Tetrahedral	FNM2	$Cl(H_2O)_6^-$
10	LiCl 13.9 H_2O	24	4.0	1.99	0.28	4	Tetrahedral	FNM1	$Cl(H_2O)_6^-$
11	LiCl 6.4 H_2O	24	8.6	2.04	0.24	4	Tetrahedral	FNM1	$Cl(H_2O)_6^-$

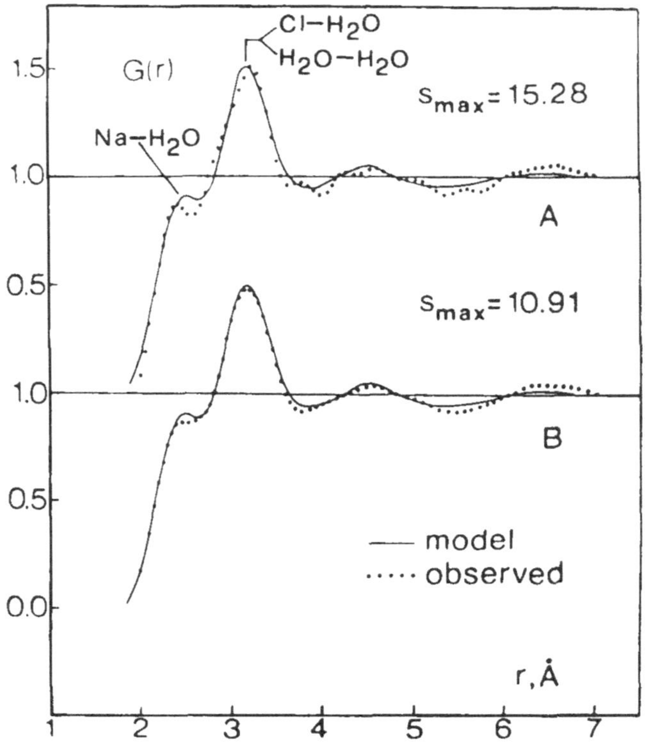

FIGURE 4. Correlation functions for a 5-*M* NaCl solution, calculated uising two different data ranges. (From Caminiti, R., Licheri, G., Piccaluga, G., and Pinna, G., *Rend. Semin. Fac. Sci. Univ. Cagliari*, 47, 19, 1977. With permission.)

also proposed in concentrated $NaNO_3$ solutions, but in this case the analysis was complicated by the existence of outer-sphere ion pairing.[30]

Some considerations suggest that this uncertainty follows not only from the basic difficulty mentioned above, but also, and perhaps prevalently, from the fact that the coordination shell of Na^+ is particularly soft and easily deformed. Proof of this is given by the comparison of the radial curves of NaCl and $CaCl_2$ solutions of similar concentration (Figures 4 and 8); the peak $Na^+ - H_2O$ at ~2.40 Å is much less resolved than the peak $Ca^{2+} - H_2O$, which falls at about the same distance. The lesser resolution does not depend on the value of s_{max}; the structure function of the NaCl solution is damped at high s, so that the correlation functions calculated using s_{max} = 15.3 $Å^{-1}$ or s_{max} = 10.9 $Å^{-1}$ look very similar, contrary to what happens in many other cases (Figure 4). As a consequence, the determination of the Na^+ coordination parameters, performed by the FNM1 method, leads to values little sensitive to the data range used in the refinements, so that different values of s_{max} are unlikely the reason of the different results reported in the various investigations.

The deformability of the Na^+ coordination shell is also indicated by the large value (0.21 to 0.27 Å) of the root mean square deviation of $Na^+ - H_2O$ distance, $\sigma_{Na^+ - H_2O}$, estimated through FNM1 or FNM2 models. Finally, if the model $Na^+ - H_2O$ distribution function is calculated from the partial structure function,[21] it becomes apparent that the first peak Na^+-H_2O is not resolved from the remaining curve, either because the values of the starting distances of the continuum are very short, or because these distances are associated with too large σ values. For this very reason, Pálinkás et al.[21] point out that the coordination number, pertaining to the geometrical model used in FNM calculation, is smaller than the number of water molecules included under the first peak of the pair distribution function up to the first minimum (called running coordination number). In our opinion, such a situation indicates, in general, a poor capability of the FNM method, which is based on rigid geometrical models, to treat cases in which the H_2O molecules do not seem rigidly bound to the central ion; the simultaneous occurrence in the same range of distances of ordered and disordered water molecules around the Na^+ ion seems a nonsense.

C. Potassium

The $K^+ - H_2O$ distance, expected on the basis of ionic radii and of results of structural studies of crystalline hydrates, is about 2.70 to 2.80 Å. Such a distance range is completely superimposed, in aqueous solution D(r)s, to the usual $H_2O - H_2O$ distances. Therefore, determinations of K^+ ion coordination are greatly affected by the not easily verifiable hypotheses made about $H_2O - H_2O$ interactions. Probably for this reason, the investigations on potassium salt solutions are few; besides the old studies by van Panthaleon van Eck et al.[5] and by Brady et al.,[14,31] only one investigation by Pálinkás et al.[21] exists, in which 2 and 4 *M* KCl solutions are examined, together with solutions of other alkali chlorides (except RbCl). The FNM2 analysis yields a $K^+ - H_2O$ distance of 2.80 Å. As regards the coordination number, the situation is similar to that mentioned before for the Na^+ ion; in fact, the six water molecules in the assumed coordination polyhedron of K^+ and the disordered (continuum) water molecules overlap, so that the "running coordination number" (see Subsection IIIB) is of about 8. This behavior again suggests that the coordination shell of K^+ is not strongly structured.

D. Cesium

Also for the Cs^+ ion the material is not abundant.[17,21,32] Only in two studies, one by Pálinkás et al.,[21] and one by Bertagnolli et al.,[32] the interpretation is based on detailed analysis of the structure function. In the first work, FNM2 succeeded in reproducing X-ray diffraction data from 2 and 4 *M* CsCl solutions through a model which sets at a distance of 3.15 Å a coordination shell of water molecules around Cs^+. Ion pairing, sometimes suggested for Cs halide solutions, has not been taken into consideration. The uncertainties mentioned above for Na^+ and K^+ coordination numbers also extend to this case.

In the second paper, which deals with very concentrated solutions of CsF (up to the composition $CsF{\cdot}2.3H_2O$), a lattice model was assumed, in which water molecules and F^- ions occupy the positions of the β-tridymite correlation lattice, while the Cs^+ cations are set in the interstitial sites of the structure, enclosed by two puckered hexagonal rings from lattice sites. The distance $Cs^+ - H_2O$ (or F^-) in this model is 3.13 Å.

E. Comments

Following the previous sections, the state of diffractometric studies of alkaline ions is certainly far from satisfactory. As the investigations are hampered in each case by basic difficulties, one should think that the present situation can unlikely be improved. However, in our opinion, this is not entirely true. In fact, if we look again at the descriptions of the single cases, we realize surprisingly that, putting aside early studies, the remaining ones are very few. In addition, some of them present very few cases, sometimes only one, which makes difficult the assessment of the results. Finally, in many studies the details about experimental conditions and data treatment are scanty (for instance, the number of accumulated counts is not indicated); probably, knowledge of these details would lower the number of studies deserving to be remembered, together with the ambiguous information. In this situation, it is impossible to ascertain how much uncertainty comes from experimental inaccuracy and how much from ambiguity in the interpretation. In the all, we believe that the potentialities of XRD in the study of alkaline cations have not been fully explored; more systematic investigation and greater experimental accuracy, adapted to the difficulties to be faced, is recommended to the researchers who intend to venture in this field.

Part of the difficulties may be the result of the weak interactions between alkali ions (except Li^+ ion) and water molecules. Several theoretical studies have pointed out the existence of many configurations of alkali ion-water molecular aggregates of similar energy. However, XRD studies have given reasonable proof of the existence of definite coordination shells around the halide ions, which, according to thermodynamic indications, interact with water no more strongly than the alkali ions. Therefore, it is not entirely convincing that the hydration shells of alkali ions should be less defined.

IV. ALKALINE EARTH IONS

A. Magnesium

This is one of the cases in which XRD meets with the most favorable conditions. In fact, in the G(r)s of $MgCl_2$ solutions (Figure 5), the ion-solvent direct distances give rise to a peak at ~2.1 Å, completely resolved from the other peaks of the radial curves.[37] In the case of Mg nitrate and sulfate solutions, also O – O distances within oxyanions contribute to the peak at 2.1 Å, but this does not give rise to particular problems, as these interactions are known in number, and the distances can be easily calculated from the known geometries of these ions. Therefore, the determination of ion-water distances and hydration numbers is not impaired by problems of resolution, and the precision of these evaluations is only conditioned by the accuracy of experimental and data treatment procedures.

Given these conditions, the studies of Mg salts solutions, which have been published in the years, almost unanimously ascribe to Mg^{2+} ions, a hydration number very close to 6 and to $Mg^{2+} - H_2O$ distance, a value of about 2.10 Å;[33-41] the only exception is an investigation by Albright[42] in which too many unfortunate factors (e.g., too short data range, use of the approximation by Warren et al., and so on) concur to indicate a coordination of Mg^{2+} ions near to 8, which surprises the author himself. In the following, only the studies will be reviewed, which fulfill the general criteria discussed in the Introduction.

Mg^{2+} is one of the species investigated by Bol et al.[35] by the isomorphous replacement method. As discussed in the previous chapter, this method aims at isolating partial radial functions describing the distribution of the various chemical species around the cation. To this end, differences of intensity curves, obtained from pairs of solutions having common anion, identical concentration (1 *M*), and cations of supposed isostructurality, must be performed. The Mg^{2+} ion was considered in $Mg^{2+} - Co^{2+}$, $Mg^{2+} - Ni^{2+}$, and $Mg^{2+} - Zn^{+2}$ pairs. To understand which advantages the method has, suffices it to observe that, in spite of the low cut-off used (s_{max} ~ 10 A^{-1}), the partial radial curves (presented in Figure 6) exhibited a perfectly isolated first peak (nearest neighbor shell of cationic coordination); besides, they provided the first unequivocal qualitative proof of the existence of a second hydration shell. However, some doubts exist about the quantitative evaluation of the hydration parameters supplied by the method in point. In fact, it is not clear how accurate the difference method is, when the ions of the pairs adopted are not perfectly isostructural. Therefore, the differences in the ionic radii of the ions in the pairs (although small) make the determination of the ion-water distance not completely reliable and direct; in particular, this distance was assumed (rather than determined) for the Mg^{2+} ion equal to 2.066 Å! The first coordination number was calculated in the range 5.7 to 6.2, while the number of second coordination should be about 12.

The most accurate data have been collected for solutions of $MgCl_2$,[37,39] $Mg(NO_3)_2$,[38] and $MgSO_4$;[40] in the studies quoted in References 37 and 38, 300,000 counts per point were stored. The experimental data were interpreted using modeling techniques and the resulting structural parameters are given in Table 2.

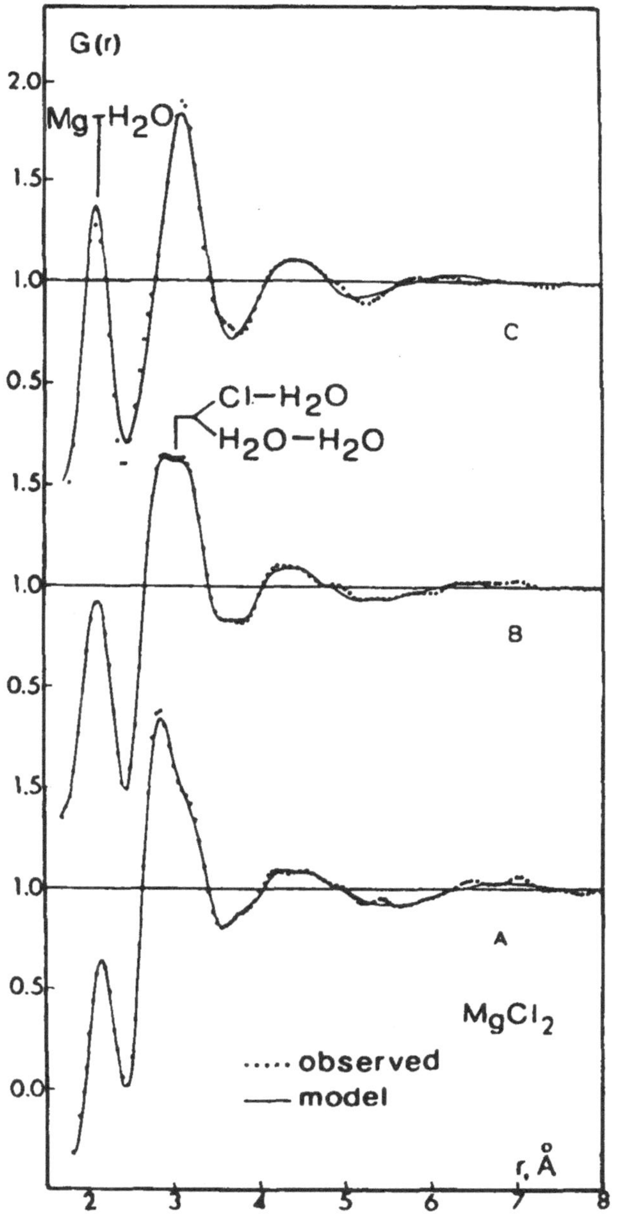

FIGURE 5. Correlation functions for $MgCl_2$ solutions: (A) 1.0 *M*; (B) 1.95 *M*; (C) 4.3 *M*. (From Caminiti, R., Licheri, G., Piccaluga, G., and Pinna, G., *J. Appl. Cryst.*, 12, 34, 1979. With permission.)

Table 2 shows the remarkable similarity of the hydration parameters of the Mg^{2+} in different investigations, within the limits of their different reliability. The first coordination number has always been taken equal to 6 in the models, and alternative coordination numbers have not been tested. This was justified, however, by the quasi-direct evaluation of first coordination number, allowed by the good resolution of the $Mg^{2+} - H_2O$ peak at ~2.1 Å. Whenever this straight evaluation is possible, the first coordination numbers are strongly reliable.

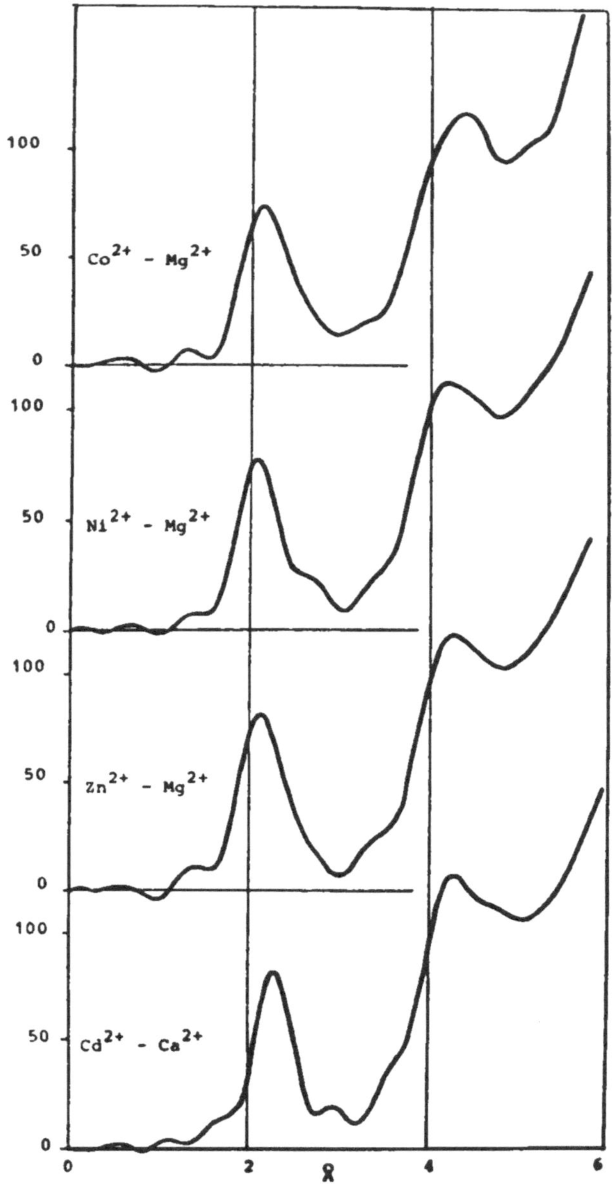

FIGURE 6. Electron density difference functions calculated for four couples of 1-*M* solutions. (From Bol, W., Gerrits, G. J. A., and van Panthaleon van Eck, C. L., *J. Appl. Cryst.*, 3, 486, 1970. With permission.)

On the contrary, the second coordination numbers are no more than a simple indication. However, calculations neglecting the contributions from the second shells result definitely worse either in the overall agreement factors of the model, or in the rationality of the entire picture of parameters. As a consequence, the existence of second shell interactions is strongly suggested by the mentioned studies, although they deal with total structure functions. Second shell contributions improve the quality of the refinements also in the most concentrated solutions, where little or no water remains, which is not engaged in the primary hydration

Table 2
COORDINATION OF Mg^{2+} ION

Sample no.	Sample	Ref.	Mg^{2+} conc.	$r_{Mg^{2+}-H_2O_I}$	$Mg^{2+}-H_2O_I$	$N_{Mg^{2+}-H_2O_I}$	Geometry	$r_{Mg^{2+}-H_2O_{II}}$	$\sigma_{Mg^{2+}-H_2O_{II}}$	$N_{Mg^{2+}-H_2O_{II}}$	$r_{H_2O_I-H_2O_{II}}$	$\sigma_{H_2O_I-H_2O_{II}}$	Method	Anion
1	$MgCl_2$ 55.5 H_2O	37	1.0*M*	2.12	0.04	6	Octahedral	4.1	0.7	12	2.77	0.01	TSM1	$Cl(H_2O)_6^-$
2	$MgCl_2$ 27.1 H_2O	37	1.9_5*M*	2.10	0.10	6	Octahedral	4.2	0.6	12	2.75	0.05	TSM1	$Cl(H_2O)_6^-$
3	$MgCl_2$ 11.5 H_2O	114	4.3*M*	2.11	0.12	6	Octahedral	4.0	0.4	6	2.69	0.07	TSM1	$Cl(H_2O)_6^-$
4	$MgCl_2$ 50.5 H_2O	39	1.1*m*	2.12	0.04	6	Octahedral	4.1	0.5	12	2.79	0.02	TSM2	$Cl(H_2O)_6^-$
5	$Mg(NO_3)_2$ 24.8 H_2O	38	2.0*M*	2.11	0.08	6	Octahedral	4.2	0.6	12	2.75	0.07	TSM1	Not hydrated
6	$Mg(NO_3)_2$ 10.8 H_2O	38	4.0*M*	2.10	0.11	6	Octahedral	4.2	0.7	12	2.78	0.11	TSM1	Not hydrated
7	$MgSO_4$ 20.5 H_2O	40	2.6*M*	2.09	0.12	6	Octahedral	4.2	0.2	12	2.81	0.09	TSM1	Hydrated

of the cations and, eventually, anions. This is the case of the 4.3 *M* $MgCl_2$ solution, for which the FNM1 was adopted as a first approach,[37] while subsequently the TSM showed to be more apt.[114] At this concentration, however, the parameters describing the second coordination shell came out different from the other cases. The structural meaning of the existence of independent (sometimes multiple) hydration shells at high concentration will be discussed at the end of this chapter.

The values in Table 2 indicate H_2O-H_2O distances shorter than the same distances in pure water (r = 2.80 to 2.85 Å), thus suggesting strong H-bonds. This hypothesis is also supported by the unusually small values of $\sigma_{H_2O_I-H_2O_{II}}$. As we shall see in the following, both findings (short $r_{H_2O_I-H_2O_{II}}$ and small $\sigma_{H_2O_I-H_2O_{II}}$) are an almost systematic result which has not been completely explained yet.

In the calculation of the contributions coming from the second coordination shell, no rigid geometry was imposed to the molecular aggregates, so that the three parameters $r_{Mg^{2+}-H_2O_I}$, $r_{Mg^{2+}-H_2O_{II}}$, $r_{H_2O_I-H_2O_{II}}$ were independently refined. The good reproducibility of the parameter values gives the opportunity to investigate the orientation of the water molecules in the first hydration shell. In fact, the distance $r_{Mg^{2+}-H_2O_{II}}$ does not depend on the distances $r_{Mg^{2+}-H_2O_I}$ and $r_{H_2O_I-H_2O_{II}}$ only, but also on the spatial arrangement of the molecule H_2O_I. If the H-bonds $O-H_I$ . . . O_{II} are nearly linear, the angle formed by the plane $O_{II}-O_I-O_{II}$ and the straight line between Mg^{2+} and O_I coincides with the angle between the bond $Mg-O_I$ and the water molecule H_2O_I (called tilt angle). Under the further assumption that the angle H$\hat{O}$H in the water molecule is 109° (that is, the average value found in crystalline hydrates),[43] a tilt angle of ~30° was calculated for solutions 1.95 *M* of $MgCl_2$ and 2.0 and 4.0 *M* of $Mg(NO_3)_2$, while an angle of 50° was proposed for the solution 1.0 *M* of $MgCl_2$. On the basis of the few cases examined, it is difficult to assess the reliability of these calculations, for which the uncertainty of the parameter $r_{Mg^{2+}-H_2O_{II}}$ is probably the critical point. However, we wish to underline that using accurate results and some reasonable assumptions, XRD studies are able to give indications also about the ion-water conformation.

The good resolving power of XRD under the favorable conditions occurring for Mg^{2+} ions is demonstrated by a study carried out on a mixed solution of $MgCl_2$ and $CaCl_2$,[44] which is one of the few examples of diffractometric studies on a solution containing two salts dissolved. The good quality of high-angle data and, consequently, the high value of s_{max} attainable (>15 $Å^{-1}$) allowed the calculation of a G(r) (Figure 7), in which two distinct maxima appear at ~2.05 and 2.40 Å, which can be attributed to direct distances Mg^{2+}-H_2O and $Ca^{2+}-H_2O$, respectively, even if the peaks are not completely separated. Also in this case, coordination models of Mg^{2+} and Ca^{2+} have been proposed, resulting in parameter values in good agreement with those reported in Tables 2 and 3.

Finally, we may observe that the Mg^{2+} hydration parameters, given in Table 2, have also been confirmed in a study of solutions of Mg acetate,[41] though in this case, on the basis of thermodynamic indications, the model assumed involves Mg^{2+} ions engaged, at least partially, in complexes with the acetate ions.

B. Calcium

The results about the coordination of Ca^{2+} ions are less conclusive than those obtained for Mg^{2+} ions. This happens because in the radial curves, the peak due to the $Ca^{2+}-H_2O$ interactions at about 2.40 Å is never completely separated from the adjacent ones. This is apparent in the radial curves of $CaCl_2$ solutions (Figure 8), though in the G(r) of the 4 *M* solution, the resolution of $Ca^{2+}-H_2O$ peak is the best ever obtained.[45] As no free water is present at this concentration, the peak $Ca^{2+}-H_2O$ is ~0.8 Å distant from the nearest peak due to Cl^--H_2O interactions. The incomplete resolution of the $Ca^{2+}-H_2O$ peak in this

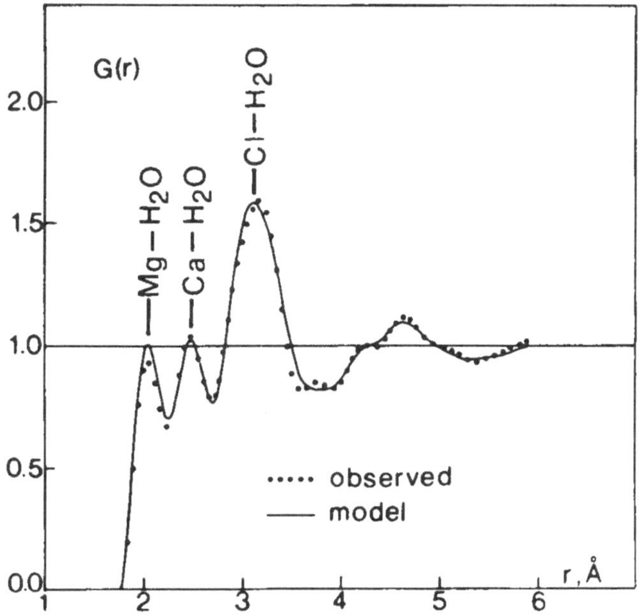

FIGURE 7. Correlation function of a $MgCl_2$-$CaCl_2$ solution. (From Caminiti, R., Licheri, G., Piccaluga, G., and Pinna, G., *Chem. Phys. Lett.*, 47, 275, 1977. With permission.)

situation suggests that the environment of Ca^{2+} ions is relatively disordered; some further support to this interpretation will be given in the following.

A good approach for Ca^{2+} solutions should be given by the isomorphous replacement method. The Ca^{2+} ion has been investigated by this method,[35] actually, but here the validity of the procedure, discussed in connection to the studies on Mg^{2+} solutions, becomes even more questionable, since the hypothesis of isostructurality for the adopted pair $Ca^{2+} - Cd^{2+}$, is even less credible. In fact, the description of the results about Ca^{2+} and Cd^{2+} coordination shows that even the direct average ion-water distances of Ca^{2+} and Cd^{2+} differ, about 0.10 to 0.15 Å. Some caution must therefore be used with the results of Bol et al., who fixed the $Ca^{2+} - H_2O$ distance at 2.330 Å and evaluated a coordination number of Ca^{2+} and Cd^{2+} of 5.8; the existence of a second hydration shell was also suggested.

The first reliable results obtained with fitting methods refer to the most concentrated solutions of $CaCl_2$.[45] In more dilute solutions of the same salt, or in solutions of $CaBr_2$,[46] the resolution of the radial functions was not enough to find out a coordination model clearly better than the others. However, if the best model for the $CaCl_2$ concentrated solutions was used in the other cases, it came out consistent with the experimental data, with minor modifications in the values of the parameters (Table 3). The same model turned out to be valid also in the case discussed above of the mixed solution of $MgCl_2 - CaCl_2$. It is worth mentioning here the high statistic accuracy of the experimental data reported in the work of Reference 45 (300,000 counts per point).

Under the same type of coordination, the values of $\sigma_{Ca^{2+}-H_2O}$ result greater than those found for other divalent ions. Besides, good fits were obtained in this case without introducing a second coordination shell in the calculations (the presence of a certain order beyond the first shell cannot be excluded, however). These observations suggest some structural disorder in the environment of Ca^{2+}, which may explain the difficulties in the formulation of one preferential model. In other words, the poor resolution of the radial curves seems to be a consequence of the natural broadening of the peaks.

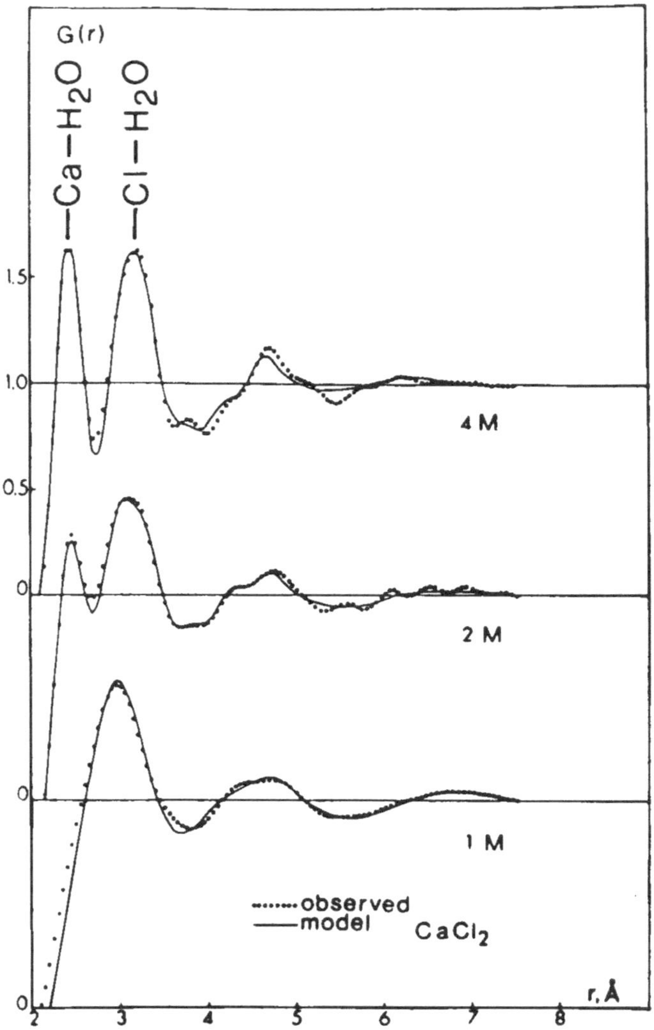

FIGURE 8. Correlation functions for $CaCl_2$ solutions. (From Licheri, G., Piccaluga, G., and Pinna, G., *J. Chem. Phys.*, 64, 2437, 1976. With permission.)

This hypothesis has been strongly supported by a recent study of a ~1 *M* solution of $CaCl_2$,[47] in which the experimental data appear in good agreement with those reported in Reference 45 for a $CaCl_2$ solution of very similar composition (see also Chapter 1, Figure 23). In this study, the conclusion was reached that, "the coordination number of Ca^{2+}, especially at lower concentrations, can still not be determined uniquely by diffraction methods." At low concentrations, the main reason of this result seems to be the very low weight of the Ca – O contribution to the structure function. However, in the same study, some small improvement of the fits was obtained when no symmetry was assumed for the cation hydration shell. This lack of symmetry suggests the existence of a static disorder, which makes the determination of the coordination number difficult. In any case, the final parameter values of this study well agree with those of Table 3 ($r_{Ca^{2+}-H_2O}$ = 2.39 Å, $\sigma_{Ca^{2+}-H_2O}$ = 0.14 Å, and $N_{Ca^{2+}-H_2O}$ = 6.9). Again, the second coordination shell around Ca^{2+} ions proved to be unnecessary for good refinements.

Table 3
COORDINATION OF Ca^{2+} ION

Sample no.	Sample	Ref.	Ca^{2+} conc. (*M*)	$r_{Ca^{2+}-H_2O}$	$\sigma_{Ca^{2+}-H_2O}$	$N_{Ca^{2+}-H_2O}$	Geometry	Method	Anion
1	$CaCl_2$ 55.8 H_2O	45	1.0	2.42[f]	0.14[f]	6	Octahedral	FNM1	$Cl(H_2O)_6^-$
2	$CaCl_2$ 26.6 H_2O	45	2.0	2.41	0.15	6	Octahedral	FNM1	$Cl(H_2O)_6^-$
3	$CaCl_2$ 12.3 H_2O	45	4.0	2.42	0.13	6	Octahedral	FNM1	$Cl(H_2O)_6^-$
4	$CaBr_2$ 44.0 H_2O	46	1.2	2.40	0.12	6	Octahedral	FNM1	$Br(H_2O)_6^-$
5	$CaBr_2$ 26.0 H_2O	46	2.0	2.44	0.15	6	Octahedral	FNM1	$Br(H_2O)_6^-$

Calcium salt solutions could be investigated by XRD under extreme conditions of concentration. In fact, the hydrated salt of $CaCl_2$ and $Ca(NO_3)_2$ melt near room temperature; superconcentrated solutions can be then obtained, which are really fused salts. In these cases, the diffraction data were interpreted describing the local order around the cation through models deduced from the crystalline structures of the hydrated salts.[48,49] The system $CaCl_2{\cdot}6H_2O$ appears particularly interesting, as the model is directly suggested by the experimental radial function (Figure 9). If we compare the G(r) function of the fused $CaCl_2{\cdot}6H_2O$ with the one of the solution of composition $CaCl_2{\cdot}12H_2O$ (~4 *M*), we see that between the peaks at 2.40 and 3.20 Å, due to the distances $Ca^{2+}-H_2O$ and $Cl^- - H_2O$, a third one appears at ~2.75 Å; moreover, meaningful correlations exist at medium range (6 to 7 Å). The structure of the crystalline calcium chloride hexahydrate might explain the three main peaks and provides the basis for the interpretation of the medium range order. In fact, in the hexagonal crystals of $CaCl_2{\cdot}6H_2O$, each cation is surrounded by three H_2O molecules at the nearest distance of about 2.35 Å, in a plane containing the Ca atom and perpendicular to the c axis, while six water molecules are at a greater distance of about 2.7 Å in positions 1/2 and $\bar{1}/2$ along the c axis. These six water molecules form $Ca-H_2O-Ca$ chains along the c axis with a Ca-Ca distance of ~4 Å. Cl^- ions are interposed between parallel chains and each of them has six water molecules at distances near 3.2 Å and three others at ~3.6 Å. It is clear that all the details of the correlation function are qualitatively in agreement with the distances existing in the structure described; a fitting procedure showed that the agreement is also quantitative. An interesting feature of this case is the presence in the model of ion-ion positional correlations ($Ca^{2+}-Ca^{2+}$), usually not taken into consideration. This entails that the models based on the independent hydration of the ions do not have absolute validity, but must be replaced, under very high concentrations, by models retaining memory of the hydrated crystal lattice structure.

C. Strontium

Besides the study by Albright,[42] which limitations have been already mentioned, only a diffractometric investigation exists on the Sr^{2+} ion, in which two $SrCl_2$ solutions were examined.[50] In the radial curves of these samples (Figure 10) two peaks indicate $Sr^{2+}-H_2O$ interactions (distance of ~2.65 Å) and $Cl^- - H_2O$ interactions (distance of ~3.20 Å). The

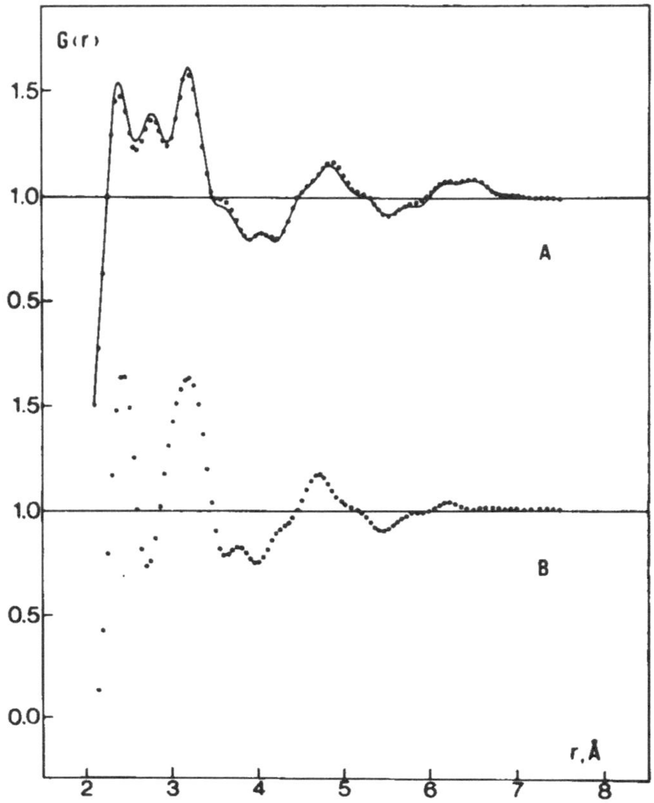

FIGURE 9. (A) Experimental (dots) and model (solid line) correlation functions for liquid $CaCl_2{\cdot}6H_2O$. (B) Experimental correlation function for the solution $CaCl_2{\cdot}12H_2O$. (From Licheri, G., Piccaluga, G., and Pinna, G., *J. Am. Chem. Soc.*, 101, 5438, 1979. With permission.)

great scattering factor of the heavy Sr^{2+} ion makes the peak due to the cationic hydration considerably more important than that due to $Cl^- - H_2O$ pairs. This fact lets us understand that it is possible in the present case to study the cationic coordination with good precision taking advantage of the heavy atom, as it is often made in crystal structure determinations. In this connection, it is worth observing that, in the medium r range, an important peak appears at ~4.75 Å. Its position and especially its height are qualitative, but direct proof that a second cation coordination shell contributes to this peak.

Since the shortest pair distance is due to the dominant term, the coordination number of the cation could be determined by high-angle analysis of the structure function, resulting in a number of $Sr^{2+} - H_2O$ contacts nearly equal to 8. A complete structural model was then proposed, which allowed a satisfactory calculation of the entire structure and correlation functions; the parameters for the cationic coordination from this calculation are given in Table 4. The best fit was obtained introducing interactions between Sr^{2+} ions and second neighbor water molecules; the refined number $N_{Sr^{2+}-H_2O_{II}}$ turned out smaller in the more concentrated solution. Needless to say, the geometry associated with the coordination number determined is not conclusive information; the square antiprismatic geometry (obtained from the simple cubic geometry by rotation of one face by 45° with respect to the opposite one) is frequently preferred in crystals with eightfold coordination, as it minimizes the ligands repulsion, and it resulted slightly preferred also in the solutions here discussed.

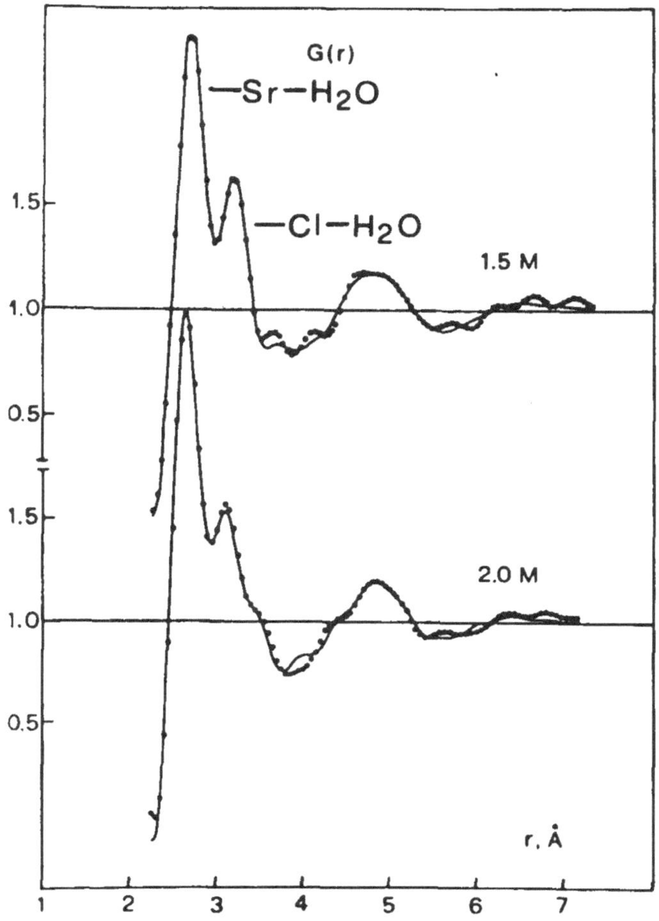

FIGURE 10.. Experimental (dots) and model (solid line) correlation functions for $SrCl_2$ solutions. (From Caminiti, R., Musinu, A., Paschina, G., and Pinna, G., *J. Appl. Cryst.*, 15, 482, 1982. With permission.)

D. Comments

The only result existing for the Sr^{2+} ion seems quite reliable, so that it is possible to examine the sequence of the results for the group of ions Mg^{2+}, Ca^{2+}, and Sr^{2+}. The intermediate ions is undoubtedly in a peculiar situation; the presence of a second hydration shell around the calcium ion, in fact, was not determined conclusively, and even the value for the first coordination number was assigned with some uncertainties. Since the experimental accuracy of the different studies was comparable and, particularly, the experimental work on $CaCl_2$ solutions was very good, a possible explanation of the result lies in a certain irregularity and indefiniteness of the hydration shell of the Ca^{2+} ion. On the other hand, it looks strange that the more voluminous Sr^{2+} ion is also more strongly hydrated than the Ca^{2+} ion, and the results can follow from the different scattering factors. Further investigations on these ions, particularly on the Ca^{2+} ion, in various conditions of concentration and counterions, are desirable to clarify this point.

Finally, we want to stress again the very satisfactory results which have been obtained for solutions containing Mg^{2+} ions. This is the first case in our review of the results (but others will follow) in which unambiguous quantitative information about structural parameters comes solely from XRD results.

Table 4
COORDINATION OF Sr^{2+} ION

Sample no.	Sample	Ref.	Sr^{2+} conc. (*M*)	$r_{Sr^{2+}\text{-}H_2O_I}$	$\sigma_{Sr^{2+}\text{-}H_2O_I}$	$N_{Sr^{2+}\text{-}H_2O_I}$	Geometry	$r_{Sr^{2+}\text{-}H_2O_{II}}$	$\sigma_{Sr^{2+}\text{-}H_2O_{II}}$	$N_{Sr^{2+}\text{-}H_2O_{II}}$	$r_{H_2O_I\text{-}H_2O_{II}}$	$\sigma_{H_2O_I\text{-}H_2O_{II}}$	Method	Anion
1	$SrCl_2$ 34.6 H_2O	50	1.5	2.64	0.14	8	Square antiprism	4.9	0.3	15	2.81	0.10	TSM1	$Cl(H_2O)_6^-$
2	$SrCl_2$ 26.6 H_2O	50	2.0	2.64	0.12	8	Square antiprism	4.9	0.3	9	2.87	0.10	TSM1	$Cl(H_2O)_6^-$

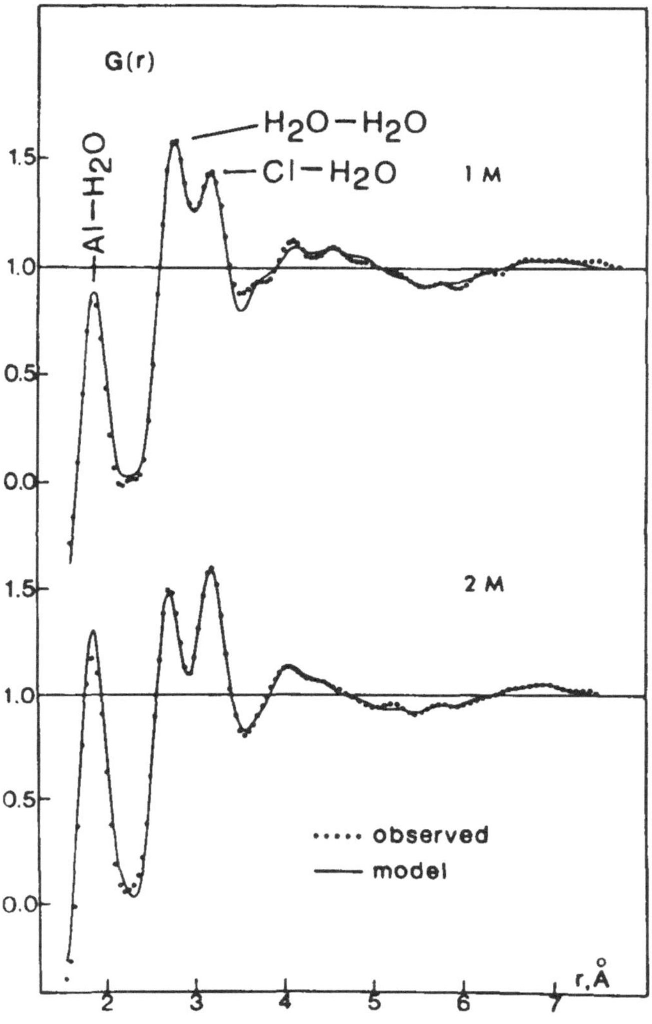

FIGURE 11. Correlation functions for $AlCl_3$ solutions. (From Caminiti, R., Licheri, G., Piccaluga, G., Pinna, G., and Radnai, T., *J. Chem. Phys.*, 71, 2473, 1979. With permission.)

V. GROUP IIIA

A. Aluminum

The correlation functions of two $AlCl_3$ solutions are given in Figure 11;[51] the first peak at ~1.90 Å, due to $Al^{3+}-H_2O$ direct distances, is perfectly isolated. In a solution of $Al(NO_3)_3$ the first peak also includes the O − O distances in the nitrate ions, but their number and value are known.[52] Therefore, a direct evaluation of the first coordination number was possible in both cases, which confirmed the sixfold coordination found in crystalline hydrates of aluminum salts.

Although in the more concentrated solution of $AlCl_3$ the amount of water is just enough to form the hydration spheres of the cations and anions, not even in this case did FNM turn out consistent with the experimental data; in all cases examined, Al^{3+} ions showed to have a second coordination shell. Full hydration parameters are given in Table 5.

While the discussion of the ionic structuring in the most concentrated solution is deferred

Table 5
COORDINATION OF Al^{3+} ION

Sample no.	Sample	Ref.	Al^{3+} conc. (*M*)	$r_{Al^{3+}\text{-}H_2O_I}$	$\sigma_{Al^{3+}\text{-}H_2O_I}$	$N_{Al^{3+}\text{-}H_2O_I}$	Geometry	$r_{Al^{3+}\text{-}H_2O_{II}}$	$\sigma_{Al^{3+}\text{-}H_2O_{II}}$	$N_{Al^{3+}\text{-}H_2O_{II}}$	$r_{H_2O_I\text{-}H_2O_{II}}$	$\sigma_{H_2O_I\text{-}H_2O_{II}}$	Method	Anion
1	$AlCl_3$ 53.9 H_2O	51	1.0	1.90	0.04	6	Octahedral	4.0	0.33	12	2.73	0.02	TSM1	$Cl(H_2O)_6^-$
2	$AlCl_3$ 23.8 H_2O	51	2.0	1.89	0.10	6	Octahedral	4.0	0.22	12	2.68	0.09	TSM1	$Cl(H_2O)_6^-$
3	$Al(NO_3)_3$ ~18 H_2O	52	2.5	1.87	0.15	6	Octahedral	4.0	0.25	12	2.70	0.09	TSM2	Hydrated

to the final section of this chapter, here we wish to point at the very short values of the distances $H_2O_I - H_2O_{II}$. This result demonstrates that, in the environment of ions with high charge density, very strong and linear H-bonds form in solution, as already found in crystalline hydrates; under these circumstances, the H_2O_I water molecules generally point the negative end of their electric dipoles towards the metal ion (the so-called trigonal orientation),[43] and the tilt angle (defined in the Section IV A of the present chapter) is null. Then, the three distances, $r_{Al^{3+}-H_2O_I}$, $r_{Al^{3+}-H_2O_{II}}$, and $r_{H_2O_I-H_2O_{II}}$, which have been independently refined, may be used to calculate the angle $\hat{HOH}$ in the molecules H_2O_I. This angle turned out to be in the range 115 to 120°, in good agreement with the supposed trigonal geometry.

An $Al^{3+} - H_2O$ partial distribution function was presented by Bol and Welzen.[53] It was calculated from the diffraction data of a 0.5 *M* $Al(NO_3)_3$ solution, removing the $H_2O - H_2O$ and $NO_3^- - H_2O$ contributions through the use of diffraction experiments on pure water and on a solution of HNO_3. This procedure assumes that the $H_2O - H_2O$ and $NO_3^- - H_2O$ pair correlation functions obtained in the studies of the simpler systems represent the respective atomic distributions also in the solution under examination. This assumption is questionable, mainly for the $NO_3^- - H_2O$ contribution, in the light of the special structural features existing in acid solutions (see Section II). Notwithstanding, the results obtained by this original method are in very good agreement with those reported in Table 5. The $Al^{3+} - H_2O$ distribution function is given in Figure 12. This curve shows the existence of two coordination shells around the cations. The quantitative analysis showed that the first shell is composed of six water molecules, 1.90 Å distant from the central ion; the evaluation of the second shell composition was more uncertain, the radial curve being consistent with the presence of 12 to 14 H_2O molecules at about 4.10 to 4.15 Å from the Al^{3+} ion.

B. Indium

The first diffractometric investigation about the environment of In^{3+} ions in aqueous solution has been performed, fairly recently, by Maeda and Ohtaki[54] on one concentrated solution of indium perchlorate. Only the composition of the first hydration sphere was determined and the following hydration parameters were proposed: $r_{In^{3+}-H_2O} = 2.15$ Å and $N_{In^{3+}-H_2O} = 6$.

Later, only two other studies had been carried out, one on an $In_2(SO_4)_3$ solution,[55] and one on $InBr_3$ and $InCl_3$ solutions.[56] In the latter case, the interest was essentially devoted to the analysis of the possible existence of a "liquid-type-quasi-close-packing" of cation complexes. The features of the cationic environment were therefore assumed, rather than determined, and it is not possible to understand the reliability of the hydration parameters adopted; indeed, some of them look suspicious (for instance, $In^{3+} - H_2O$ distances fixed at 2.32 Å in the $InCl_3$ solution and at 2.60 Å in the $InBr_3$ one).

The solution of indium sulfate, on the contrary, was examined by the usual modeling technique. In the model of cationic coordination, besides the first coordination shell, a second shell was introduced and also inner-sphere ionic complexation was accounted for. This model is directly suggested by the experimental correlation function (Figure 13), in which the pair distances involving the heavy In^{3+} ion are clearly observable. In fact, besides the peak at 1.5 Å due to S − O distances within sulfate ions and the peak at 2.85 Å due to the various O − O distances always present in aqueous systems, other details are apparent in the G(r): (1) a big peak at 2.15 Å (direct distance $In^{3+} - O$); (2) an important peak at 4.15 Å (second cationic coordination shell); and (3) a small peak at ~3.35 Å, which, according to the crystal structure of $InH(SO_4)_2 \cdot 5H_2O$, is probably to be ascribed to In^{3+}-S distances in the anionic complexes. The most meaningful structural parameters of the final model are given in Table 6.

Hexacoordination with O atoms was imposed to the In^{3+} ion, while the number of In^{3+} −

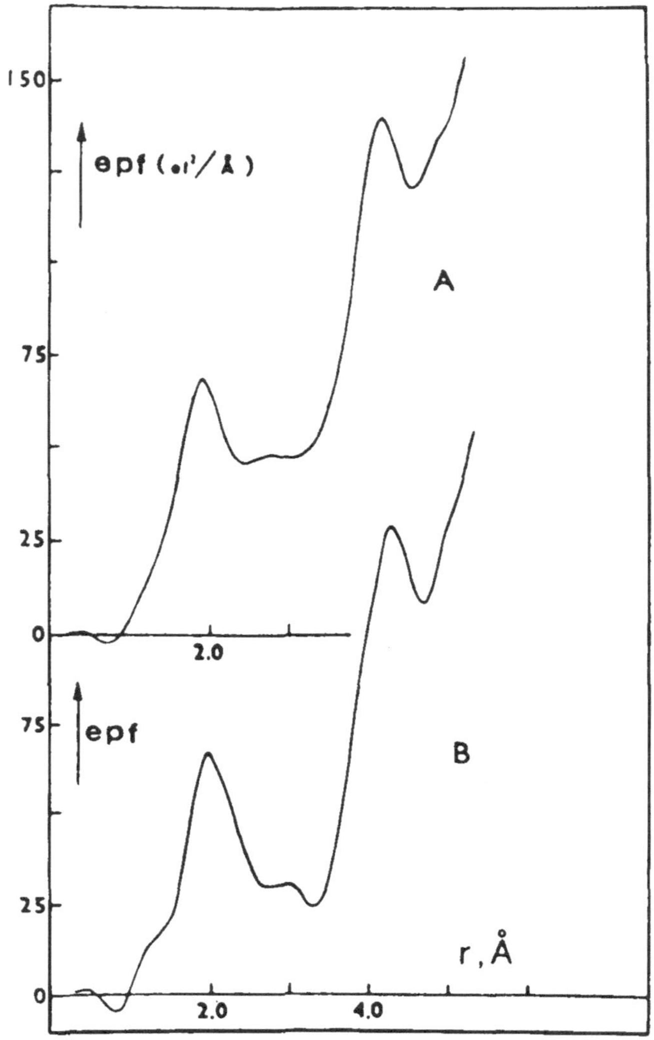

FIGURE 12. Partial cation-water distribution function for the aluminum(III) nitrate (A) and chromium (III) nitrate (B) solutions. (From Bol, W. and Welzen, T., *Chem. Phys. Lett.*, 49, 189, 1977. With permission.)

S contacts was refined; 0.88 oxygen atoms, out of the six enveloping the In^{3+} ion, come from the sulfate ion (the model is the same that we will describe for $Fe_2(SO_4)_3$ solutions). The In − O distance of 2.16 Å was kept for all the six distances in the octahedron InO_6. The In^{3+} − S distance was not refined, but was deduced from the crystal structure mentioned above.

As in Al^{3+} ion solutions, the parameter $r_{H_2O_I-H_2O_{II}}$ has a very small value and suggests the same interpretation, that is, that the water molecule H_2O_I, in dipole orientation towards the cation, forms with external water molecules strong and linear H-bonds. Under this assumption, in this case as well, the distances $r_{In^{3+}-H_2O_I}$, $r_{In^{3+}-H_2O_{II}}$, and $r_{H_2O_I-H_2O_{II}}$ may be used for the calculation of the HÔH angle in the molecule H_2O_I, which, according to the assumed trigonal orientation, turned out to be very near to 120°.

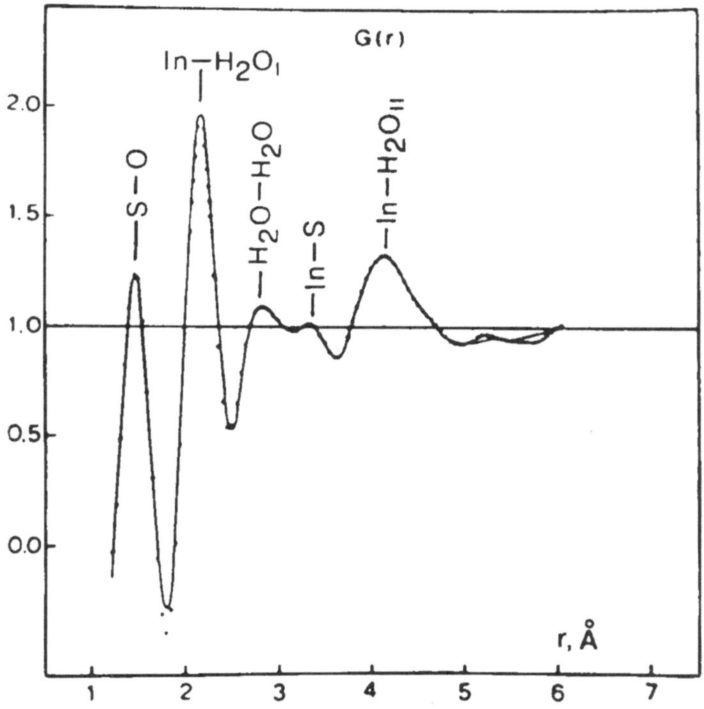

FIGURE 13. Experimental (dots) and model (solid line) correlation functions for a 1.7-*M* indium sulfate solution. (From Caminiti, R. and Paschina, G., *Chem. Phys. Lett.*, 82, 487, 1981. With permission.)

C. Thallium

1. Hydration

The only study of the environment of the aquo Tl^{3+} ion is due to Glaser and Johansson,[57] who examined two solutions of Tl(III) perchlorate about 1 and 2 *M*, acidified with $HClO_4$. The coordination of the cations was determined by analyzing the high-angle region of the structure functions in terms of intramolecular interactions only (in the ions ClO_4^- and in the groups $Tl(H_2O)_N{}^{3+}$). Even if some uncertainty remains in the evaluation of N, the results support the existence in solution of ions $Tl(H_2O)_6{}^{3+}$, with distance $Tl^{3+}-H_2O$ of 2.24 Å and regular octahedral geometry, similarly to that found in the crystal of $Tl(ClO_4)_3{\cdot}6H_2O$.

2. Halide Complex Formation

The main object of the mentioned study,[57] and of one that followed soon after,[58] was the study of the structures of Tl(III) halo-complexes, which are among the strongest metal-ion halide complexes known in aqueous solution. These studies show example of one of the most rational procedures to be adopted in this kind of study. Various investigations (thermodynamic, spectroscopic, crystallographic, etc.) point out at the existence of different halo-complexes; some controversy exists about the details of their formation (for instance, about the existence of polynuclear complexes). Under these circumstances, the structure of the single complexes can be determined by means of diffraction methods, if it is possible to prepare solutions, in which a single complex is the dominant species. On the other hand, this is possible if the stability constants of the various species are known with sufficient precision, and if their values suggest that the different thallium-halide complexes have well-separated regions of existence. Obviously, this problem cannot be tackled by XRD without the help of complementary techniques.

Table 6
COORDINATION OF In^{3+} ION

Sample no.	Sample	Ref.	In^{3+} conc. (*M*)	$r_{In^{3+}-H_2O_I}$	$\sigma_{In^{3+}-H_2O_I}$	$N_{In^{3+}-H_2O_I}$	$r_{In^{3+}-H_2O_{II}}$	$N_{In^{3+}-H_2O_{II}}$	$r_{H_2O_I-H_2O_{II}}$	$\sigma_{H_2O_I-H_2O_{II}}$	$r_{In^{3+}-S}$	$N_{In^{3+}-S}$	Method	Free anion
1	$In_2(SO_4)_3$ 58.9 H_2O	55	1.7	2.16	0.11	6—0.88	4.2	10	2.67	0.06	3.34[f]	0.88	TSM1 + CF	Hydrated

Therefore, diffraction measurements were preceded by a study of Tl-205 NMR of thallium(III) halide complexes, by which the stability constants of the various complexes were determined for the same solutions examined by the diffraction technique.[59] Following these indications, solutions were prepared, in which the large excess of just one type of halo-complex was guaranteed and, furthermore, reasonable estimates of the percentages of the minor complexes were also possible. In these conditions, the structures of the complexes $TlBr_2(H_2O)_4{}^+$, $TlBr_3(H_2O)_2$, $TlBr_4{}^-$, $TlCl_4{}^-$, and $TlCl_6{}^{3-}$ were resolved.

Going into greater detail, the procedure was based once more on the analysis of the high-angle parts of the intensity curves, including the most important intramolecular interactions (Tl − X and X − X, X = Br, and Cl), and accounting for the minor Tl(III) complexes in the final calculations. In Tables 7 and 8 the parameters describing the dominant interactions are given for the various solutions examined. For sample 1 of Table 8, Cl − Cl interactions contribute very little and could not be refined. The frequency factors $N_{Tl^{3+}-X}$, mainly for the bromide solutions, explicitly indicate the type of halo-complex formed. In dubious cases, the complete picture of parameters is of help, as it also gives information on the coordination geometry of the complexes. For example, when $N_{Tl^{3+}-X}$ is close to 4, the ratio $r_{X-X}/r_{Tl^{3+}-X}$ near the value 1.633 expected for a regular tetrahedral arrangement, allows to confirm the coordination numbers and to draw indications about the geometry through the distance parameters, which are more reliable than the frequency factors. On the same grounds, linear $TlBr_2{}^+$ complexes in sample 1 of Table 7, planar triangular complexes $TlBr_3$ in sample 2 of Table 7, and octahedral complexes $TlCl_6{}^{3-}$ in sample 2 of Table 8 have been deduced. Also, the numbers of halide-halide interactions per metal atom, N_{X-X}, are in good agreement with the complexes proposed. The overall view of the results clearly shows the success of these investigations. It is worth observing that, at high X^-/Tl^{3+} ratios, the hexachloro complex forms to a great extent, while the hexabromo complex is a minor species. Besides, in no case is there evidence of polynuclear complexes.

Finally, in the lower Br^- complexes, the dominating contribution of the anions to the scattering prevented the estimation of water present in the coordination shells. For a picture of the whole coordination polyhedra, the knowledge of the crystalline structures of various halo-complexes and the comparison of Tl-NMR chemical shifts for solids and solutions were exploited in these cases, thus suggesting the structures sketched in Figure 14. The trigonal-bipyramidal shape of the complex $TlBr_3(H_2O)_2$ is noteworthy.

VI. GROUP IVA (Sn,Pb)

The hydration of the ions coming from the metals of the group IVA have hardly ever been studied by XRD. The Swedish School has directed its efforts to the characterization of the polynuclear aggregates formed in Sn and Pb salt solutions by alkaline hydrolysis phenomena.[60,61] Only one ~3.3 *M* solution of $Sn(ClO_4)_2$ has been investigated in acid medium which prevents hydrolysis;[61] experimental data from this study have been reexamined in an investigation, in which EXAFS spectroscopy was applied to a solution of very similar composition.[62]

The interest for this solution followed from crystal structures determinations, in which the 5s electron pair of Sn(II) in the ground state has revealed a stereochemically active role, since it forms a nonspherical charge distribution around the element, giving rise to very distorted coordination polyhedra. It was then interesting to check whether analogous phenomena take place in solution as well. The answer obtained is not very clear. In fact, a peak is present in the radial curve at ~2.3 Å, which, besides O − O distances in the ClO_4^- ions, includes the shortest $Sn(II) - H_2O$ distances. The quantitative analysis of this contribution indicate that the number of H_2O molecules at this distance is not high (in the range 2 to 3 in the evaluation by Johansson et al.,[61] in the range 3 to 4 according to the

Table 7
PARAMETERS FOR Tl-Br AND Br-Br INTERACTIONS IN THALLIUM(III) BROMIDE COMPLEXES

Sample no.	Sample	Ref.	Tl^{3+} conc. (*M*)	Br/Tl ratio	$r_{Tl^{3+}-Br^-}$	$\sigma_{Tl^{3+}-Br^-}$	$N_{Tl^{3+}-Br^-}$	$r_{Br^--Br^-}$	$\sigma_{Br^--Br^-}$	$N_{Br^--Br^-}$	$r_{Br^--Br^-}/r_{Tl^{3+}-Br^-}$	Method
1	$TlBr_2ClO_4$ 15.1 H_2O	57	2.7	2.0	2.48	0.06	2.0[f]	4.94	0.14	1[f]	1.99	HAA
2	$TlBr_3$ 16.8 H_2O	57	2.6	3.0	2.51	0.06	2.7	4.38	0.22	3[f]	1.74	HAA
3	$TlBr_3$ 1.1 LiBr 14.4 H_2O	57	2.6	4.1	2.56	0.06	3.8	4.16	0.22	6	1.62	HAA
4	$TlBr_3$ 1.5 LiBr 46.8 H_2O	57	1.0	4.5	2.57	0.07	4.0	4.18	0.22	8	1.63	HAA
5	$TlBr_3$ 9.3 LiBr 33.0 H_2O	57	1.1	12.3	2.58	0.04	4.0	4.22	0.22	7	1.64	HAA

Table 8
PARAMETERS FOR Tl-Cl AND Cl-Cl INTERACTIONS IN THALLIUM(III) CHLORIDE COMPLEXES

Sample no.	Sample	Ref.	Tl^{3+} conc. (*M*)	Cl/Tl ratio	$r_{Tl^{3+}-Cl^-}$	$\sigma_{Tl^{3+}-Cl^-}$	$N_{Tl^{3+}-Cl^-}$	$r_{Cl^--Cl^-}$	$\sigma_{Cl^--Cl^-}$	$N_{Cl^--Cl^-}$	$r_{Cl^--Cl^-}/r_{Tl^{3+}-Cl^-}$	Method
1	$TlCl_3$ 1.8 LiCl 37.2 H_2O	58	1.3	4.8	2.44	0.10	4.6					HAA
2	$TlCl_3$ 10.4 LiCl 45.7 H_2O	58	0.9	13.4	2.59	0.09	5.5	3.69	0.20	13	1.42	HAA
3	$TlCl_3$ 1.1 LiCl 15.5 H_2O	58	2.6_5	4.1	2.43	0.09	4.2	3.96	0.20	5	1.63	HAA

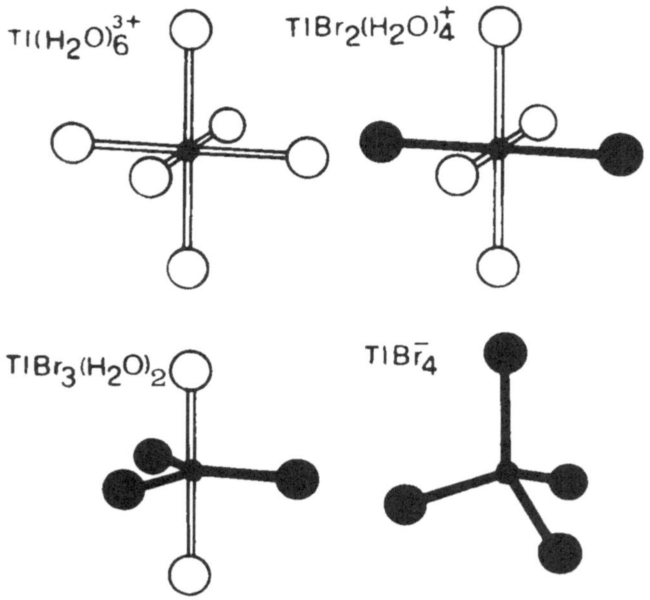

FIGURE 14. Suggested structures for the $TlBr_N(H_2O_M)^{3-N}$ complexes in aqueous solution. Dark spheres represent Br and light spheres, H_2O.

conclusions by Yamaguchi et al.[62]). These results agree with a general view of the stereochemically active Sn(II) ion. However, unfortunately, possible slightly longer $Sn-H_2O$ distances (at about 2.8 Å) could not be identified unequivocally, as various contributions involving the solvent fall in this r range. The omission in the interpretation of the longer $Sn(II)-H_2O$ term, therefore, does not compromise the modeling calculations and, as a consequence, it is impossible to understand if longer, discrete $Sn-H_2O$ distances are absent or unobservable. We can anticipate that EXAFS analysis does not shed more light on this subject.

VII. RARE EARTH IONS

It is known that the elements named rare earth (or lanthanides) are characterized by extreme likeness of chemical-physical properties, since the variation of the number of 4f electrons from La to Lu involves the filling of a shell strongly screened by electron shells with a higher principal quantic number. Likewise, it is known that homogeneity in their behavior is not absolute. Sophisticated systematic investigations of electrolyte solutions have collected a great deal of thermodynamic data (solubility, viscosity, free energies, and others mentioned in Reference 63), showing that the expected monotonous variations of these properties along the series break off in going from Nd to Tb, and then start again in the last group of ions. This phenomenon is often called "the anomaly in the rare earth series".

Searching for an explanation of this behavior, the hypothesis was put forth that the anomaly between Nd and Tb arises from a difference in the number of water molecules forming the first coordination sphere of light and heavy rare earth ions, exciting great interest for the structural studies of these systems. It surely is a challenge for the scholars of hydration phenomena. In fact, the electronic structure of these elements makes the interaction between a rare earth ion and water molecules predominantly ion-dipole and electrostatic in character. The study of hydration along the series should provide invaluable information on the dependence of this interaction from parameters varying slightly and continously, like the ionic radius, the charge density, and so on.

Therefore, several XRD studies have been performed,[63-70] but the following discussion will be almost entirely dedicated to three papers by Habenschuss and Spedding.[63-65] The first reason for this choice lies in the fact that only the studies of the authors mentioned are rigorously systematic, as required if possible small differences of some property along the series are looked for. However, we cannot conceal a second reason, that is, the unreliability of a second group of papers on this subject,[66-70] thoroughly discussed in Reference 64. In the investigations of Habenschuss and Spedding, ten solutions of rare earth chlorides were examined near the saturation concentration. Every effort was made to treat the solutions uniformly in terms of sample preparation, intensity measurements, data reduction, and resolution of the radial distribution functions. Object of these precautions was to maximize the accuracy and precision of the results and especially to minimize changes in the systematic errors from element to element, so that weak trends along the series could be identified. Besides, the various lanthanides (light, intermediate, and heavy elements) were actually run in random sequence, in order to ''minimize the chance that accumulating operator experience would correlate with a sequential examination of the rare earth series''.

The normalized radial correlation functions, G(r), for the ten solutions are given in Figure 15. This comprehensive view of results illustrates the uniformity of the radial functions. The changes from element to element are small and subtle, and only cumulative differences along the series are apparent. Three main peaks appear in the G(r)s in the distance ranges of 2.3 to 2.6, 3.1 to 3.2, and 4.7 to 5.0 Å. The positions of the first peak agree with the cation-water distances found in the various crystalline hydrates, clearly showing that this peak is due to cation-water interactions. The second peak includes H_2O-H_2O and Cl^--H_2O interactions, while the third peak is ascribed to indirect ion-water distances, as well as to ion$-Cl^-$ outer-sphere ion pairs, which are likely to exist at the concentrations examined. Significant amounts of ion$-Cl^-$ inner-sphere ion pairs, observed in hydrated crystals, were ruled out by Habenschuss and Spedding, because of the lack of structural details in the radial curves at about 2.7 Å and on the grounds of various spectroscopic and thermodynamic evidences.

Even in absence of inner-sphere ion pairs, the evaluation of direct cationic coordination is not easy. This is why the rare earth ions are voluminous and the ion-water distances are close to H_2O-H_2O (and Cl^--H_2O) ones, so that the first peak of the radial curves is not well separated from the following ones; the situation worsens progressively going from the smallest (Lu^{3+}) to the biggest (La^{3+}) ion. Thus, the analysis of the first peak of the radial curves by the ideal peak shapes, though not requiring complete geometrical models for the hydrated ions, yet requires the simulation of the adjacent peaks. This can be accomplished in various ways, for instance, independently refining all the parameters introduced, or imposing some restriction to certain parameters (coordination number of Cl^-, Cl^--H_2O distance, and so on), in order to treat the different cases with the utmost homogeneity. However, luckily, the hydration parameters of the cations turned out to be relatively independent from these different treatments, particularly when the ion-water distances have the smallest values. In Table 9 the values of the hydration parameters of the cations are given; the procedure adopted does not allow to determine other structural parameters (e.g., Cl^--H_2O parameters). Here the distance values are reported with three-decimal digits and the coordination numbers with two digits (as in the original papers) to make the comparison of the various cases clear. The regularity of the sequence shows that the uniformity of the treatment of the various samples has lead to very good relative accuracy; the absolute accuracy, however, is similar to that of all the other studies presented.

From Table 9, it is apparent that light rare earths (with great radius) have a coordination number around 9, the heavy ones around 8, with a transition from one value to the other,

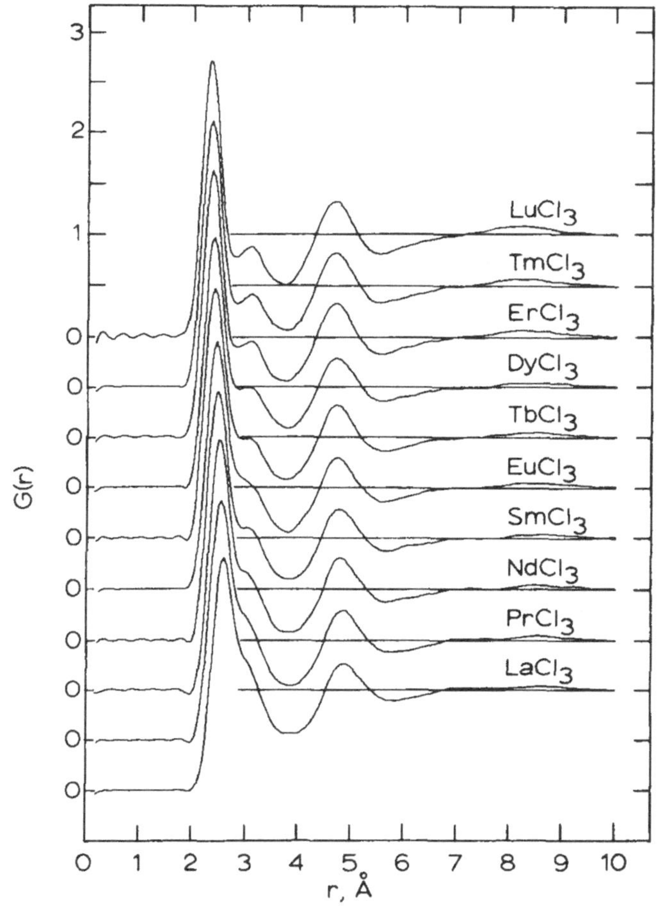

FIGURE 15. Correlation functions for ten concentrated rare earth chloride solutions. (From Habenschuss, A. and Spedding, F. H., *J. Chem. Phys.*, 73, 442, 1980. With permission.)

which takes place between Sm^{3+} and Eu^{3+} ions. Correspondingly, a break exists in the relation between the ion $-$ H_2O distance and the ionic radius of the cation, since this distance depends not only on the length of the radius of the ions along the series, but also on the variation of coordination of the ions, which occurs in the middle of the series. This is better visualized in Figure 16, where the ion-water distances and the coordination numbers are plotted against the ionic radii. In the same figure, also the values of ion-water distances in the crystals are given and they show a break in a range different from that found in the solutions; this is not surprising, as in the crystals the ions are surrounded by water molecules and by Cl^- ions as well.

In conclusion, even if the absolute determinations can be not completely exact owing to resolution problems, the uniformity of the experiments permits to affirm that a variation in the hydration takes place ''not abruptly'' along the series of rare earth ions.

An ion not investigated by Habenschuss and Spedding, that is, Ce^{3+}, has been studied afterwards in nitrate medium.[71] While the $Ce^{3+} - H_2O$ distance proposed (2.55 Å) interposes itself correctly between the $La^{3+} - H_2O$ and the $Pr^{3+} - H_2O$ ones, the coordination number (~8) is 1 unit lower than that of the nearest ions in the group. The origin of this difference is not clear. Actually, in the study of the $Ce(NO_3)_3$ solution, an interpretation method (analysis

Table 9
COORDINATION OF RARE EARTH IONS

Sample no.	Sample	Ref.	Metal ion conc.(m)	$r_{Me^{3+}\text{-}H_2O}$	$N_{Me^{3+}\text{-}H_2O}$	Method
1	$LaCl_3$ 14.6 H_2O	64	3.8	2.580	9.13	PSA
2	$PrCl_3$ 14.6 H_2O	64	3.8	2.539	9.22	PSA
3	$NdCl_3$ 16.5 H_2O	64	3.4	2.513	8.90	PSA
4	$SmCl_3$ 17.2 H_2O	65	3.2	2.474	8.80	PSA
5	$EuCl_3$ 17.2 H_2O	65	3.2	2.450	8.34	PSA
6	$TbCl_3$ 15.9 H_2O	63	3.5	2.409	8.18	PSA
7	$DyCl_3$ 16.9 H_2O	63	3.3	2.396	7.93	PSA
8	$ErCl_3$ 15.7 H_2O	63	3.5	2.369	8.19	PSA
9	$TmCl_3$ 15.3 H_2O	63	3.6	2.358	8.12	PSA
10	$LuCl_3$ 15.3 H_2O	63	3.6	2.338	7.97	PSA

of the whole structure function by means of geometrical models) completely different from that used by Habenschuss and Spedding was applied, so that the absolute determinations are likely to be not homogeneous; besides, an inner-sphere ion pairs model was used for $Ce(NO_3)_3$, which entails a different structural arrangement of the near environment of the ion in nitrate and chloride solutions, so that the differences in the coordination numbers might be true.

Finally, La, Sm, Tb, and Er have been investigated in perchlorate solutions (absence of complex formation) and in selenate solutions (presence of complex formation).[72] In this work the interpretation of the experimental data is particularly elaborate. To separate the cation – O distributions from the O – O ones, a method of isomorphous substitutions was adopted for the smallest ions (Tb and Er), using Y(III) as the ion of supposed isostructurality. For the biggest ions (La and Sm) the separation of the cation-water contributions required an even more questionable subtraction of the partial contributions of the lightest atoms to the radial curves, as deduced from the Y(III) solutions and considered equal in the rare earth solutions. For admission of the authors themselves, the differences in coordination polyhedra of Y(III) and rare earths affect the precision of the determination (always critical in difference methods), particularly in the cases of La and Sm. The results obtained confirm the values of cation-water distances proposed by Habenschuss and Spedding, but do not show any variation of the coordination number along the series (the value proposed was 8 in all the cases). This result, partially not clear because of the complex analysis involved, does not seem to convey much information, as, with different evaluation procedures used for small and big ions, the criterion of uniformity has been lost, which is essential to estimate variations in the range of 10 to 15% in the coordination numbers along the series.

VIII. GROUP VIB

The metals of the group VIB (chromium, molybdenum, and tungsten) exhibit the oxidation state + 6, which is practically the only interesting one for the two heavier elements. However,

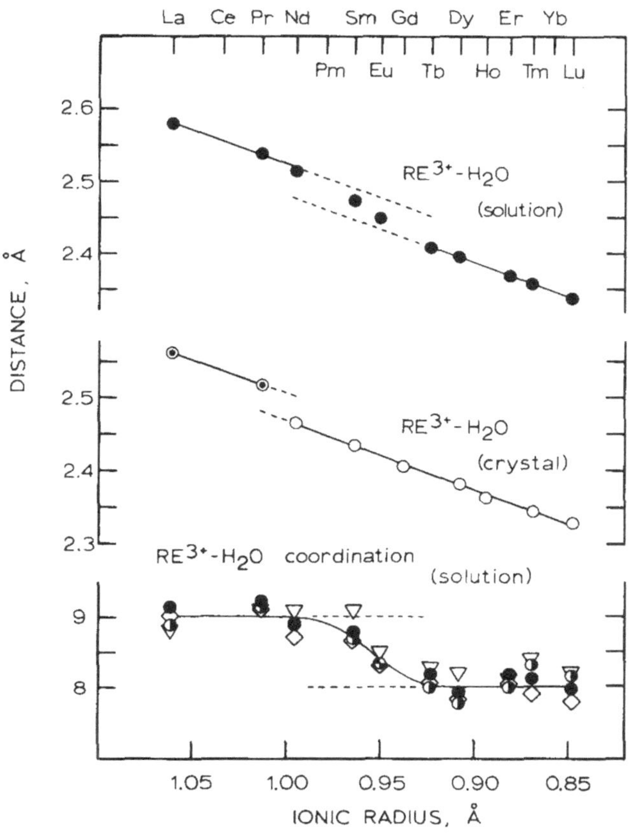

FIGURE 16. (Top) average ion-water distances in concentrated rare earth chloride solutions. (Middle) average ion-water distances in crystalline rare earth chloride hydrates. (Bottom) average water coordination of the rare earth ions in concentrated chloride solutions; the different symbols indicate different calculation methods.[63-65] (From Habenschuss, A. and Spedding, F. H., *J. Chem. Phys.*, 73, 442, 1980. With permission.)

in this state, the metals form oxygenated species (cromates, molybdates), and in solution they are present in anionic groups. The molybdenum oxide, in particular, dissolves in alkaline medium, giving rise to a complicated series of polymeric oxyanions; some of these solutions have been investigated by XRD by Johansson et al.[73-76]

Only Cr(III) is stable in water as monoatomic cation; Cr(II), on the contrary, shows reducing properties. This paragraph will therefore deal exclusively with aqueous solutions of Cr(III) salts.

A. Chromium

1. Hydration

The studies of Cr(III) salt solutions hold a prominent position in the outline of XRD applications to liquid systems. Since 1950, a series of experiments using ^{18}O as a tracer showed that the $Cr(H_2O)_6^{3+}$ ion exchanges its hydration water quite slowly, the half-life being about 40 hr.[77] The existence of ion-water hydration complexes with defined structural and kinetic identities, supported by several experimental proofs, induced XRD studies to search for possible order phenomena beyond the first cationic hydration shell.

A solution of chromium(III) chloride was investigated first, which was obtained by dis-

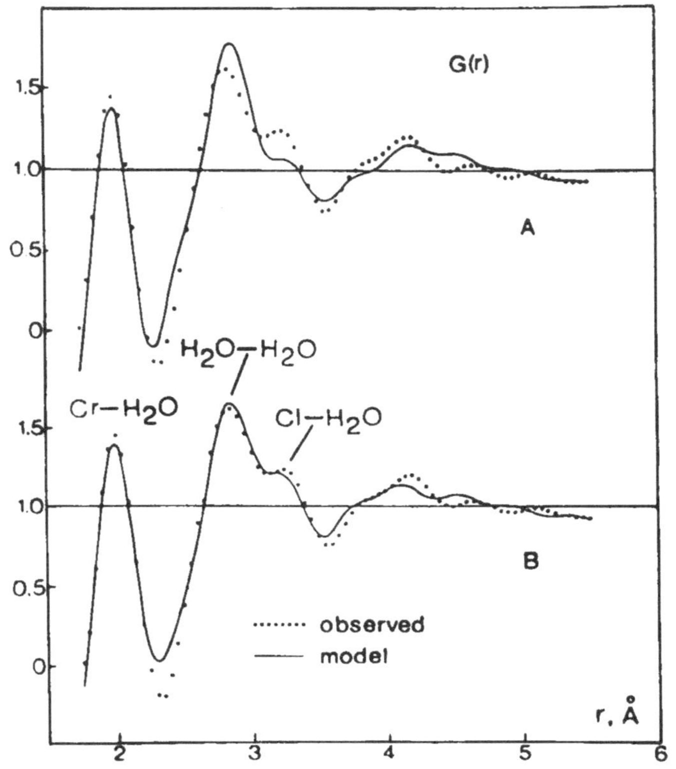

FIGURE 17. Observed (dots) and model (solid line) correlation functions of a 0.85-*m* solution of $[Cr(H_2O)_6]Cl_3$. The species assumed in the model are $Cr(H_2O)_6{}^{3+}$ and $Cl(H_2O)_6{}^{-}$ without a second hydration shell around the cation (curve A) and with a second hydration shell of 12 water molecules around the cation (curve B). (From Caminiti, R., Licheri, G., Piccaluga, G., and Pinna, G., *J. Chem. Phys.*, 65, 3134, 1976. With permission.)

solving the violet salt $[Cr(H_2O)_6]Cl_3$, properly prepared from the commercial green salt $[Cr(H_2O)_4Cl_2]Cl{\cdot}2H_2O$.[78] The presence of the Cr^{3+} ions as fully hydrated species was checked through careful titrations of the free Cl^- anion.

The analysis of XRD data led to the conclusion that the first neighbor model is inadequate to describe the structure of the solution considered. The data were only consistent with a model in which the hydrated ions $Cr(H_2O)_6{}^{3+}$ interact strongly with a second shell of water molecules. The results of the simulations through FNM and TSM are given in Figure 17, which shows clearly that TSM is the better choice. This study represents the first case in which a single scattering experiment supported unambiguously the existence of an ordered second coordination shell around a cation. The same conclusion was reached in a subsequent study on two $Cr(NO_3)_3$ solutions.[79]

These investigations pointed essentially at verifying the decisive contribution of a second coordination shell for a satisfactory simulation of the experimental data, and no particular meaning was attached to the parameters describing it. Only later, the analysis was devoted to a better understanding of the interactions between hydration water (H_2O_I) and external water (H_2O_{II}).[80]

Two facts encouraged this analysis : (1) the observation that the H-bonds connecting H_2O_I and H_2O_{II} molecules around trivalent hexacoordinated ions in crystalline hydrates are very short and linear, falling in the category which Brown[81] defines as strained; and (2) the observation, made by Friedman and Lewis,[82] that around the same ions, the water molecules tend to assume trigonal orientation (that is, they point the negative end of their electric dipole vector towards the metal ion).

Two models of Cr^{3+} hydration were then compared. In the first (used in the two early studies of $CrCl_3$ and $Cr[NO_3]_3$), the water molecules H_2O_I near the Cr^{3+} ions were arranged in pyramidal orientation (i.e., pointing a lone pair towards the ion), so that each of them could H-bond three external water molecules; the angle $H-\hat{O}_I-H$ was fixed at 109°, which is the average value found in hydrated crystals,[43] so that the two O—H bonds and the two lone pairs form a regular tetrahedron. In the second model, the H_2O_I molecule was set on a plane including the Cr^{3+} ion; owing to the linearity of the bonds $O_I-H....O_{II}$, this plane also includes the O_{II} atoms of the external water molecules. In this case, as the trigonal orientation is probably associated with sp^2 hybridation of the O atom, which could give rise to a widening of the $H-\hat{O}_I-H$ angle beyond the value of 109°,[43] the angular value was not kept fixed in the model; its value was calculated *a posteriori* from the final parameters $r_{Cr^{3+}-H_2O_I}$, $r_{H_2O_I-H_2O_{II}}$, and $r_{Cr^{3+}-H_2O_{II}}$, which were independently refined. The same procedure was later adopted in the study of Al^{3+} hydration, as described above.

The second model provided a fit slightly, but meaningfully, better than the first one. From the Figures 18 and 19, it is apparent that the peak at ~4.0 Å, where the $Cr^{3+}-H_2O_{II}$ distances fall, is better described when the trigonal model is used. The parameters from the best fits are given in Table 10. Obviously, first coordination numbers different from 6 were not tried in this case. The reproducibility of parameter values in the various cases is noteworthy, as well as the expected short values of the $H_2O_I-H_2O_{II}$ distances are. Under the hypotheses mentioned above, the angle of the H_2O_I molecule was calculated and it turned out very close to 120°, thus confirming the trigonal geometry with sp^2 hybridation of the O atom.

The values of the parameters in Table 10 are in very good agreement with those proposed in a study of a solution of $Cr_2(SO_4)_3$, acidified with H_2SO_4;[83] small diffrences are quite expected, since the formation of inner-sphere chromium-sulfate complex was accounted for in the model. These results will be discussed later, together with other similar cases.

A $Cr^{3+}-H_2O$ partial distribution function was reported by Bol and Welzen[53] in a paper mentioned above, in which results on Al^{3+} hydration were also given (Figure 12). While referring the reader to section V A for description of the method used, we recall here that the quantitative results obtained are in very good agreement with the ones reported above, the only meaningful difference concerning the $Cr^{3+}-H_2O_{II}$ distance, for which a value slightly longer than the one reported in Table 10 was proposed (4.2 to 4.25 Å against 4.05 to 4.10 Å).

2. *Chloride Complex Formation*

Solutions of chromium chloride were also investigated with the aim of studying chloro-complex formation.[84] Therefore, they were prepared directly from the commercial green salt, since it is known that, dissolving this salt in water, the dark green color due to the ion $Cr(H_2O)_4Cl_2^+$ turns pale as consequence of the following equilibria:

$$\underset{\text{II}}{Cr(H_2O)_4Cl_2^+} + Cl^- + 2H_2O = \underset{\text{I}}{Cr(H_2O)_5Cl^{2+}} + 2Cl^- + H_2O = \underset{\text{O}}{Cr(H_2O)_6^{3+}} + 3Cl^-$$

Besides, dissolving the salt in concentrated HCl, another Cr^{3+} complex, the neutral pale green $Cr(H_2O)_3Cl_3$(III) was identified. Two solutions in water and one solution in HCl were investigated (the symbols O, I, II, III will be used in the following to designate the complexes and to indicate the number of halogen atoms present in their coordination shells). Obviously, the diffractometric measurements of these samples started when the equilibrium was reached.

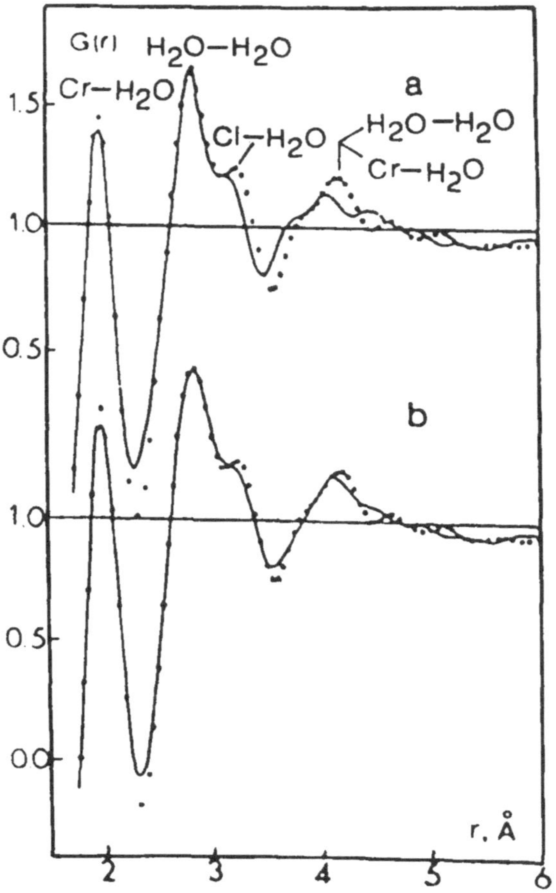

FIGURE 18. Experimental (dots) and model (solid line) correlation functions obtained for the $CrCl_3$(violet) solution with a trigonal orientation of H_2O_I (b) and with a pyramidal orientation (a). (From Caminiti, R., Licheri, G., Piccaluga, G., and Pinna, G., *J. Chem. Phys.*, 69, 1, 1978. With permission.)

The radial distribution curves, given in Figure 20, clearly show the existence of chlorocomplexation. In fact, between the $Cr^{3+}-H_2O_I$ peak at 2.0 Å and the H_2O-H_2O and Cl^--H_2O peak at 2.9 to 3.1 Å, a third one appears at about 2.3 Å which can be ascribed to the pair $Cr^{3+}-Cl^-$, on the basis of ionic radii and crystallographic data. This peak is more evident in the most concentrated solution, as expected, and even more prominent in the solution with atomic ratio $Cl^-/Cr^{3+} > 3$.

The following step of the analysis was the evaluation of the average number of Cl^- ions bound to the metal atom, $N_{Cr^{3+}-Cl^-}$ (called N_{Cl} for brevity). N_{Cl} results from the concentrations of the various chromo-complexes present in each solution. For each sample, therefore, several theoretical structure functions were calculated by assuming different sets of complexes which concentrations were treated as variables to be refined, while their sum was kept constant and equal to the total metal concentration. Only for the acid solution, the neutral complex $Cr(H_2O)_3Cl_3$ was assumed to be present in accordance with elution test results in column resin. While further details will be given in the discussion of this chapter, we can anticipate that the method is accurate for the determination of the average complexation number, but is rather poor in discriminating between sets of complexes with the

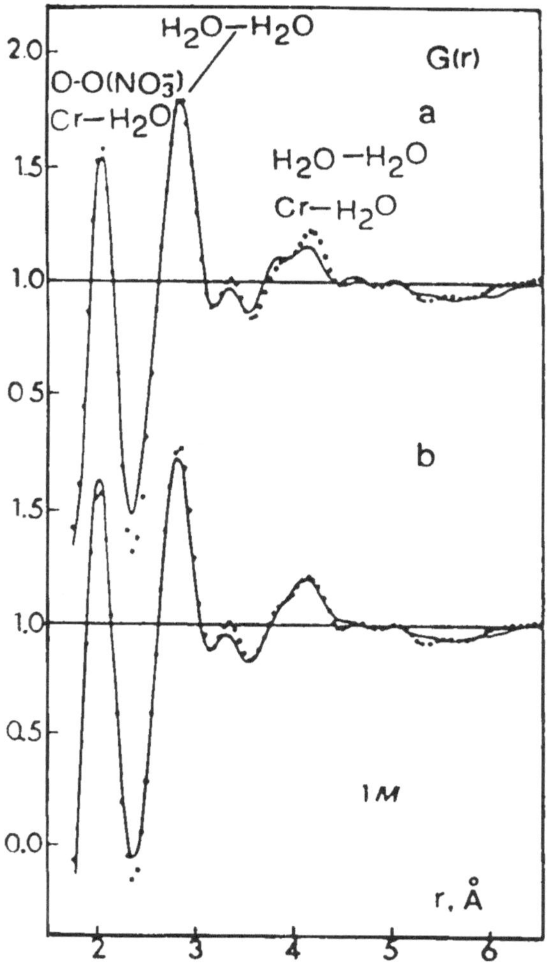

FIGURE 19. Experimental (dots) and model (solid line) correlation functions obtained for a 1-*M* solution of $Cr(NO_3)_3$ with a trigonal orientation of H_2O, (b) and with a pyramidal orientation (a). (From Caminiti, R., Licheri, G., Piccaluga, G., and Pinna, G., *J. Chem. Phys.*, 69, 1, 1978. With permission.)

same N_{Cl} value. The final simulation of the structure functions was based on the observation that the N_{Cl} values determined were in good agreement with those inferred from stability constants, although the latter ones had been measured in chemical conditions (metal concentration, ionic medium, and so on) very far from those of the cases investigated by XRD. Thus, the stability constants provided the preliminary information to divide N_{Cl} into different possible contributions, omitting some complexes which should be present in very small amounts (see Figure 21).

In the final calculations, interatomic distances and root mean square deviations were refined; the best values of the cationic coordination parameters are given in Table 11 (A) and the percentages of the complexes considered are also reported (B). The sixfold coordination of the Cr^{3+} ion was assumed throughout these investigations.

Parameters r and σ of Table 11 are average values which are valid for all the complexes; clearly, the method cannot give a more precise description of single entities when more than one complex is present. Furthermore, the parameters describing the second coordination

Table 10
HYDRATION OF Cr^{3+} ION

Sample no.	Sample	Ref.	Cr^{3+} conc.	$r_{Cr^{3+}-H_2O_I}$	$\sigma_{Cr^{3+}-H_2O_I}$	$N_{Cr^{3+}-H_2O_I}$	Geometry	$r_{Cr^{3+}-H_2O_{II}}$	$\sigma_{Cr^{3+}-H_2O_{II}}$	$N_{Cr^{3+}-H_2O_{II}}$	$r_{H_2O_I-H_2O_{II}}$	$\sigma_{H_2O_I-H_2O_{II}}$	Method	Anion
1	$CrCl_3$(violet) 0.6 HCl 64.8 H_2O	80	0.8_5 *m*	1.99	0.07	6	Octahedral	4.0_5	0.28	12	2.68	0.01	TSM1	$Cl(H_2O)_6^-$
2	$Cr(NO_3)_3$ 50.8 H_2O	80	1.0 *M*	2.00	0.08	6	Octahedral	4.1	0.26	12	2.72	0.06	TSM1	Not hydrated
3	$Cr(NO_3)_3$ 24.4 H_2O	80	2.0 *M*	1.99	0.10	6	Octahedral	4.1	0.27	12	2.70	0.07	TSM1	Not hydrated

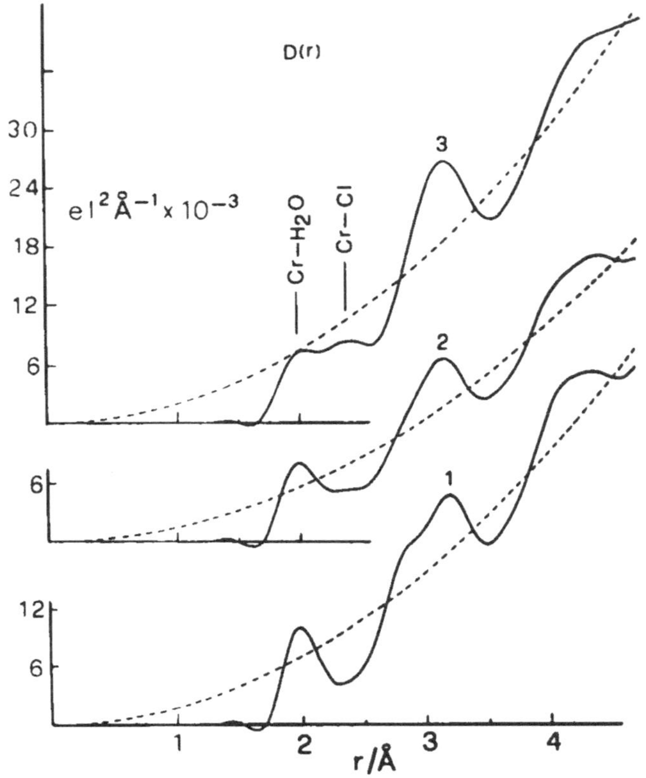

FIGURE 20. Radial distribution functions, D(r), for three solutions of $[Cr(H_2O)_4Cl_2]Cl{\cdot}2H_2O$ (solid lines). Dashed lines are the corresponding $4\pi r^2\rho_0$ functions. (From Magini, M., *J. Chem. Phys.*, 73, 2499, 1980. With permission.)

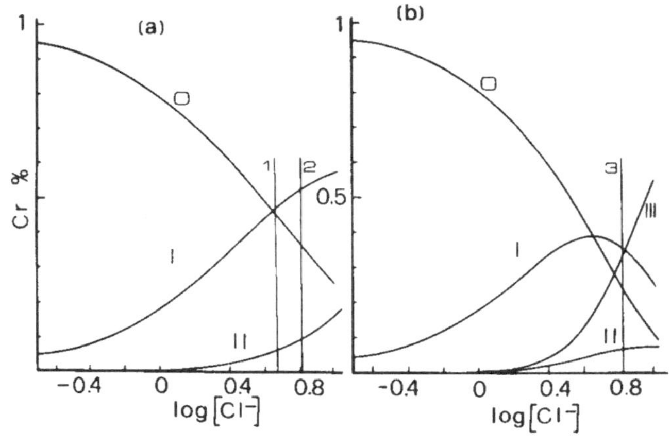

FIGURE 21. Fractions of chromium bonded in the different chloride complexes (indicated by O to III, see the text) as a function of the chloride concentration. The calculated complex distributions for the solutions examined, assuming the equilibrium constant to be valid, are indicated by vertical lines. For the equilibrium constants used, see Reference 84. (From Magini, M., *J. Chem. Phys.*, 73, 2499, 1980. With permission.)

Table 11
COORDINATION OF Cr^{3+} ION IN $CrCl_3{\cdot}6H_2O$ (GREEN) SOLUTION

A. Structural Parameters

Sample no.	Sample	Ref.	Cr^{3+} conc. (*M*)	Cl/Cr ratio	$r_{Cr^{3+}-H_2O_I}$	$\sigma_{Cr^{3+}-H_2O_I}$	$N_{Cr^{3+}-H_2O_I}$	$r_{Cr^{3+}-Cl^-}$	$\sigma_{Cr^{3+}-Cl^-}$	$N_{Cr^{3+}-Cl^-}$	$N_{Cr^{3+}-Cl^-}$ from k_{eq}	Method	Free anion
1	$CrCl_3$ 26.5 H_2O	84	2.0	3.0	1.97	0.07	6—0.65	2.32	0.09	0.65	0.60	TSM1 + CF	$Cl(H_2O)_6$
2	$CrCl_3$ 17.9 H_2O	84	2.8	3.0	1.98	0.07	6—1.1	2.32	0.09	1.1	0.73	TSM1 + CF	$Cl(H_2O)_6^-$
3	$CrCl_3$ 1.9 HCl 24.5 H_2O	84	1.9	4.9	1.96	0.08	6—1.5	2.30	0.10	1.5	1.57	TSM1 + CF	$Cl(H_2O)_6^-$

B. Percentages of the Complexes Used

Sample no.	Ref.	Cr^{3+} conc. (*M*)	Cr^{3+}	$CrCl^{2+}$	$CrCl_2^+$	$CrCl_3$
1	84	2.0	35	65		
2	84	2.8	20	50	30	
3	84	1.9	25	37.5		37.5

shell have not much meaning, as the presence of Cl^- in the first (and, likely, also in the second) coordination sphere implies many slightly different distances contributing to the second shell peak, which is in fact quite broad. The corresponding parameters, therefore, are not given in the table. However, it is worth noticing that even in this case, the $H_2O_I - H_2O_{II}$ distances turned out very short, (2.68 to 2.70 Å in the neutral solutions and 2.58 Å in the acid solution). The last value is probably ascribable to the contribution of very short O....O distances around free H_3O^+ ions (see Section II).

IX. GROUP VIIB

A. Manganese

This is the only species of the group for which interesting, even if not abundant, information exists.[85-90]

1. Hydration

The characteristics of the first coordination shell of the fully hydrated Mn^{2+} have been determined by Ohtaki et al.[86] in an investigation dealing with the structure of divalent first-row transition metal ions in aqueous solution. Mn(II) was examined in a perchlorate solution. The $Mn^{2+} - H_2O_I$ distances give the main contribution to the peak at about 2.2 Å in the D(r), which also includes contributions from the O – O distances within the ClO_4^- ions. Deducing the structural parameters of the perchlorate ions by literature results, the first coordination parameters of Mn^{2+} were determined both through the ideal peak shape analysis of the radial curve, and by simulation of the high angle part of the i(s) function. The two calculations converged, within small uncertainty, to the coordination number of 6 and $Mn^{2+} - H_2O_I$ distance of 2.20 Å.

2. Complex Formation

a. With Chloride Ions

Three solutions of $MnCl_2$ have been examined.[90] Similar to the case of green chromium(III) chloride solutions, the radial curves (Figure 22) show evidence of Cl^- ions in the first cationic coordination shell. The first peak of the D(r)s appears in fact doubly composed; the first component, centered at the distance found for the $Mn^{2+} - H_2O$ pair, lowers at increasing concentration, while the second component heightens. The latter is clearly ascribable to $Mn^{2+} - Cl^-$ distances. Unfortunately, reliable values of formation constants for chloro-complexes of Mn(II) are not available. Therefore, although the average complexation number, $N_{Mn^{2+}-Cl^-}$, could be obtained through the procedures already described for $CrCl_3$ solutions, the indications about type and percentage of complexes reported in Table 12, together with the complete set of coordination parameters, have only orientative meaning.

b. With Sulfate and Nitrate Ions

Solutions of $MnSO_4$ have been examined in two studies;[87,88] a ~2-*M* solution was examined in both studies, so that an interesting comparison may be done of results obtained in two separate experiments on a practically identical sample (an early investigation by Shapovalov et al.[85], indeed, also examined a solution of $MnSO_4$; however the values proposed for the ion-water distances were in general incorrect).

The model used for the interpretation of the radial curves assigns hexacoordination to the Mn^{2+} ion and implies formation of inner complexes with the anions. Complexation, which is indeed in agreement with indications of equilibrium constants, is also strongly suggested by the radial curves themselves (Figure 23), which display, besides the obvious peaks $Mn^{2+} - H_2O_I$, $H_2O - H_2O$, and $Mn^{2+} - H_2O_{II}$, a peak at about 3.5 Å, that should result from

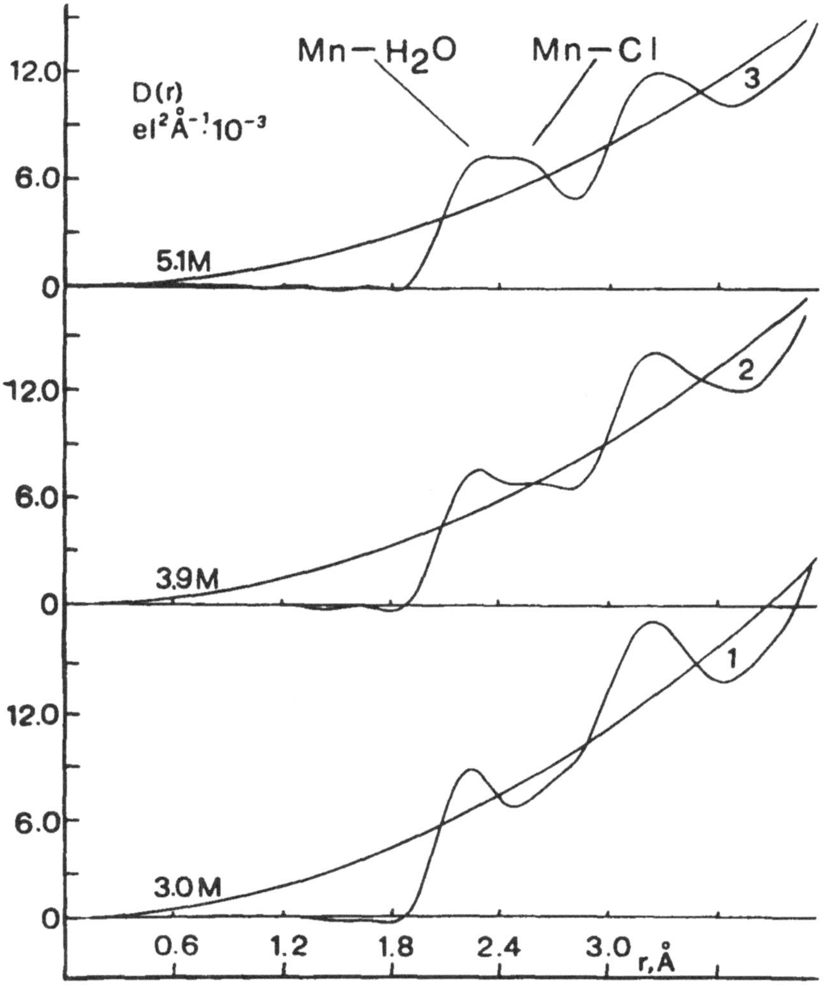

FIGURE 22. Radial distribution functions, D(r), for three solutions of $MnCl_2$. The parabolic curves are the corresponding $4\pi r^2\rho_0$ functions.

Mn – S contacts and is more evident in the saturated solution. Moreover, a model excluding sulfate complex formation gave rise to definitely poorer results.[88] The parameters from the best fits are reported in Table 13.

Noteworthy, in Table 13, is the remarkable agreement between parameters coming from the two experiments on the 2-*M* solution, although the fitting procedures were carried out in a somehow different way (for instance, the Mn – S distance was refined in Reference 88, while it was deduced from the crystal structure of $MnSO_4{\cdot}5H_2O$ in Reference 87). So, in this case, quality of the experimental data and amount of the phenomenon investigated led to qualitative evidence of sulfate complex formation and to meaningful quantitative results.

A similar comparison cannot be made for Mn nitrate solutions, since in this case just one solution about 4.4 *M* has been examined.[89] Also, for this solution, following the indications of Raman spectroscopy, a model involving inner-sphere Mn^{2+} – ONO_2^- complexes was adopted. The final parameters are also given in Table 13; an even surprising agreement of coordination parameters in nitrate and sulfate solutions is apparent.

Table 12
COORDINATION OF Mn^{2+} ION IN $MnCl_2$ SOLUTIONS: STRUCTURAL PARAMETERS AND PERCENTAGES OF THE COMPLEXES USED

Sample no.	Sample	Ref.	Mn^{2+} conc. (*M*)	$r_{Mn^{2+}-H_2O_I}$	$\sigma_{Mn^{2+}-H_2O_I}$	$N_{Mn^{2+}-H_2O_I}$	$r_{Mn^{2+}-Cl^-}$	$\sigma_{Mn^{2+}-Cl^-}$	$N_{Mn^{2+}-Cl^-}$	Mn^{2+}	$MnCl^+$	$MnCl_2$	Method	Free anion
1	$MnCl_2$ 17.0 H_2O	90	3.0	2.21	0.11	6—0.9	2.54	0.11	0.9	25	58	17	TSM1 + CF	$Cl(H_2O)_6^-$
2	$MnCl_2$ 12.6 H_2O	90	3.9	2.21	0.10	6—1.1	2.53	0.10	1.1	25	40	35	TSM1 + CF	$Cl(H_2O)_6^-$
3	$MnCl_2$ 9.3 H_2O	90	5.1	2.20	0.09	6—1.3	2.51	0.09	1.3	10	45	45	TSM1 + CF	$Cl(H_2O)_6^-$

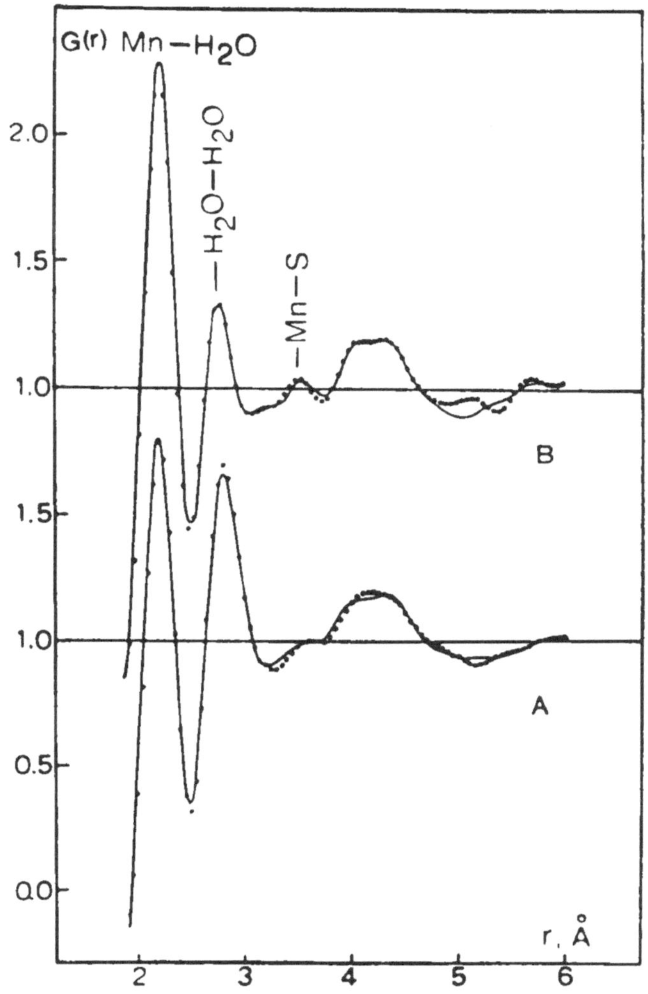

FIGURE 23. Experimental (dots) and model (solid lines) correlation functions for $MnSO_4$ solutions: (A) 2.0 *M*; (B) 3.8 *M*. (From Caminiti R., Marongiu, G., and Paschina, G., *Z. Naturforsch.*, 37a, 581, 1982. With permission.)

X. GROUP VIII (Fe, Co, Ni)

A. Iron(II)

Few solutions containing Fe(II) salts have been examined by XRD. In a systematic study on hydrated divalent transition metal ions in aqueous solution, Ohtaki et al.[86] examined a solution of Fe(II) perchlorate. Preparation, chemical analysis, and diffractometric measurements were performed under nitrogen gas, in order to prevent oxidation of Fe^{2+} ions. Calculations of theoretical peak shapes and analysis of the high angle part of the structure function concordantly indicated that the Fe^{2+} ion is hexahydrated, with the $Fe^{2+}-H_2O$ distance of about 2.12 Å.

The structure of bromo-iron(II) complexes in concentrated aqueous solutions of $FeBr_2$ under N_2 atmosphere was investigated by Ichihashi et al.[91] Qualitative evidence of $Fe^{2+}-Br^-$ direct interactions is given by the presence in the radial curves of a peak at ~2.7Å, which becomes blurred in the less concentrated solution. The quantitative evaluation of Fe(II) coordination parameters was carried out by simulation of the high-angle part of the si(s)s

Table 13
COORDINATION OF Mn^{2+} IONS IN SULFATE AND NITRATE SOLUTIONS

Sample no.	Sample	Ref.	Mn^{2+} conc. (*M*)	$r_{Mn^{2+}-H_2O_I}$	$\sigma_{Mn^{2+}-H_2O_I}$	$N_{Mn^{2+}-H_2O_I}$	$r_{Mn^{2+}-H_2O_{II}}$	$N_{Mn^{2+}-H_2O_{II}}$	$r_{H_2O_I-H_2O_{II}}$	$\sigma_{H_2O_I-H_2O_{II}}$	$r_{Mn^{2+}-X}$ (X = S,N)	$N_{Mn^{2+}-X}$	Method	Free anion
1	$MnSO_4$ 13.3 H_2O	87	3.8	2.22	0.12	6—0.92	4.2	9.8	2.74	0.03	3.47'	0.92	TSM1 + CF	Hydrated
2	$MnSO_4$ 26.8 H_2O	87	2.0	2.20	0.10	6—0.75	4.2	10.4	2.78	0.04	3.47'	0.75	TSM1 + CF	Hydrated
3	$MnSO_4$ 26.7 H_2O	88	2.0	2.20	0.10	6—0.50	4.3	10.7	2.73	0.05	3.5	0.50	TSM1 +CF	Hydrated
4	$Mn(NO_3)_2$ 10.3 H_2O	89	4.4	2.19	0.13	6—0.60	4.2_5	8.9	2.71	0.06	3.1	0.60	TSM1 + CF	Hydrated

and the results are reported in Table 14. The values of $N_{Fe^{2+}-H_2O}$ and $N_{Fe^{2+}-Br^-}$ suggest that Fe(II) is prevailingly involved in the hexacoordinated $Fe(H_2O)_6^{2+}$ and $Fe(H_2O)_5Br^+$ complexes.

B. Iron(III)

1. Hydration

The structure of Fe(III) aquo-complexes has been studied in nitrate and perchlorate medium. Two of the first studies examined almost exclusively the first coordination shell of the cation.[92,93] The concentration range explored in nitrate solutions extended from 1.5 *M* (H_2O/Fe^{3+} = 32.5) to 4.9 *M* (H_2O/Fe^{3+} = 6.5).[92] The most concentrated solutions were prepared by directly melting the salt $Fe(NO_3)_3{\cdot}9H_2O$ and evaporating some water from the melt. The radial functions of the nitrate solutions are reported in Figure 24; they look very similar, and small variations in the short range order of the system through the wide concentration range explored were suggested. The peak at ~2.05 Å comes from ion-water distances; the incomplete resolution on the right side is due to overlapping by O – O distances within the NO_3^+ ion. However, as the value of these distances is known and their number is fixed by the stoichiometry, the evaluation of ion-water distance and hydration number is not hampered. Perchlorate solutions are in a similar situation.[93]

Calculations of $Fe-H_2O$ peak shapes and high-angle analysis of si(s)s suggested the existence of the aquo-complex $Fe(H_2O)_6^{3+}$; $Fe^{3+}-H_2O$ distances turned out to be 2.03 Å in nitrate and 2.00 Å in perchlorate solutions. In both cases, anionic structuring does not give rise to chracteristic peaks in the radial curves (apart from intraionic distances) and cannot be quantitatively analyzed. However, a shoulder appears at ~5.0 Å in the perchlorate radial curves, which is better evidenced when distribution curves of perchlorate and nitrate solutions of close composition are compared (Figure 25). The 5-Å peak in the difference curve is probably due to Fe – Cl distances in outer sphere complexes $Fe(H_2O)_6^{3+}-ClO_4^-$. Analogous Fe – N interactions should exist in nitrate solutions, particularly in the most concentrated ones; the lack of evidence in the difference curves may be explained with their light weight. The difference peaks at 3.4 to 3.7 Å in Figure 25 were tentatively interpreted as due to anion hydration.

Three solutions out of the Fe(III) nitrate ones were reexamined more thoroughly,[94] following the same method adopted for Al^{3+} and Cr^{3+}. Structural parameters were obtained from a model in which the complex $Fe(H_2O)_6^{3+}$ was surrounded by a second shell of water molecules. Preliminary calculations suggested that the trigonal orientation of H_2O_I molecules, with angles H–Ô–H of 120°, allows the calculation of synthetic structure functions in very good agreement with the experimental ones. Keeping the angular value fixed and considering $O_I-H\ldots O_{II}$ bonds linear, only two independent parameters, $r_{Fe^{3+}-H_2O_I}$ and $r_{H_2O_I-H_2O_{II}}$, had to be refined for a complete description of the hydration structure of Fe^{3+}. The best parameters are given in Table 15. It is worth noting the rather short value obtained for $H_2O_I-H_2O_{II}$ distances.

2. Complex Formation

a. With Chloride Ions

Fe(III) chloride solutions have been repeatedly examined during the years; here we have an example of an apparent entanglement of contrasting interpretations, which can be easily sorted out by critical examination of the results.

The early studies unaminously indicated a high degree of complexation with chloride ions, but were not in agreement about the type of complexes existing in solution.[95-97] Later, Wertz et al.[98] questioned even the high degree of chloro-complexation and affirmed that, soon after the preparation, halo-complexes exist in the solutions obtained from anhydrous $FeCl_3$, but they are absent in the solutions obtained from the hydrated salt $FeCl_3{\cdot}6H_2O$. One sample obtained from the hydrated salt was reexamined by Wertz et al.[99] after 14 months

Table 14
COORDINATION OF Fe^{2+} IONS IN $FeBr_2$ SOLUTIONS

Sample no.	Sample	Ref.	Fe^{2+} conc. (*M*)	$r_{Fe2+\text{-}H_2O}$	$\sigma_{Fe2+\text{-}H_2O}$	$N_{Fe2+\text{-}H_2O}$	$r_{Fe2+\text{-}Br^-}$	$\sigma_{Fe2+\text{-}Br^-}$	$N_{Fe2+\text{-}Br^-}$	Method
1	$FeBr_2$ 17.8 H_2O	91	2.7	2.12	0.09	5.5	2.60	0.11	0.3	HAA
2	$FeBr_2$ 10.2 H_2O	91	4.5	2.12	0.08	5.1	2.62	0.12	0.7_5	HAA

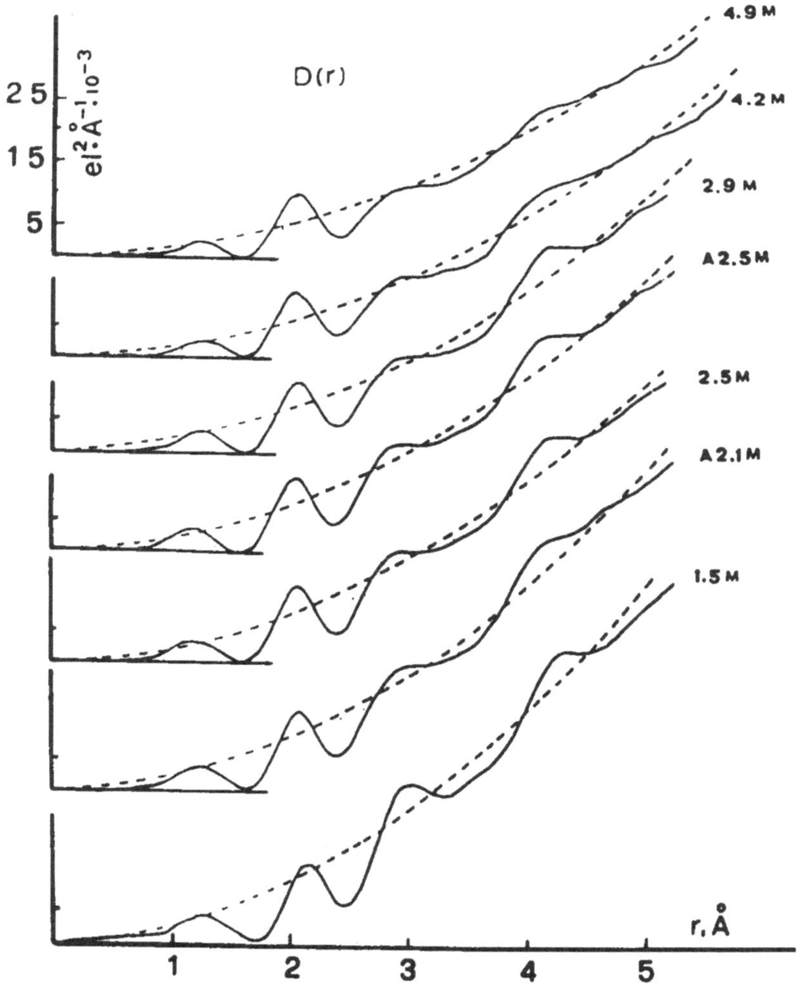

FIGURE 24. Radial distribution functions (solid lines) and $4\pi r^2\rho_0$ functions (dashed lines) for $Fe(NO_3)_3$ solutions. (From Magini, M. and Caminiti, R., *J. Inorg. Nucl. Chem.*, 39, 91, 1977. With permission.)

from its preparation, and the conclusion was reached that some of the inner-sphere water molecules of the former complex $Fe(H_2O)_6^{3+}$ had been replaced by chloride ions.

However, all these studies are lacunose. Chemical equilibria, in fact, were not accounted for in the first group of investigations, so that the analyses did not consider all kinds of complexes which may exist at the working conditions. As far as the conclusions of Wertz et al. are concerned, they are likely the result of some experimental accident. In fact, the presence of solely $Fe(H_2O)_6^{3+}$ ions in concentrated solutions of $FeCl_3 \cdot 6H_2O$ not only contradicts an impressive mass of experimental data, but was also excluded by an accurate repetition of the diffractometric work.[100] The following solutions were indeed investigated by X-ray: (1) a 3.7 *M* neutral solution, prepared from $FeCl_3 \cdot 6H_2O$ and examined as soon as prepared (H_R°); (2) a solution as above, independently prepared and examined by another diffractometer (H_C°); (3) the solution H_R°, reexamined after 8 months (H_R'); and (4) a 3.65 *M* neutral solution, prepared from the anhydrous $FeCl_3$ salt and examined as soon as prepared using the same diffractometer as for collections 1 and 3 (A_R°). The resulting radial distribution curves are shown in Figure 26. The coincidence in both position of the first maximum and

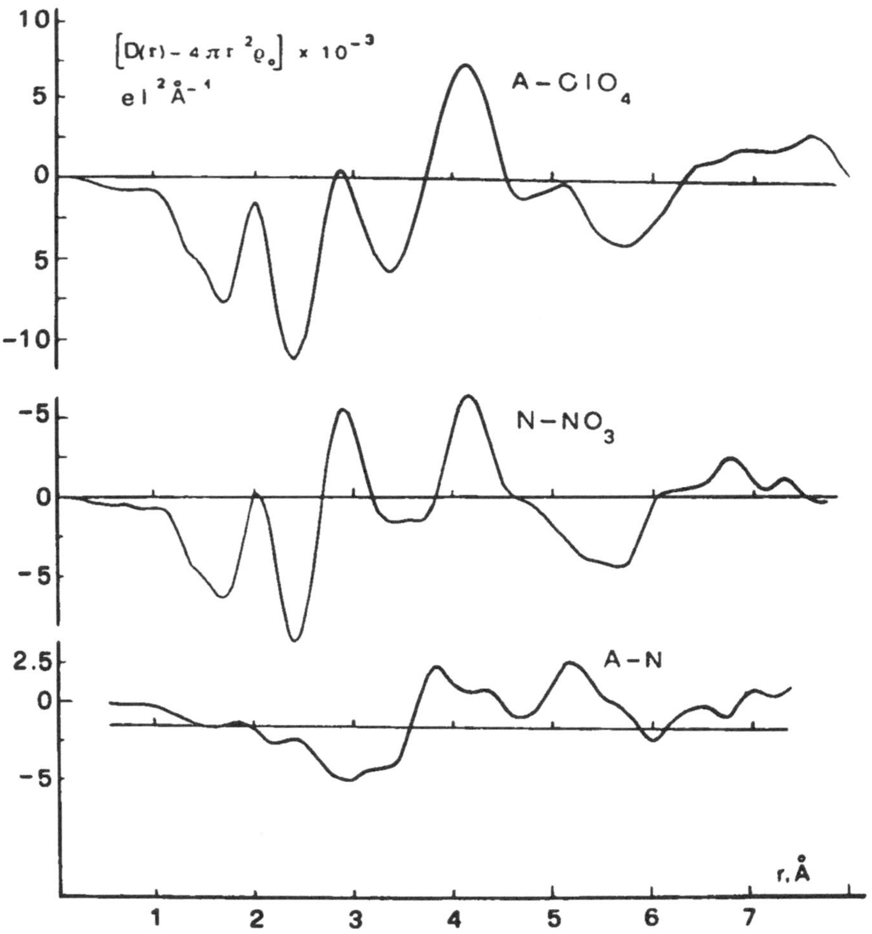

FIGURE 25. Radial distribution functions in differential form, $D(r) - 4\pi r^2\rho_0$, for a Fe(III) perchlorate solution (A-ClO_4) and for a Fe(III) nitrate solution (N − NO_3). The lower curve shows the difference between the curves A and N. The intramolecular contributions within ClO_4^- and NO_3^- ions have been subtracted from the respective curves. (From Magini, M., *J. Inorg. Nucl. Chem.*, 40, 43, 1978. With permission.)

general shape of the four curves is very good. On one hand, the similarity of $H°_C$ and $H°_R$ curves warrants the correctness and reproducibility of the results; on the other hand, the coincidence of all curves demonstrates that no important difference exists between solutions prepared either from $FeCl_3$ or $FeCl_3{\cdot}6H_2O$, and between fresh and aged solutions. The maximum is always centered at 2.25 Å, and this position clearly indicates a strong chloro-complexation. In fact, while $Fe^{3+} - H_2O$ distances fall at ~2.0 Å, $Fe^{3+} - Cl^-$ distances measured in various crystalline compounds are in the range 2.2 to 2.4 Å.

Therefore, chloro-complexes form in aqueous solution of iron(III) chloride and the point is a correct determination of the extent of this phenomenon. Reliable information can be found in a study of five solutions of $FeCl_3{\cdot}6H_2O$, neutral or acidified with HCl.[101] The radial curves are shown in Figure 27 and appear very similar. The first peak is centered at ~2.25 Å; a small shift from 2.22 Å in the most dilute solution to 2.27 to 2.29 Å in concentrated or acidified solutions probably indicates increasing chloro-complexation.

The changes with concentration and/or acidity, and hence indication about the nature of the main species present in solution, may be better followed through the differences between

Table 15
HYDRATION OF Fe^{3+} IONS

Sample no.	Sample	Ref.	Fe^{3+} conc. (*M*)	$r_{Fe^{2+}\text{-}H_2O_I}$	$\sigma_{Fe^{3+}\text{-}H_2O_I}$	$N_{Fe^{3+}\text{-}H_2O_I}$	Geometry	$r_{F^{3+}\text{-}H_2O_{II}}$	$\sigma_{F^{3+}\text{-}H_2O_{II}}$	$N_{F^{3+}\text{-}H_2O_{II}}$	$r_{H_2O_I\text{-}H_2O_{II}}$	$\sigma_{H_2O_I\text{-}H_2O_{II}}$	Method	Anion
1	$Fe(NO_3)_3$ 68.0 H_2O	94	0.8	2.04	0.08	6	Octahedral	4.1[f]	—	12	2.72	0.06	TSM1	Not hydrated
2	$Fe(NO_3)_3$ 32.4 H_2O	94	1.5	2.05	0.08	6	Octahedral	4.1[f]	—	12	2.70	0.05	TSM1	Not hydrated
3	$Fe(NO_3)_3$ 0.2 HNO_3 18.1 H_2O	94	2.5	2.04	0.09	6	Octahedral	4.1[f]	—	12	2.70	0.04	TSM1	Not hydrated

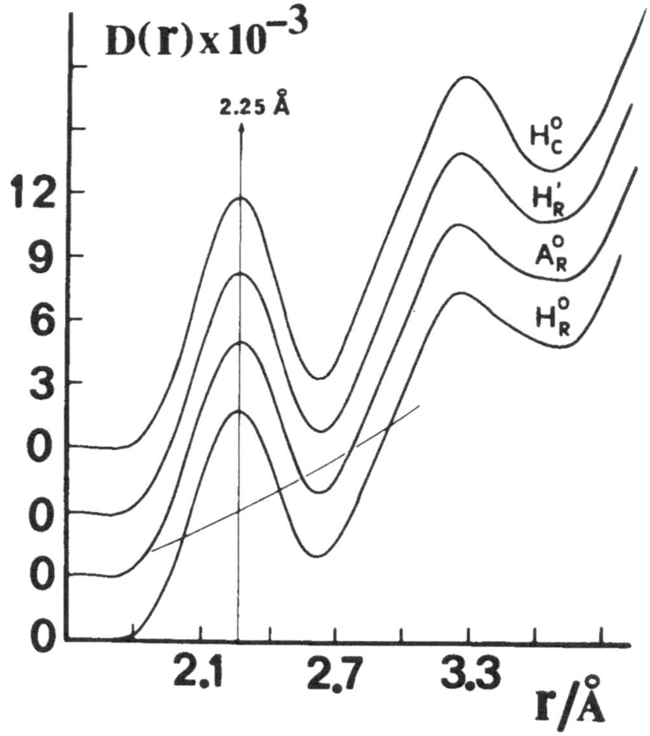

FIGURE 26. Radial distribution functions, D(r), for four solutions of $FeCl_3$ (strong solid lines) and the $4\pi r^2\rho_0$ function drawn on the H^o_R curve (weak solid line). The meaning of the labels is given in the text. (From Giubileo, G., Magini, M., Licheri, G., Paschina, G., Piccaluga, G., and Pinna, G., *Inorg. Chem.*, 22, 1001, 1983. With permission.)

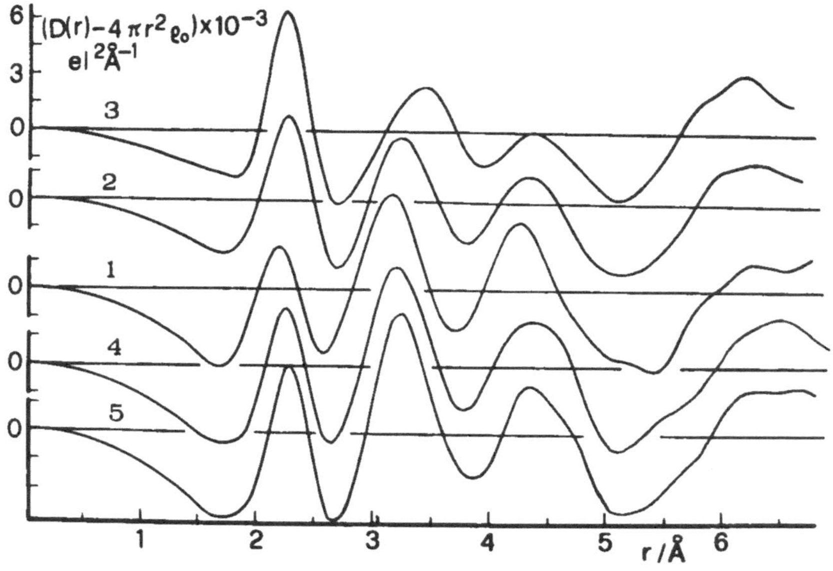

FIGURE 27. Radial distribution functions in the differential form, $D(r) - 4\pi r^2\rho_0$, for five solutions of $FeCl_3$. (From Magini, M. and Radnai, T., *J. Chem. Phys.*, 71, 4255, 1979. With permission.)

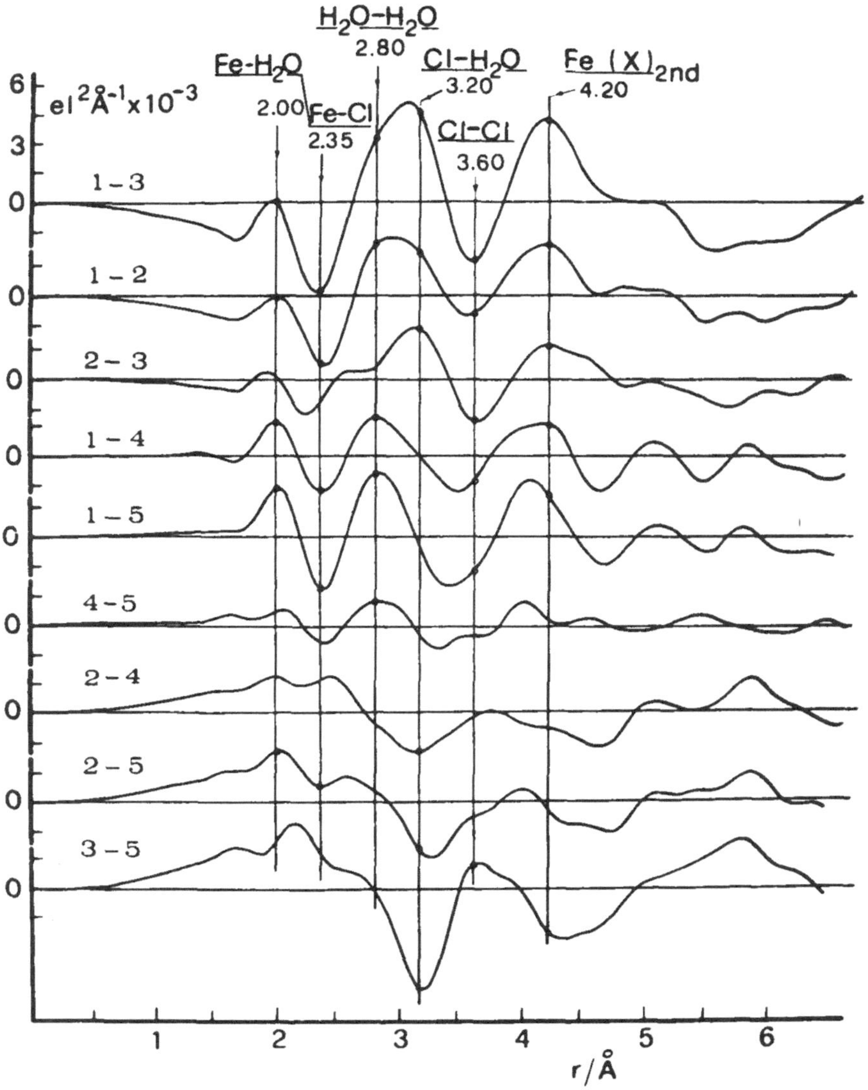

FIGURE 28. Differences curves between the D(r) — $4\pi r^2\rho_0$ functions for pairs of $FeCl_3$ solutions. The positions of expected contributions from pair interactions in solution are indicated by vertical weak solid lines. (From Magini, M. and Radnai, T., *J. Chem. Phys.*, 71, 4255, 1979. With permission.)

the distribution curves for pairs of solutions (Figure 28). In these curves, positive peaks indicate interatomic distances more numerous in the subtrahend D(r), while negative peaks indicate distances more frequent in the subtracted D(r). For instance, the positive $Fe^{3+} - H_2O$ peak in $D(r)_1$—$D(r)_3$ suggests more extensive hydration of Fe(III) in the more dilute solution and, correspondingly, the negative Fe − Cl peak in the same curve indicates higher halo-complexation in the more concentrated solution. Among the various details, the peak at 3.6 Å is particularly interesting, as it is likely due to Cl − Cl interactions in $FeCl_4^-$ tetrahedra, which form at high chloride concentration. Qualitatively, these observations confirm the indications of equilibrium constants that point to the formation of a continuous series of complexes, $FeCl^{2+}$ (I), $FeCl_2^+$ (II), $FeCl_3$ (III) and $FeCl_4^-$ (IV), with distribution changing with the solution composition.

The first step of the quantitative analysis, based on the calculation of a synthetic structure function, was the determination of the average chloride and water numbers bonded to one iron atom, $N_{Fe^{3+}-Cl^-}$ and $N_{Fe^{3+}-H_2O}$ (from now on called N_{Cl} and N_{H_2O} for brevity). To this end, for each data set, combinations of iron-complexes were introduced and tested; their concentrations were refined, obviously keeping their sum fixed and equal to the total iron concentration. Preliminary calculations and indications from total and difference radial curves allowed some complexes to be neglected as nonexistent or in strong minority. For example, hexa-aquo complexes were never considered, the mono-chloro complex was introduced only in the most dilute solution, and so on. In the calculations, octahedral geometry was ascribed to $Fe(H_2O)_5Cl^{2+}$, $Fe(H_2O)_4Cl_2^+$, and $Fe(H_2O)_3Cl_3$, while $FeCl_4^-$ was described as tetrahedral. One distance value was assumed for the same interaction in different octahedral complexes, while the Fe – Cl distance in $FeCl_4^-$ was refined separately. In the final fittings, second-shell interactions around Fe^{3+} ions were also introduced, although second-sphere parameters are almost meaningless when different complexes with several chloride ions in the first shell are accounted for (see Cr^{3+}).

The parameters of cationic coordination are given in Table 16 which also includes the percentages of the various complexes used. Ion-water and, to a lesser extent, ion-chloride direct distances show some variations along the series which are probably unphysical. It may be due to the description of octahedral complexes with equal parameters, but it certainly depends on the unresolved contributions under the envelope of the first peak in the radial curves. With the exception of the most dilute solution, the value of $N_{Cl} + N_{H_2O} < 6$ reveals the presence of a high percentage of the tetrachloro complex. As far as the distribution of the species is concerned, we have seen (see Cr^{3+}) that discrimination is not provided by the scattering data. Despite this limit, least square refinements provide some indications about the dominant species in solution. $Fe(H_2O)_4Cl_2^+$, $FeCl_4^-$, and $Fe(H_2O)_3Cl_3$ complexes are the prevailing species in the dilute neutral solution, in the concentrated neutral solution, and in the other solutions, respectively. Higher chloro-complexes (for instance, $FeCl_6^{3-}$) are probably absent.

N_{Cl} turned out a little greater than 3 in two cases. For the solution $FeCl_3 \cdot 6.7H_2O$ this result can be explained only invoking some sharing of chloride ions between the simple species described above. Possible models involving the dimeric form Fe_2Cl_6, made up by two tetrahedra sharing an edge were tested, and they gave rise to neither an improved nor a worse fit of the structure function. The solution $FeCl_3 \cdot 6.7H_2O$ has a composition very near to the hydrated salt; however, the structural units found in the crystal are *trans* $Fe(H_2O)_4Cl_2^+$, while in the solution, the dominant complex is $FeCl_4^-$, thus suggesting that deep structural rearrangements take place with melting.

In a following study,[102] the existence of a small amount of a slightly distorted dimeric Fe_2Cl_6 complex was inferred from the comparison of the D(r) of the sample $FeCl_3 \cdot 6.7H_2O$ with those of two reference solutions with the same average electronic density and obtained by addition of LiCl or HCl to $FeCl_3$ solutions. In the difference radial curves, a small peak was detected at 2.75 to 2.80 Å, a plausible distance for Fe–Fe interaction in the dimeric complex. The small entity of this difference peak explains why Fe–Fe distances are not "eye evident" in the total radial curves.

b. With Sulfate Ions

The formation of inner-sphere $Fe^{3+} - SO_4^{2-}$ complexes is suggested by the three radial curves reported in Figure 29, obtained in a study of $Fe_2(SO_4)_3$ solutions.[103] In addition to the peaks arising from hydration of Fe(III) and from SO_4^{2-} ions (probably hydrated), a peak appears at 3.3 Å that is deemed real from its entity and occurrence in all the cases examined. Crystal structures of several iron(III) sulfate compounds show that 3.3 Å is a typical Fe – S distance within $Fe(H_2O)_{6-N}(OSO_3)_N^{+3-2N}$ complexes.

Table 16
COORDINATION OF Fe^{3+} IONS IN CHLORIDE SOLUTIONS

A. Structural Parameters

Sample no.	Sample	Ref.	Fe^{3+} conc. (*M*)	Cl/Fe ratio	$r_{Fe^{3+}-H_2O_I}$	$\sigma_{Fe^{3+}-H_2O_I}$	$N_{Fe^{3+}-H_2O_I}$	$r_{Fe^{3+}-Cl^-}$ octah.	$\sigma_{Fe^{3+}-Cl^-}$ octah.	$r_{Fe^{3+}-Cl^-}$ tetrah.	$\sigma_{Fe^{3+}-Cl^-}$ tetrah.	$N_{Fe^{3+}-Cl^-}$	Method	Free anion
1	$FeCl_3$ 21.5 H_2O	101	2.3	3.0	2.08	0.14	4.2	2.30	0.17	—	—	1.8	TSM1 + CF	$Cl(H_2O)_6^-$
2	$FeCl_3$ 10.7 H_2O	101	4.2	3.0	2.00	0.12	2.3	2.32	0.13	2.26	0.12	3.0	TSM1 + CF	Not present
3	$FeCl_3$ 6.7 H_2O	101	5.9	3.0	2.05	0.12	1.7	2.35	0.15	2.24	0.07	3.3	TSM1 + CF	Not present
4	$FeCl_3$ 1.0 HCl 20.6 H_2O	101	2.3	4.0	1.94	0.15	2.6	2.30	0.15	2.24	0.05	3.0	TSM1 + CF	$Cl(H_2O)_6^-$
5	$FeCl_3$ 1.7 HCl 20.9 H_2O	101	2.2	4.7	1.97	0.12	2.1	2.37	0.17	2.26	0.04	3.3	TSM1 + CF	$Cl(H_2O)_6^-$

B. Percentages of the Complexes Used

Sample no.	Ref.	Fe^{3+} conc. (*M*)	Fe^{3+}	$FeCl^{2+}$	$FeCl_2^+$	$FeCl_3$	$FeCl_4^-$
1	101	2.3		24	76		
2	101	4.2			33	34	33
3	101	5.9			20	30	50
4	101	2.3			20	60	20
5	101	2.2				70	30

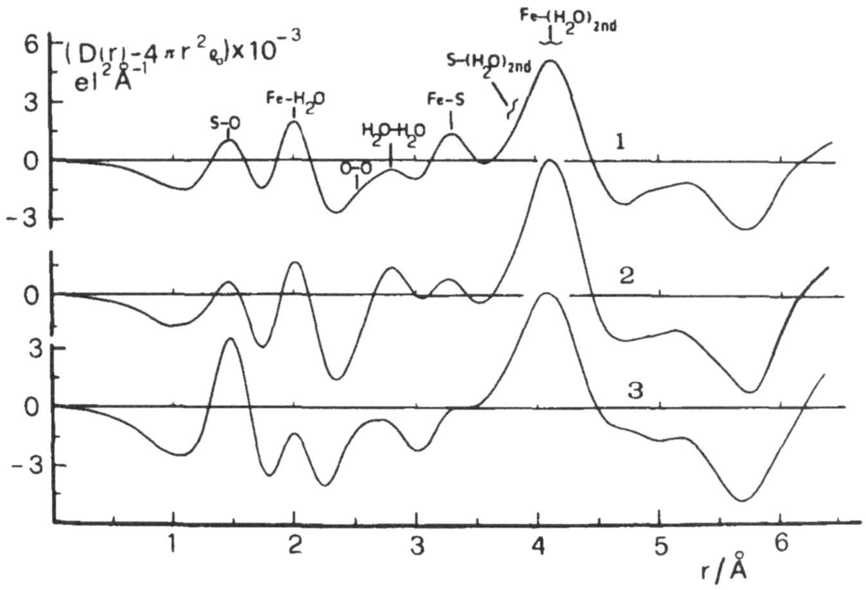

FIGURE 29. Radial distribution functions in the differential form, $D(r) - 4\pi r^2\rho_0$, for three solutions of $Fe_2(SO_4)_3$. (From Magini, M., *J. Chem. Phys.*, 70, 317, 1979. With permission.)

A clearer evidence of Fe – S interactions and a first evaluation of the number of Fe – S contacts was achieved by comparison of radial curves of $Fe_2(SO_4)_3$ and $Fe(ClO_4)_3$ solutions having the same Fe(III) concentration. The difference radial curve (Figure 30) exhibits the peak Fe – S at 3.3 Å (the negative region at about 5 Å should be due to outer-sphere Fe – ClO_4 complexes, as previously discussed). Peak shape analysis of the 3.3 Å peak suggested that the number of bonded sulfate groups per iron(III) atom is greater than 1 (~1.4).

As final analysis, the scattering curves for the $Fe(H_2O)_{6-N}$ $(OSO_3)_N{}^{+3-2N}$ units were calculated, refining geometrical parameters and the quantity N. Interactions from cationic second shells and hydrated anions were also accounted for. With regard to iron-sulfate complexes, the idea was to start from structural units which form in the crystalline state. Preliminary calculations indicated the model in Figure 31 as the most promising one. Once the reciprocal orientation between SO_4 tetrahedron and FeO_6 octahedron was fixed, the distances inside the units could be calculated from r_{S-O} and $r_{Fe^{3+}-O}$ distances, and from the α angle $Fe-\hat{O}-S$.

The parameters from the final fits are given in Table 17. Although the method (TSMl + CF) used is equivalent to the one used for $MnSO_4$ and $In_2(SO_4)_3$ solutions, second-shell parameters have not been reported in Table 17 for the present case. In fact, continuum terms were not introduced in the refinements and parameters describing medium-range discrete distances are not reliable (in any case, they are not homogenous with those given in previous and following tables). From mean values of α and $r_{Fe^{3+}-O}$, the mean distance Fe – S comes out at 3.24 Å.

N_{Fe-S} values are in excellent agreement with those calculated from equilibrium constants, although the latter were measured at low Fe(III) concentration. Obviously, N_{Fe-S} is an average value and, in general, it follows from the presence of various complexes; in particular, $N_{Fe-S} > 1$ implies presence of complexes with $N_{Fe-S} = 2$. However, only an "average" complex was introduced in the analysis, on account of the very similar fits obtained from different sets of complexes under the same average N_{Fe-S} value.

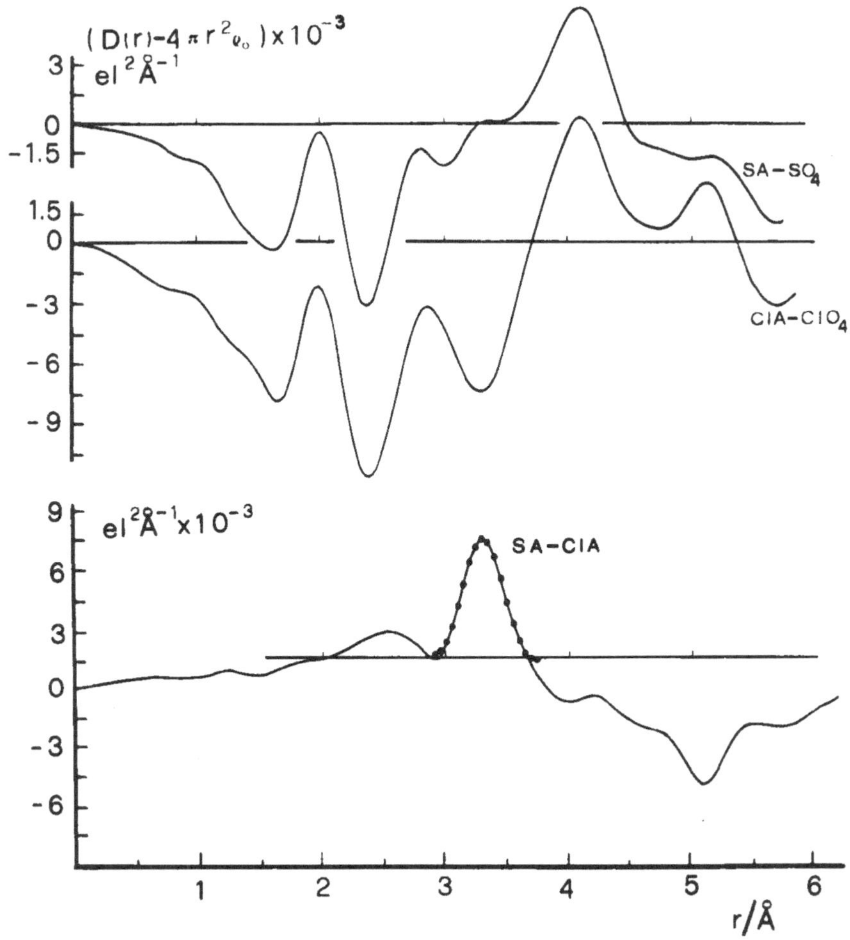

FIGURE 30. Radial distribution functions in the differential form, $D(r) - 4\pi r^2\rho_0$, for a Fe(III) sulfate solution ($SA-SO_4$) and for a Fe(III) perchlorate solution ($ClA-ClO_4$). The intramolecular contributions within sulfate and perchlorate ions have been subtracted from the respective curves. The lower curve shows the difference between the curves SA and ClA. (From Magini, M., *J. Chem. Phys.*, 70, 317, 1979. With permission.)

C. Cobalt

1. Hydration

The few results published about the hydration of Co^{2+} ions may be considered quite reliable. In fact, the peak $Co^{2+}-H^2O$ appears in the radial curve at 2.1 Å, well separated from the nearest H_2O-H_2O peak and, at worse, mixed with intramolecular peaks of oxy-anions of known structure. Therefore, the existence in aqueous solutions of species $Co(H_2O)_6{}^{2+}$, with $Co^{2+}-H_2O$ direct distance of about 2.08 to 2.09 Å, was clearly demonstrated by Bol et al.[35] in nitrate, and by Ohtaki et al.[86] in perchlorate solutions. In the work of Bol et al., the existence of a second hydration shell around Co^{2+} ions was also inferred.

A complete set of parameters describing the hydration of Co^{2+} ions in a $Co(ClO_4)_2$ solution was presented by Magini et al.,[104] and is reported in Table 18. As Fe(III) perchlorate solutions, the $Co(ClO_4)_2$ solution shows a weak trace of possible outer-sphere complex formation with the perchlorate ion, which occurrence is suggested by comparison of radial curves of $Co(ClO_4)_2$ and $CoCl_2$ solutions (Figure 32). Also $Co^{2+}-ClO_4{}^-$ outer-sphere interactions were accounted for in the final calculation of the synthetic structure function.

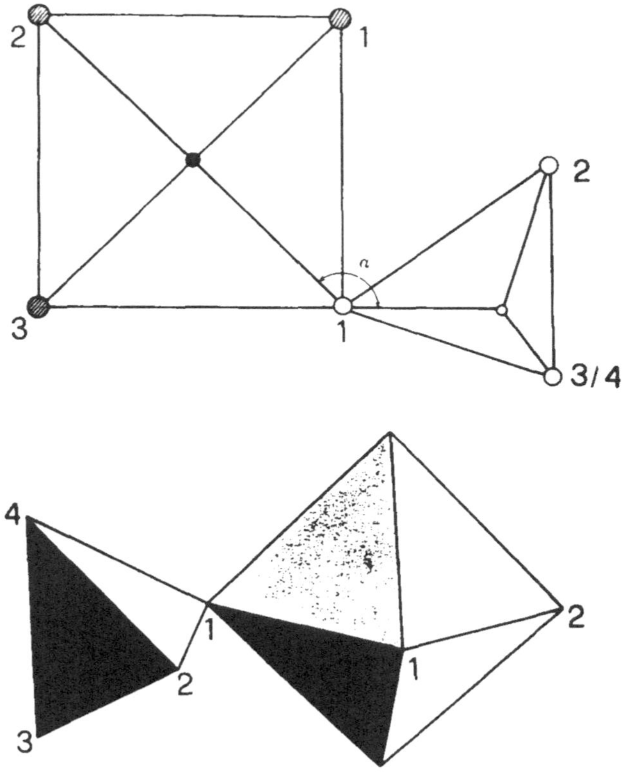

FIGURE 31. The best model for the Fe(III)-sulfate units: projection and perspective views. (From Magini, M., *J. Chem. Phys.*, 70, 317, 1979. With permission.)

2. Complex Formation

a. With Chloride Ions

Much interest has been devoted to the study of Co(II) chloro-complex formation.[11,105-108] Both stoichiometric solutions (Cl^-/Co^{2+} = 2) and solutions in which higher ratios Cl^-/Co^{2+} have been obtained through the addition of suitable electrolytes (LiCl, HCl) have been examined at Co^{2+} concentration ranging from ~0.5 to ~4.0 *M*, and Cl^-/Co^{2+} ratio from 2 to ~22! In addition to structural purposes, these investigations aimed at finding possible differences in chloro-complexing trend of Ni^{2+} and Co^{2+} and at correlating these differences with the selective extraction of the two ions from chloride medium.

The tendency of the Co^{2+} ion to bind chloride ions is not as strong as that of Cr^{3+}, Fe^{3+}, and, even less, as that of Zn^{2+} or Cd^{2+}. Satisfactory fits of the first peak in the radial curves were achieved in stoichiometric solutions with only interactions within the aquo-complexes $Co(H_2O)_6{}^{2+}$. The incomplete resolution of the peak in point (contrary to what happens in the radial curves of $MgCl_2$ solutions, with the same $Me^{2+}-H_2O$ distance) was noticed for the first time in ~3 *M* solution of $CoCl_2$ and ascribed to $Co^{2+}-Cl^-$ contacts with mean distance 2.45 Å.[106] The best fit of the experimental i(s) was obtained introducing equal amounts of $Co(H_2O)_6{}^{2+}$ and $Co(H_2O)_5Cl^+$ complexes.

In this model, one out of twelve ligand atoms (~8%) is a Cl^- ion and this amount accounts for the difficulty of XRD in resolving $Co^{2+}-Cl^-$ contacts. Some help in this direction is offered by the examination of series of solutions. In fact, comparisons of results can allow the evolution of the system to be observed and the formation of first chloro-complexes to be put in evidence. In this connection, an enlightening example is found in a study of three

Table 17
COORDINATION OF Fe^{3+} IONS IN SULFATE SOLUTIONS

Sample no.	Sample	Ref.	Fe^{3+} conc. (*M*)	$r_{Fe^{3+}-O}$	$\sigma_{Fe^{3+}-O}$	$N_{Fe^{3+}-O}$	α_{Fe-O-S}	$N_{Fe^{3+}-S}$	$N_{Fe^{3+}-S}$ from k_{eq}	Method
1	$Fe_2(SO_4)_3$ 28.7 H_2O	103	3.4	2.02	0.09	6	135°	1.2	1.4	TSM1 + CF
2	$Fe_2(SO_4)_3$ 40.0 H_2O	103	2.5	2.02	0.07	6	134°	1.3	1.4	TSM1 + CF
3	$Fe_2(SO_4)_3$ $3H_2SO_4$ 47.7 H_2O	103	1.9	2.01	0.08	6	135°	1.0	0.9—1.1	TSM1 + CF

Table 18
HYDRATION OF Co_2^+ IONS

Sample no.	Sample	Ref.	Co^{2+} conc. (*M*)	$r_{Co^{2+}-H_2O_I}$	$\sigma_{Co^{2+}-H_2O_I}$	$N_{Co^{2+}-H_2O_I}$	Geometry	$r_{Co^{2+}-H_2O_{II}}$	$\sigma_{Co^{2+}-H_2O_{II}}$	$N_{Co^{2+}-H_2O_{II}}$	$r_{H_2O_I-H_2O_{II}}$	$\sigma_{H_2O_I-H_2O_{II}}$	Method	Anion
1	$Co(ClO_4)_2$ 15.0 H_2O	104	2.9	2.10	0.12	6	Octahedral	4.2	0.29	6.3	2.79	0.04	TSM1	Not hydrated

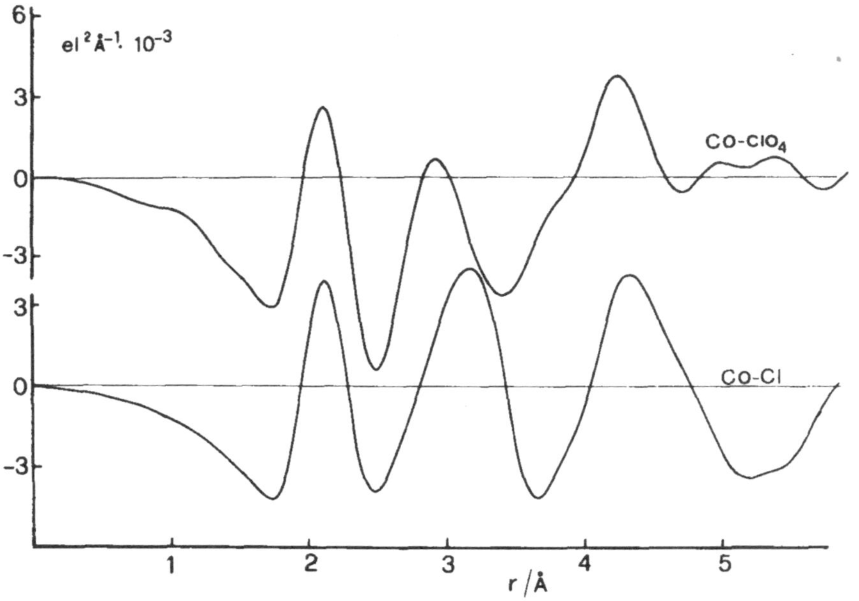

FIGURE 32. Radial distribution functions of two cobalt solutions in perchlorate (Co-ClO_4) and chloride (Co-Cl) medium. The intramolecular contributions within the ClO_4^- group have been subtracted from the perchlorate distribution curve for a clearer comparison. (From Magini, M., and Giubileo, G., *Gazz. Chim. Ital.*, 111, 449, 1981. With permission.)

$CoCl_2$ solutions,[108] in which the concentration of Co^{2+} was kept fixed at ~1*M* while the ratios Cl^-/Co^{2+} were 4, 6, and 7, respectively (hence the names Co4, Co6, and Co7 given to the samples). The first peak of the correlation functions, reported in Figure 33, is perfectly resolved and symmetric only in the sample Co4. In this case, it is highly probable that only water molecules form the Co^{2+} environment. On the contrary, the minimum between the first peak and following ones rises progressively in Co6 and Co7 G(r)s. The obvious explanation of this behavior lies on formation of Co^{2+}-Cl^- contacts in increasing amounts. A convincing proof of this hypothesis was obtained by subtracting the radial curve of Co4 from those of Co6 and Co7. The result of this procedure is shown in Figure 34; the minima in the difference curves indicate correlation distances more frequent in Co4, the maxima put distances more abundant in Co6 and Co7 into evidence. Then, the minima at 2.10 and 2.80 Å denote the progressive decreasing of $Co^{2+}-H_2O$ and H_2O-H_2O contacts when the ionic concentration of the solutions increases, the maxima at 2.40 and 3.20 Å indicate the increasing of $Co^{2+}-Cl^-$ and Cl^--H_2O contacts. The formation of chloro-complexes of Co(II) in Co6 and Co7 solutions is directly and unambiguously demonstrated. Clearly, a region of transition has been explored, from a state where only $Co(H_2O)_6^{2+}$ exists to a state planation of this behavior lies on formation of $Co^{2+}-Cl^-$ contacts in increasing amounts. A convincing proof of this hypothesis was obtained by subtracting the radial curve of Co4 solutions (~0.5 to 0.6 *M*) having high ratios Cl^-/Co^{2+} (~21 to 22), so that practically only the extreme chloro-complex $CoCl_4^-$ was present. Dichloro- and trichloro-complexes have never been investigated, as they have too narrow a concentration range of existence. The results of all the quantitative analyses are summarized in Table 19.

As in the cases already described of Cr^{3+} and Fe^{3+} halo-complexation, the average numbers of $Co^{2+}-Cl^-$ contacts per Co^{2+} ion, $N_{Co^{2+}-Cl^-}$ (from now on called N_{Cl} for brevity), are quite reliable. In accordance with obvious expectations, at a given concentration, N_{Cl} increases with the Cl^-/Co^{2+} ratio, and, at a given Cl^-/Co^{2+} ratio, it goes up with

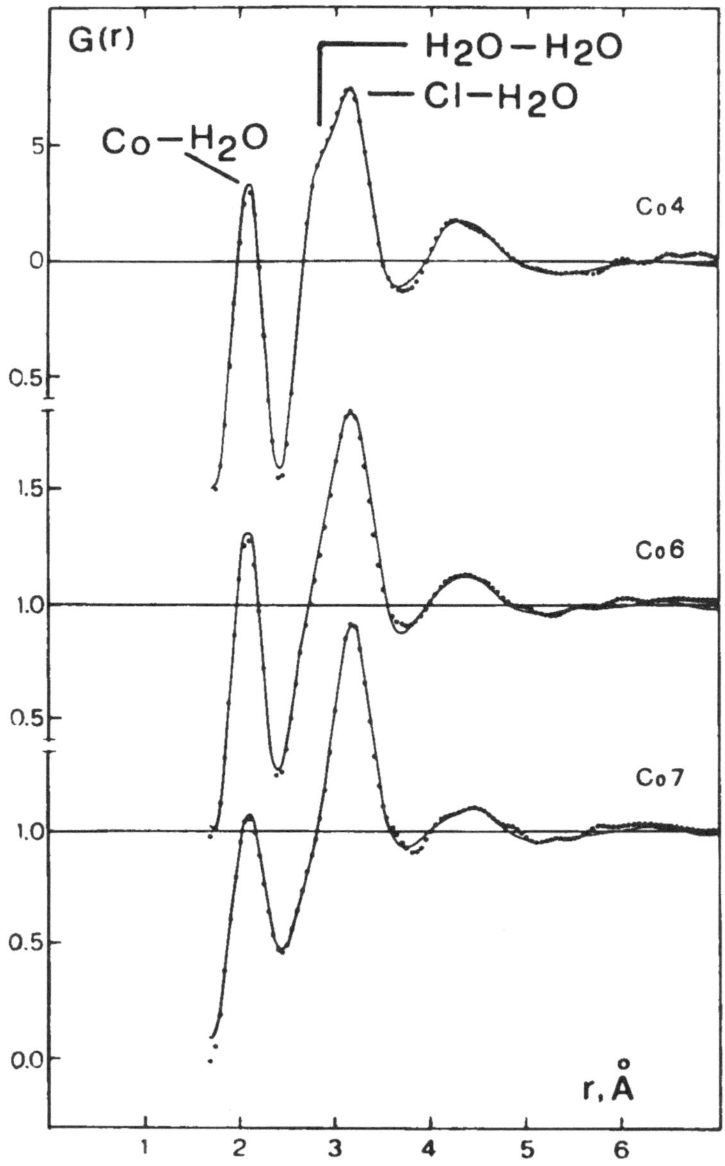

FIGURE 33. Experimental (dots) and model (solid lines) correlation functions for three $CoCl_2$-LiCl solutions. (From Musinu, A., Paschina, G., Piccaluga, G., and Magini, M., *J. Chem. Phys.*, 80, 2772, 1984. With permission.)

increasing concentration. On the contrary, the evaluations of complex percentages are reliable only in the extreme situations of full hydration (N_{Cl} = 0) or full chloro-complexation (N_{Cl} = 4); since in the other cases the degree of complexation is relatively low, the amounts of complexes could not be estimated uniquely and are not given in the table. Because of these difficulties, different strategies were adopted in the fitting procedures (use of K_{eq} indications, introduction of several complexes or an "average" complex, and so on). These differences are in part reflected on the values of the geometrical parameters, some of which are not in complete accordance. This is the case in particular of $r_{Co^{2+}-Cl^-}$ octahedral distance, which uncertainty is also determined by the small weight of $Co^{2+}-Cl^-$ interactions. In any case,

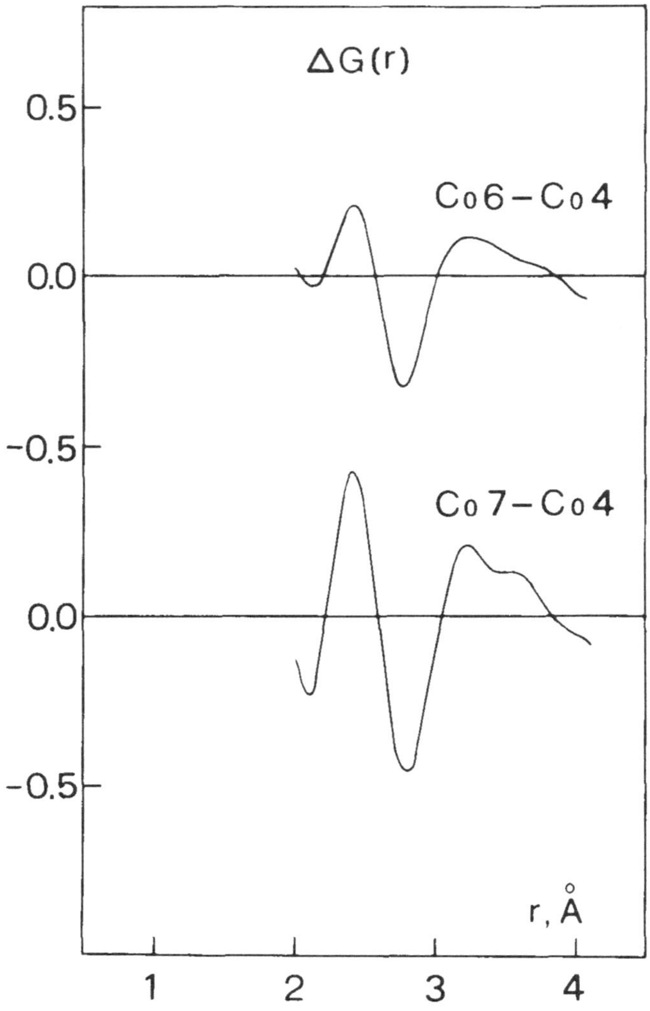

FIGURE 34. Difference curves of the experimental correlation functions given in Figure 33. (From Musinu, A., Paschina, G., Piccaluga, G., and Magini, M., *J. Chem. Phys.*, 80, 2772, 1984. With permission.)

$Co^{2+}-Cl^-$ tetrahedral distance turns out definitely shorter than the octahedral one. As far as the distance $Co^{2+}-H_2O_I$, the first of the two values from Reference 11 has been derived from the Ohtaki et al.[86] description of the complex $Co(H_2O)_6{}^{2+}$; the second value refers to the refined parameter for the complex, $Co(H_2O)_5Cl^+$. In the other cases, the same parameter value has been used in the two complexes. The highest Cl^-/Co^{2+} ratios give rise to a further difficulty, as also coordination models for the additional species must be proposed. Different choices (for example, different models for the hydration of Li^+ or H_3O^+ ions) might have some influence on the Co^{2+} coordination parameters. However, Lee et al.[11] have shown that various descriptions of H_3O^+ structuring have secondary effects on the parameters given in Table 19 (sample 4).

b. With Bromide Ions

Recently, a research on $CoBr_2$ solutions has been added to the investigations on $CoCl_2$ solutions,[91] and to an old study on a $CoBr_2$ solution.[109] As in the cases of $FeBr_2$ and $NiBr_2$

Table 19
COORDINATION OF Co^{2+} IONS IN CHLORIDE SOLUTIONS

Sample no.	Sample	Ref.	Co^{2+} conc. (*M*)	Cl/Co ratio	$r_{Co^{2+}-H_2O_I}$	$\sigma_{Co^{2+}-H_2O_I}$	$N_{Co^{2+}-H_2O_I}$	$r_{Co^{2+}-Cl^-}$ octa	$\sigma_{Co^{2+}-Cl^-}$ octa	$r_{Co^{2+}-Cl^-}$ tetra	$\sigma_{Co^{2+}-Cl^-}$ tetra	$N_{Co^{2+}-Cl^-}$	Method	Free anion
1	$CoCl_2$ 17.4 H_2O	106	3.0	2.0	2.10	0.10	6—0.5	2.47	0.10			0.5	TSM1 + CF	$Cl(H_2O)_6^-$
2	$CoCl_2$ 1.4 LiCl 23.2 H_2O	11	2.1	3.4	2.08—2.14	0.09—0.10	6—0.6	2.35	0.10			0.6	HAA + PS	Hydrated
3	$CoCl_2$ 20.1 LiCl 64.3 H_2O	11	0.6	22.1						2.29	0.06	4	HAA + PS	Hydrated
4	$CoCl_2$ 19.1 HCl 82.0 H_2O	11	0.5	21.1						2.29	0.08	4	HAA + PS	Hydrated
5	$CoCl_2$ 2.2 LiCl 26.0 H_2O	107	2.0	4.2	2.10	0.12	6—1.0	2.49	0.16			1.0	TSM1 + CF	$Cl(H_2O)_6^-$
6	$CoCl_2$ 2.1 LiCl 55.5 H_2O	108	0.9	4.1	2.08	0.11	6					0	TSM1	$Cl(H_2O)_6^-$
7	$CoCl_2$ 4.1 LiCl 50.4 H_2O	108	1.0	6.1	2.09	0.10	6—0.2_5	2.45[f]	0.13			0.2_5	TSM1 + CF	$Cl(H_2O)_6^-$
8	$CoCl_2$ 5.2 LiCl 49.9 H_2O	108	1.0	7.2	2.09	0.12	6—1.0	2.41	0.12			1.0	TSM1 + CF	$Cl(H_2O)_6^-$

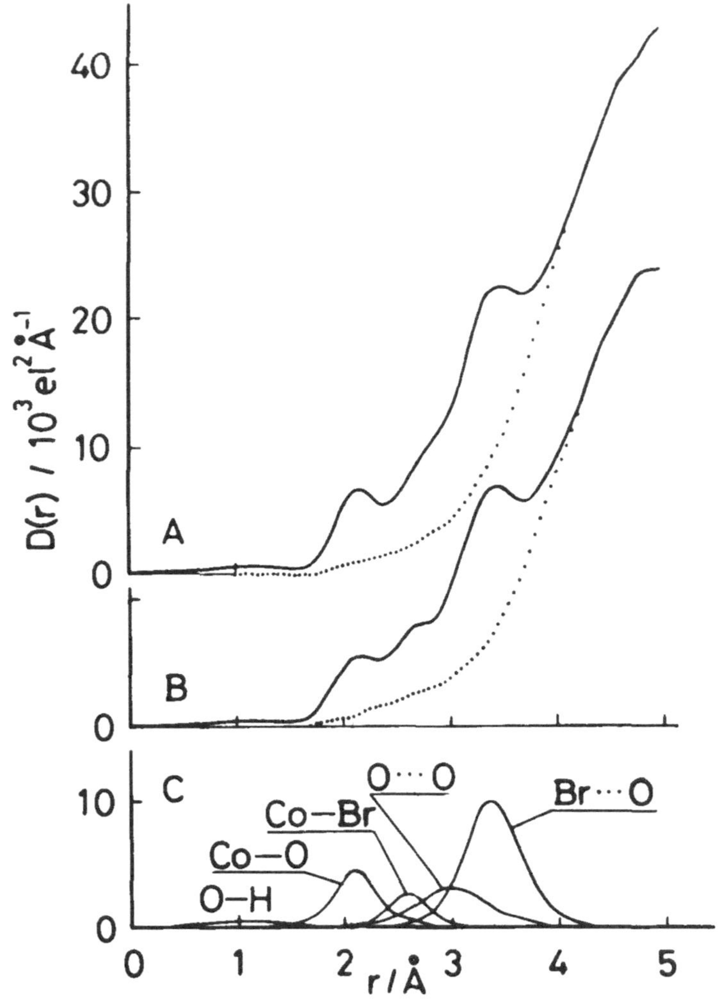

FIGURE 35. Radial distribution functions, D(r), for $CoBr_2$ solutions (solid lines): (A) 2.8 *M*; (B) 4.2 *M*. Dotted curves are the residual D(r)s after the subtraction of the theoretical short-range peaks, indicated in the lower part of the figure. (From Ichihashi, M., Wakita, H., and Masuda, I., *J. Solution Chem.*, 13, 505, 1984. With permission.)

solutions, the existence of $Co^{2+} - Br^-$ discrete distances can be deduced simply from the inspection of the experimental radial curves (Figure 35). In fact, the peak at ~2.7 Å, appearing between the maximum $Co^{2+} - H_2O$ at 2.1 Å and the $Br^- - H_2O$ one at ~3.4 Å, receives a strong contribution from the $Co^{2+} - Br^-$ term. The quantitative evaluation of the complexation degree has been performed by the high-angle simulation of the structure function and it has led to the structural parameters given in Table 20.

D. Nickel

1. Hydration

The study of the first hydration shell of Ni^{2+} by XRD is not difficult in theory, since the direct $Ni^{2+} - H_2O$ distance expected (and actually found in all the cases examined) is well separated from the distances occurring in aqueous solutions of simple monoatomic ions, or it is, at most, entangled with known intramolecular distances within poliatomic anions.

Table 20
COORDINATION OF Co^{2+} IONS IN $CoBr_2$ SOLUTIONS

Sample no.	Sample	Ref.	Co^{2+} conc. (*M*)	$r_{Co^{2+}-H_2O}$	$\sigma_{Co^{2+}-H_2O}$	$N_{Co^{2+}-H_2O}$	$r_{Co^{2+}-Br^-}$	$\sigma_{Co^{2+}-Br^-}$	$N_{Co^{2+}-Br^-}$	Method
1	$CoBr_2$ 18.2 H_2O	91	2.8	2.10	0.09	6.0	2.58	0.13	0.3	HAA
2	$CoBr_2$ 11.0 H_2O	91	4.2	2.09	0.08	5.3	2.59	0.12	0.6	HAA

Therefore, even in the earliest investigations,[33,110] were the ion-water distance (~2.05 Å) and the coordination number (~6.0) evaluated with good precision. The careful work by Bol et al.[35] and Ohtaki et al.[86] on a series of Me^{2+} nitrate and perchlorate solutions, respectively, must be remembered for the valuable information about nickel(II) hydration.

More detailed investigations on nickel-water interactions have come out lately. Indeed, great interest has been devoted to Ni salt solutions, mainly $NiCl_2$ solutions, but structural problems different from the hydration have been actually dealt with. In fact, Enderby et al., around 1975, on the basis of neutron diffraction studies of $NiCl_2$ solutions in D_2O, suggested the existence of a correlation $Ni^{2+} - Ni^{2+}$, involving long range order in a "quasi-lattice" structure.[111,112] Since then, the main object of structural studies on that subject became the research of a confirmation of this solute-connected medium-range lattice. Indeed, XRD studies soon verified that order phenomena in $NiCl_2$ solutions are similar to those occurring in the other electrolyte solutions of bi- and trivalent (transition and nontransition) metal ions. Meanwhile, the common object of these investigations has taken the attention off the "trivial" short-range order phenomena. As a consequence, only for a $Ni(ClO_4)_2$ solution, a complete set of structural parameters has been obtained through a modeling procedure,[104] and are reported in Table 21. The same comments made about the $Co(ClO_4)_2$ solution hold here.

In other cases, XRD does not definitely clarify whether the environment of Ni^{2+} ions is formed exclusively by water molecules or by some anionic group as well (for example, in cases of medium concentration $NiCl_2$ solutions or of $NiSO_4$ solutions). It is convenient to discuss these cases together, in the section on complex formation, even if sometimes the hypothesis of full hydration seems the most realistic.

2. *Complex Formation*

a. With Chloride Ions

The existence of Cl^- ions in the first coordination shell of Ni^{2+} in aqueous solutions of $NiCl_2$ has been the focus of much debate. First, XRD data seemed to indicate full hydration of the Ni^{2+} ions in the entire range of possible concentrations.[113,114] The symmetric shape of the $Ni^{2+} - H_2O$ peak and the lack of evidence of $Ni^{2+} - Cl^-$ contributions suggested this interpretation, which, furthermore, was supported by satisfactory calculations of structure and correlation functions based on the TSM1 method (Figure 36). Later, one of the authors observed that the best fit of XRD data from a ~3 *M* $NiCl_2$ solution was obtained when equal amounts of the complexes $Ni(H_2O)_6{}^{2+}$ and $Ni(H_2O)_5Cl^+$ were considered;[106] the same complex distribution was also proposed for Co^{2+} in a $CoCl_2$ solution. Following this indication, the experimental data of the 4 *M* $NiCl_2$ solution of Figure 36 were reexamined and it was verified that a model, in which equal quantities of the mentioned complexes were introduced, was in even better agreement with the experimental data than the model only based on $Ni(H_2O)_6{}^{2+}$ ions.[115] The improvement concerned the small shoulder around 2.6 Å, which is real (contrary to the early opinion) and clearly due to $Ni^{2+} - Cl^-$ interactions in the first shell, even if it falls outside the first peak envelope.

Table 21
HYDRATION OF Ni^{2+} IONS

Sample no.	Sample	Ref.	Ni^{2+} conc. (*M*)	$r_{Ni^{2+}-H_2O_I}$	$\sigma_{Ni^{2+}-H_2O_I}$	$N_{Ni^{2+}-H_2O_I}$	Geometry	$r_{Ni^{2+}-H_2O_{II}}$	$\sigma_{Ni^{2+}-H_2O_{II}}$	$N_{Ni^{2+}-H_2O_{II}}$	$r_{H_2O_I-H_2O_{II}}$	$\sigma_{H_2O_I-H_2O_{II}}$	Method	Anion
1	$Ni(ClO_4)_2$ 15.0 H_2O	104	2.9	2.05	0.10	6	Octahedral	4.1	0.21	5.5	2.77	0.10	TSM1	Not hydrated

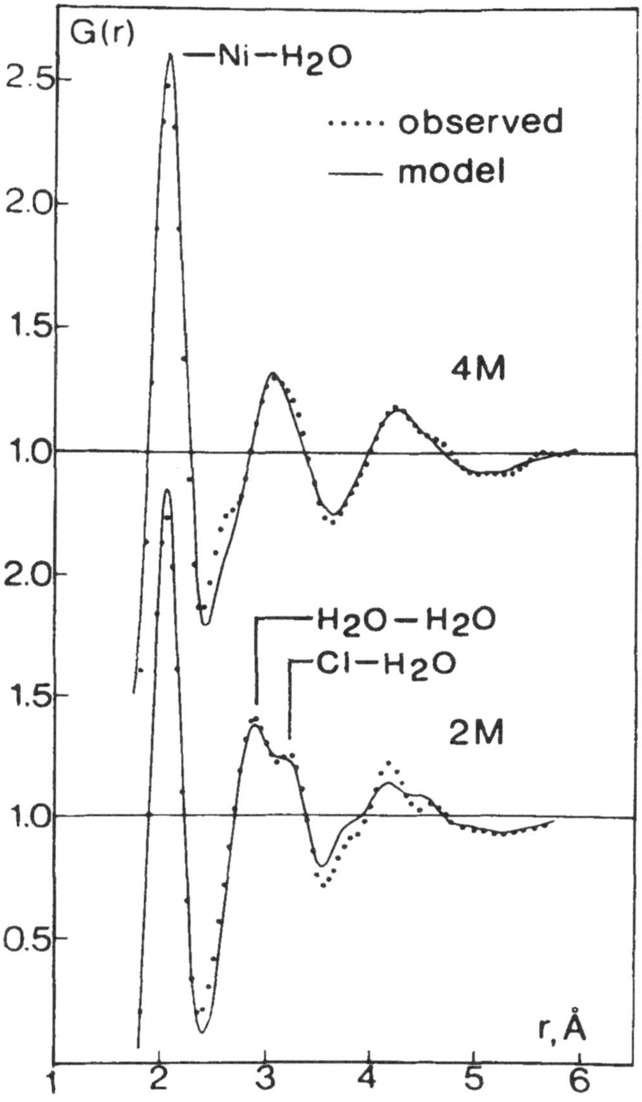

FIGURE 36. Observed (dots) and model (solid line) correlation functions for $NiCl_2$ solutions. (From Caminiti, R., Licheri, G., Paschina, G., Piccaluga, G., and Pinna, G., *Z. Naturforsch.*, 35a, 1361, 1980. With permission.)

To strengthen this interpretation, Ni(II) solutions of about 2 *M* were studied in presence of a large excess of Cl^- achieved by addition of LiCl or HCl.[116] In the reference symbols used (Ni4, NiA4,and NI5), the numbers 4 and 5 represent the ratio Cl^-/Ni^{2+} and the letter A denotes the solution acidified with HCl. It was observed that, when the ratio Cl^-/Ni^{2+} increases, the shoulder evolves into a maximum centered at ~2.50 Å, clearly arising from the increasing number of $Ni^{2+}-Cl^-$ contacts (Figure 37). The quantitative analysis was performed by evaluating approximately the mean complexation numbers, $N_{Ni^{2+}-Cl^-}$, through the simulation of high-angle structure functions, deducing from these values the kind of complexes to be considered, and, finally, carrying out the complete simulation of the structure functions.

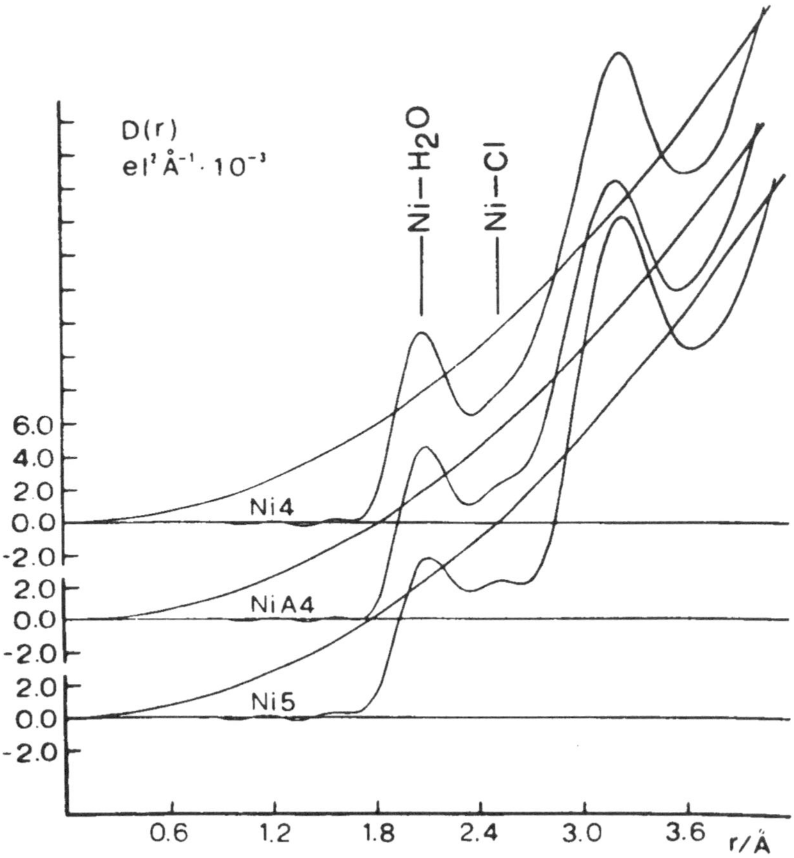

FIGURE 37. Radial distribution functions, D(r), for three solutions of $NiCl_2$ at atomic ratio $Cl^-/Ni^{2+}>2$. The parabolic curves represent the $4\pi r^2\rho_0$ functions. (From Magini, M., Paschina, G., and Piccaluga, G., *J. Chem. Phys.*, 76, 1116, 1982. With permission.)

The final parameters for all the $NiCl_2$ solutions examined are given in Table 22; as regards their reliability, the usual caution, recalled in all the cases in which halo-complexation has been discussed, must be used. The comparison of Tables 19 and 22 shows that, at concentrations > 2 *M* and under the same ratio Cl^-/Me^{2+}, Ni^{2+} and Co^{2+} have very similar complexation degrees. Unfortunately no attempt has been made to study Ni^{2+} solutions in the same conditions in which $CoCl_2$ gives rise to formation of $CoCl_4^{2-}$.

b. With Bromide Ions

The difficulty in the observation of chloro-complexation certainly arises from the fact that, at $N_{Ni^{2+}-Cl^-} \sim 1$, the frequency of $Ni^{2+}-Cl^-$ distances in the coordination polyhedra is much smaller than the one of $Ni^{2+}-H_2O$ interactions, and the contribution of Cl^- to the total scattered intensity is not so high as to counterbalance the weight of predominant interactions. Ni^{2+}-halogen interactions should be more easily revealed and characterized, if the Cl^- ions were replaced by heavier anions. For this reason, the interest of XRD investigations was recently moved towards $NiBr_2$ solutions.[117-119] The results indicate that, even in stoichiometric solutions, $Ni^{2+}-Br^-$ interactions give rise to structural details clearly observable in the total radial distribution functions. In Figure 38, the D(r)s obtained by some of the authors for two solutions of $NiBr_2$ are reported. The results of the quantitative analyses are given in Table 23 for the various solutions examined in References 118 and 119. These

Table 22
COORDINATION OF Ni^{2+} IONS IN CHLORIDE SOLUTIONS

Sample no.	Sample	Ref.	Ni^{2+} conc. (*M*)	Cl/Ni ratio	$r_{Ni^{2+}-H_2O_I}$	$\sigma_{Ni^{2+}-H_2O_I}$	$N_{Ni^{2+}-H_2O_I}$	$r_{Ni^{2+}-Cl^-}$	$\sigma_{Ni^{2+}-Cl^-}$	$N_{Ni^{2+}-Cl^-}$	Method	Free anion
1	$NiCl_2$ 27.1 H_2O	114	2.0	2	2.06	0.12	6	—	—	—	TSM1	$Cl(H_2O)_6^-$
2	$NiCl_2$ 17.8 H_2O	106	2.9_5	2	2.07	0.09	6—0.5	2.44	0.10	0.5	TSM1 + CF	$Cl(H_2O)_6^-$
3	$NiCl_2$ 13.3 H_2O	115	3.9	2	2.06	0.11	6—0.5	2.40	0.19	0.5	TSM1 + CF	$Cl(H_2O)_6^-$
4	$NiCl_2$ 2 LiCl 24.5 H_2O	116	2.0	4	2.06	0.09	6—0.9	2.47	0.14	0.9	TSM1 + CF	$Cl(H_2O)_6^-$
5	$NiCl_2$ 2 HCl 24.8 H_2O	116	2.0	4	2.06	0.08	6—1.2	2.43	0.14	1.2	TSM1 + CF	$Cl(H_2O)_6^-$
6	$NiCl_2$ 3 LiCl 23.2 H_2O	116	2.0	5	2.07	0.09	6—1.4	2.44	0.13	1.4	TSM1 + CF	$Cl(H_2O)_6^-$

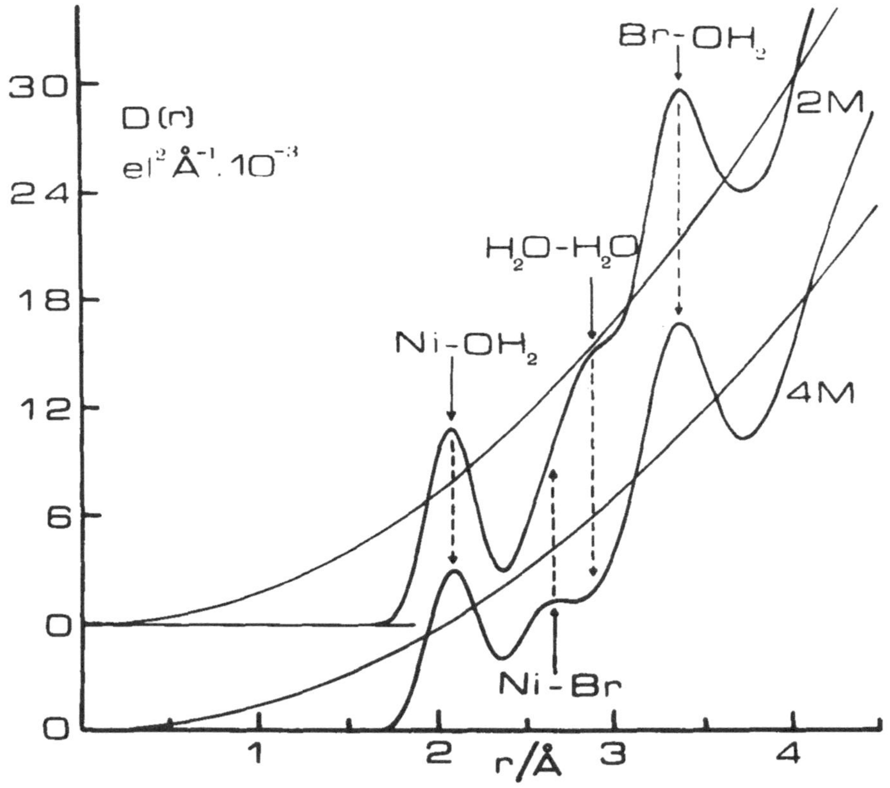

FIGURE 38. Radial distribution functions, D(r), for $NiBr_2$ solutions; the parabolic curves represent the $4\pi r^2\rho_0$ functions. (From Magini, M., de Moraes, M., Licheri, G., and Piccaluga, G., *J. Chem. Phys.*, 83, 5797, 1985. With permission.)

results confirm that bromo-complexation of the cation takes place in stoichiometric solutions; even in the 2 *M* solution, the complete omission of $Ni^{2+} - Br^-$ interactions yields to a worse simulation of the experimental structure function.

The $N_{Ni^{2+}-Br^-}$ values, under the same concentration, are in good agreement; besides, they indicate complexation degrees similar to those found for the stoichiometric $NiCl_2$ solutions. The last observation is in line with thermodynamic indications. In fact, stability constants proposed by various authors, although different according to the different determination method used, concordantly indicate that the Ni^{2+} tend to bind Cl^- and Br^- ions almost to the same extent.

In the third paper about $NiBr_2$ solutions mentioned above,[117] the results presented, in clear disagreement with those given in Table 23 and with thermodynamic indications, suggest a complexation degree much greater for $Ni^{2+} - Br^-$ than for $Ni^{2+} - Cl^-$ pair; furthermore, also to some other structural parameters odd values are ascribed. The origin of the discrepancy in this case is surely in the experimental data, since data treatment procedures and interpretation methods in References 117 and 119 are the same.

c. With Sulfate and Phosphate Ions

No definite information about the formation of complexes with oxyanions can be found in literature. In a solution ~2 *M* of $NiSO_4$, the introduction in the model of complexes $Ni(H_2O)_{6-x}(OSO_3)_x$ produces little effect on the final agreement between experimental and calculated data.[88] This means that $Ni - SO_4$ contacts cannot be clearly evidenced, neither can they be considered incompatible with XRD data. However, it is interesting to note that,

Table 23
COORDINATION OF Ni^{2+} IONS IN BROMIDE SOLUTIONS

Sample no.	Sample	Ref.	Ni^{2+} conc. (*M*)	$r_{Ni^{+2}-H_2O_I}$	$\sigma_{Ni^{2+}-H_2O_I}$	$N_{Ni^{2+}-H_2O_I}$	$r_{Ni^{2+}-Br^-}$	$\sigma_{Ni^{2+}-Br^-}$	$N_{Ni^{2+}-Br^-}$	Method	Free anion
1	NiBr 24.7 H_2O	119	2.1	2.07	0.08	6—0.3	2.61	0.13	0.3	TSM1 + CF	$Br(H_2O)_6^-$
2	$NiBr_2$ 11.9 H_2O	119	4.1	2.08	0.10	6—0.4	2.62	0.13	0.4	TSM1 + CF	$Br(H_2O_6^-$
3	$NiBr_2$ 26.5 H_2O	118	2.0	2.04	0.09	5.7	2.58	0.13	0.2	HAA	Hydrated
4	$NiBr_2$ 11.8 H_2O	118	4.1	2.05	0.09	5.5	2.53	0.13	0.5	HAA	Hydrated
5	$NiBr_2$ 9.6 H_2O	118	4.8	2.04	0.08	5.4	2.52	0.12	0.7	HAA	Hydrated

in confirmation of the small weight of eventual $Ni-SO_4$ interactions, the parameters describing $Ni^{2+}-H_2O$ interactions are remarkably concordant in two different studies, independently from the consideration of ion pairs in the model adopted.[40,88]

The formation of complexes with oxyanions seems more convincingly demonstrated in a solution ~1 *M* of Ni phosphate.[120] In the experimental G(r) an important peak at ~3.5 Å indicate Ni-P correlations straightforwardly; the high concentration of H_3PO_4 in the solution was probably responsible for this event. A model including anionic complexes turned out consistent with the experimental data and led to the structural parameters given in Table 24. The Ni^{2+} ions turned out to be surrounded by 6 O atoms, 5.25 of which are provided on the average by water molecules, while 0.75 belong to phosphate ions. The Ni – O distance of 2.03 Å was kept fixed for all six positions of the polyhedron NiO_6. Ni – P distance was not refined independently, but its value was calculated from the parameters $r_{Ni^{2+}-O}$, r_{P-O} (in the phosphate ion) and from the angle $Ni-\hat{O}-P$.

XI. GROUP VIII (Rh, Pt)

A. Rhodium

The hydration of Rh(III) has been studied in a solution of $Rh(ClO_4)_3$, acidified with $HClO_4$.[121] The case is interesting because the relatively high scattering power of the ion allowed a quantitative study of the cationic hydration at metal concentrations lower than those usually explored. It was also made easier by the small ionic radius of Rh^{3+} ions, which results in a $Rh^{3+}-H_2O$ peak in the radial curves at about 2.05 Å, well separated from the following peaks. The analysis of the $Rh^{3+}-H_2O$ peak in the D(r) was attempted through simulation of its backtransform and provided a value of the hydration number very near to 6 (6.2). A complete structural model followed, with simulation of the whole structure function and evaluation of the structural parameters which are given in Table 25. An environment very similar to that found for other trivalent cations was suggested.

B. Platinum

Diffractometric studies of Pt compounds solutions have been performed with the aim of identifying the hexahalogeno-complexes of Pt(IV), already found in solid state. In this line, concentrated solutions (~2.7 to 2.8 *M*) of the hexachloro and hexabromo platinic acids have been investigated.[122]

In the radial curves a big peak appears at distances observed for Pt-halogen pairs in solid hexahalogen platinates (2.33 Å for the Pt – Cl pair and 2.47 Å for the Pt – Br one). From peak areas and from high-angle analysis of the structure functions, the number of halogen atoms surrounding the Pt atom turned out to be very close to 6.

The analysis of the two following peaks in the D(r)s clearly indicated the geometry of PtX_6 units. In fact, the maxima fell at values expected for X – X *cis* and *trans* distances in PtX_6 octahedra, that is, $2^{1/2}$ and 2 times the direct r_{Pt-X} distance. Obviously, thanks to the higher scattering factor of Br, this indication is very clear for the hexabromo platinic acid. It is worth underlining that the evidence of the structural indications in the case discussed ensues from the presence of a heavy atom in the solution, an advantage well known to crystallographers, but not much exploited in studies of solutions.

XII. GROUP IB

A. Copper

Cu(II) is the only oxidation form examined, since in aqueous solution Cu(I) is unstable.

1. Hydration

Special attention has been devoted to Cu(II) salt solutions. In fact, Cu^{2+} is the species

Table 24
COORDINATION OF Ni^{2+} IONS IN PHOSPHATE SOLUTION

Sample no.	Sample	Ref.	Ni^{2+} conc. (*M*)	$r_{Ni^{2+}-H_2O_I}$	$\sigma_{Ni^{2+}-H_2O_I}$	$N_{Ni^{2+}-H_2O_I}$	$r_{Ni^{2+}-H_2O_{II}}$	$N_{Ni^{2+}-H_2O_{II}}$	$r_{H_2O_I-H_2O_{II}}$	$\sigma_{H_2O_I-H_2O_{II}}$	$r_{Ni^{2+}-P}$	$N_{Ni^{2+}-P}$	Method	Free anion
1	$Ni_3(PO_4)_2$ 10.0 H_3PO_4 15.2 H_2O	120	1.0	2.03	0.11	6—0.75	4.14	7.3	2.70	0.06	3.38	0.75	TSM1 + CF	Hydrated

Table 25
COORDINATION OF Rh^{3+} IONS

Sample no.	Sample	Ref.	Rh^{3+} conc. (*M*)	$r_{Rh^{3+}-H_2O_I}$	$\sigma_{Rh^{3+}-H_2O_I}$	$N_{Rh^{3+}-H_2O_I}$	Geometry	$r_{Rh^{3+}-H_2O_{II}}$	$\sigma_{Rh^{3+}-H_2O_{II}}$	$N_{Rh^{3+}-H_2O_{II}}$	$r_{H_2O_I-H_2O_{II}}$	$\sigma_{H_2O_I-H_2O_{II}}$	Method	Free anion
1	$Rh(ClO_4)_3$ 1.0 $HClO_4$ 14.5 H_2O	121	0.35	2.04	0.10	6	Octahedral	4.02	0.21	13.7	2.69	0.06	TSM1	Not hydrated

with d^9 electronic configuration, for which the greatest number of expected Jahn-Teller distortions has been found in solid crystalline compounds. Different kinds of distortion have also been identified in CuL_6 complexes and it was interesting to see whether these distortions still maintain in solution.

Even an early study showed that the peak of the radial curves due to direct $Cu^{2+} - H_2O$ distances could not be explained on the basis of a regular coordination polyhedron.[85] The first study, that adds to qualitative evidence of distortion a quantitative analysis, is due to Ohtaki et al.[123] In the radial curve D(r) of a concentrated (~3.55 *M*) solution of $Cu(ClO_4)_2$ two peaks appear at ~2.0 and 2.4 Å, between the peak at 1.5 Å (Cl – O distance in the ClO_4^- group) and that at 2.9 Å ($H_2O - H_2O$ distances) (Figure 39a). While the peak at 2.0 Å is certainly due to $Cu^{2+} - H_2O$ distances, the peak at 2.4 Å can be, at least partially, the result of O – O distances within the ClO_4^- ion. However, in other solutions of perchlorates, these O – O distances do not give rise to distinct peaks, so that also longer Cu^{2+} distances are likely contributing to the peak at issue. This turned out evident by subtracting from the total radial curve the assumed contributions of ClO_4^- ions; the difference curve (Figure 39b) shows evidence of the peak at ~2.4 Å on the right side of the peak at 2.0 Å. At a qualitative level, the Jahn-Teller distortion was thus demonstrated. The quantitative analysis showed that the experimental curves were consistent with four H_2O molecules at 1.94 Å and two at 2.43 Å from the central Cu^{2+}, suggesting that the kind of distorted polyhedron should be the most commonly found also in solid state, that is, the tetragonally distorted elongated octahedron. The same interpretation (with minor differences in distance values) was put forward by Ohtaki et al.[86] for an almost saturated (~1.37 *M*) solution of $CuSO_4$, although the evidence of distortion in the radial curve here is less than in the previous case.

These results have been substantially confirmed in two more recent studies performed by some of the authors,[124,125] in which, in addition, the quantitative analysis was carried out so as to describe the whole structure existing in the solutions; as for the other bi- and trivalent metal ions, the interpretation implies the presence of a second coordination shell. The most meaningful result is still the demonstration of the distortion in the first coordination shell of Cu(II). In the case of perchlorate solutions, the comparison of the radial curves of $Cu(ClO_4)_2$ solutions with those of $Co(ClO_4)_2$ of the same concentration is self explaining (Figure 40).[124] The only peak, $Me - H_2O$ at 2.1 Å, in the solution of Co(II) splits up in the solutions of $Cu(ClO_4)_2$, unequivocally indicating the occurrence of two $Cu^{2+} - H_2O$ distances. The similarity of the radial curves in Figure 40 with those presented by Ohtaki et al. (Figure 39a) is worthy of notice.

The comparative examination of Cu(II) and Co(II) perchlorate radial curves can also provide quantitative indications. In fact, subtracting from the D(r) of the $Cu(ClO_4)_2$ solution the one of the $Co(ClO_4)_2$ solution, and taking into account the contribution of complexes $Co(H_2O)_6^{2+}$ to the difference function (calculated according to the hydration parameters of Co^{2+} given in Table 18), a curve is obtained, which represents at low r values the radial distribution around the Cu(II) (Figure 41, dots). The sequence of the peaks indicated by arrows is what is expected if distorted octahedral units are occurring in solution. Peak sizes are also consistent with such units. The full line in Figure 41, in fact, gives the contribution calculated (but not refined) for units $Cu(H_2O)_{4,eq}(H_2O)_{2,ax}$, and it is consistent both in position and size with the peaks drawn by the dotted line.

The $CuSO_4$ solution shows with less evidence the Jahn-Teller distortion (again confirming the results by Ohatki et al.); in fact, no maximum appears in the D(r) function at ~2.4 Å.[125] However, if the radial curves of equally concentrated solutions of $CuSO_4$ and $ZnSO_4$ are compared, the distortion of the Cu(II) coordination polyhedron shows up (Figure 42). The peak at 2.1 Å, which in $ZnSO_4$ solution is diagnostic of $Zn^{2+} - H_2O$ pairs, is in fact higher than the peak at 2.0 Å of $CuSO_4$, thus indicating that the number of the shortest $Cu^{2+} - H_2O$ contacts included in the first peak is lower than the number of direct $Zn^{2+} - H_2O$ contacts.

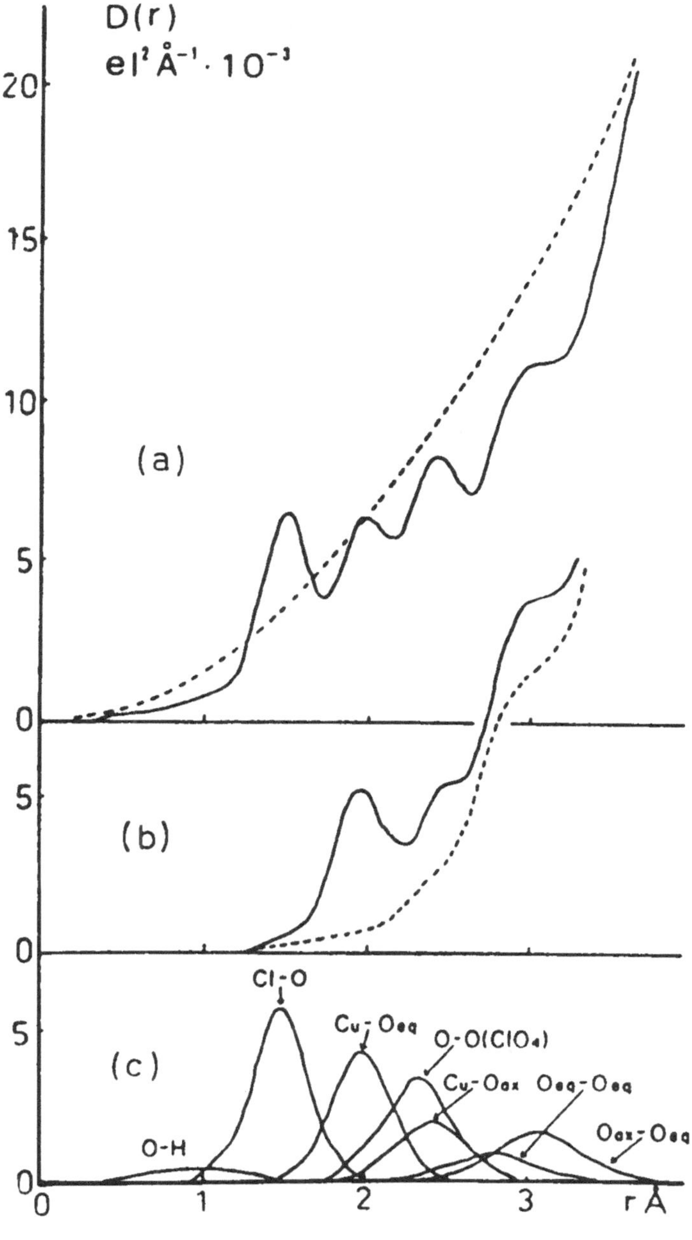

FIGURE 39. (a) The radial distribution curve, D(r), for a 3.55-*M* $Cu(ClO_4)_2$ solution; the broken curve denotes $4\pi r^2\rho_0$. (b) The solid curve gives the difference between D(r) and the calculated peak shapes for the Cl – O and O – O atom pairs within ClO_4^-; the broken curve gives the residual distribution function obtained by subtraction of the theoretical peak shapes shown in (c) from the D(r) curve. (From Ohtaki, H. and Maeda, M., *Bull. Chem. Soc. Jpn.*, 47, 2197, 1974. With permission.)

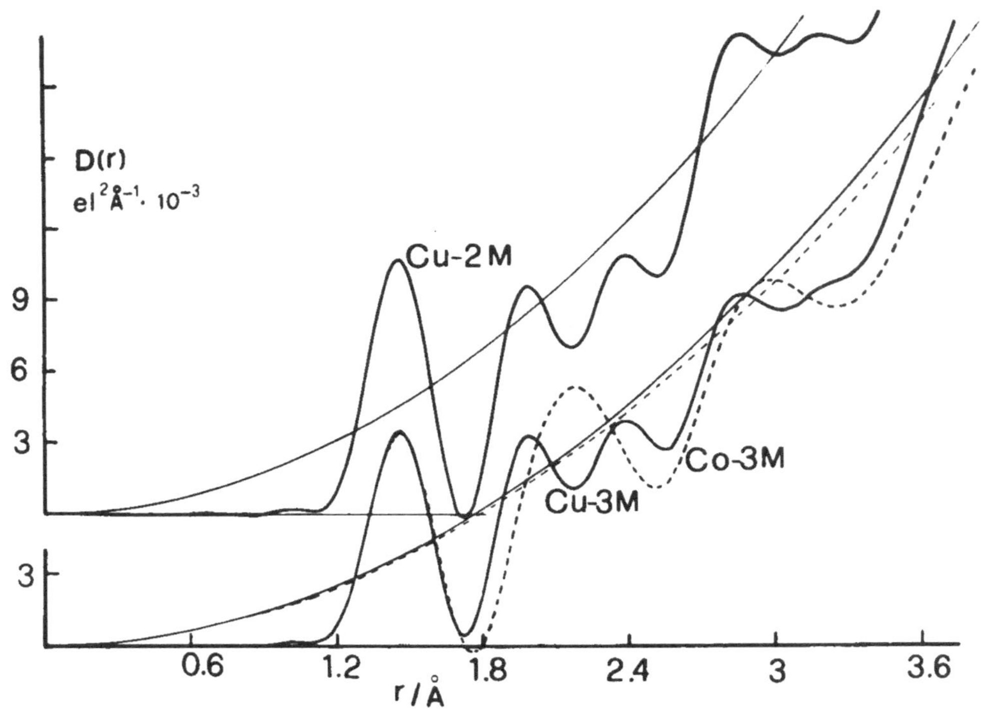

FIGURE 40. Radial distribution functions, D(r), for two copper perchlorate solutions. The distribution curve of a 3-*M* cobalt perchlorate solution is also shown by dashes for comparison. The parabolic curves are the $4\pi r^2\rho_0$ functions. (From Magini, M., *Inorg. Chem.*, 21, 1535, 1982. With permission.)

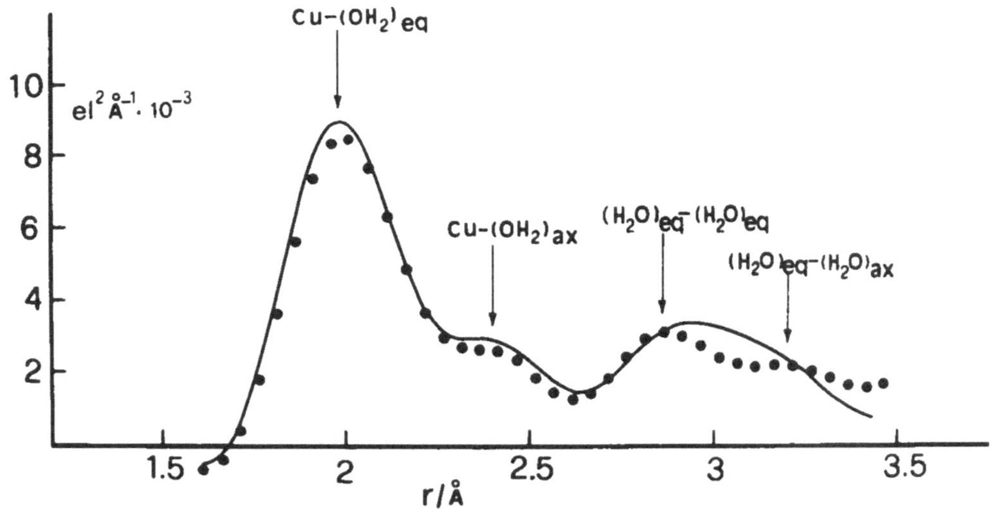

FIGURE 41. The partial Cu(II) – H_2O residual distribution curve, obtained from the experimental D(r) by a subtraction procedure described in the text, is shown by dots. The solid line has been calculated for an octahedral tetragonally distorted $Cu(H_2O)_{4,eq}(H_2O)_{2,ax}$. (From Magini, M., *Inorg. Chem.*, 21, 1535, 1982. With permission.)

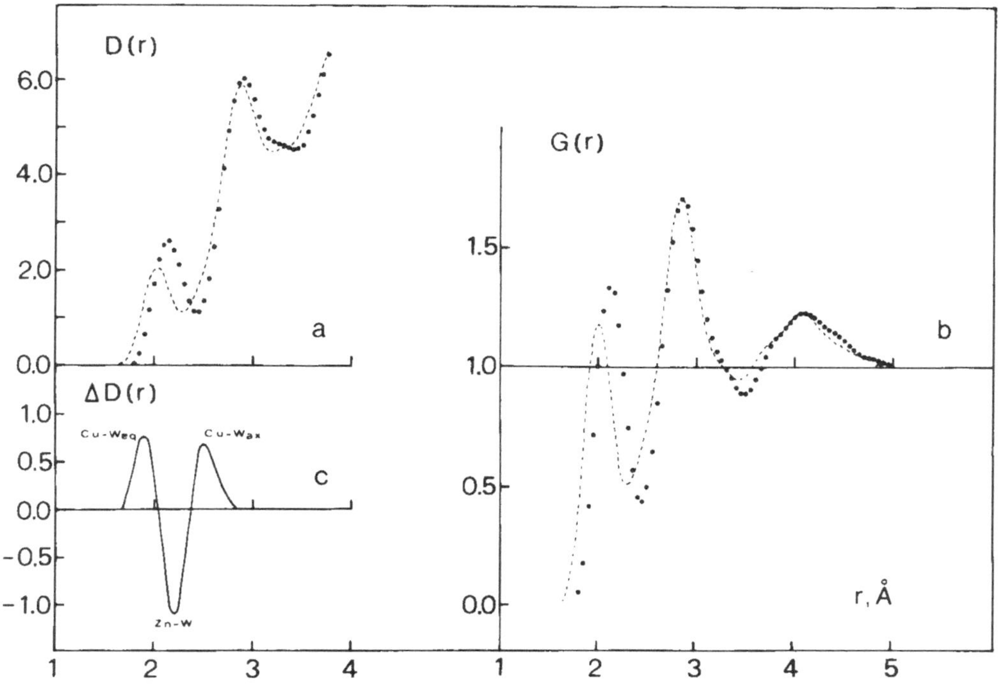

FIGURE 42. (a) Experimental radial distribution functions of $CuSO_4$ (— — —) and $ZnSO_4$(· · ·) solutions. (b) Experimental correlation functions of $CuSO_4$ (— — —) and $ZnSO_4$ (· · ·) solutions. (C) $\Delta D(r) = D(r)_{CuSO4} - D(r)_{ZnSO4}$. (From Musinu, A., Paschina, G., Piccaluga, G., and Magini, M.,*Inorg. Chem.*, 22, 1184, 1983. With permission.)

However, the $CuSO_4$ radial curve is higher at about 2.4 Å than that of $ZnSO_4$. As O−O interactions in the sulfate ion are the same, it means that longer $Cu^{2+}-H_2O$ interactions are present. This is even more evident in the difference curves $\Delta D(r) = D(r)_{CuSO4} - D(r)_{ZnSO4}$ reported in Figure 42c; two peaks clearly appear at distance values that correspond to $r_{Cu^{2+}-H_2O_{eq}}$ and $r_{Cu^{2+}-H_2O_{ax}}$, as evaluated in the $Cu(ClO_4)_2$ solutions; the minimum in the middle corresponds to $Zn^{2+}-H_2O$ distances.

The values of the parameters describing the whole structure of the hydrated Cu^{2+} ions are given in Table 26. It is interesting to observe that the structural parameters confirm the difference between perchlorate and sulfate solutions. A meaningful example is given by the distribution of $Cu^{2+}-H_2O_{I,ax}$ distances as described by the $\sigma_{Cu^{2+}-H_2O_{I,ax}}$ value, which turns out much broader in $CuSO_4$ than in $Cu(ClO_4)_2$; this suggests that a greater disorder exists around Cu^{2+} ions in aqueous $CuSO_4$, which accounts for the lesser evidence of $Cu^{2+}-H_2O_{I,ax}$ distances in the $CuSO_4$ radial curve. This increasing disorder may be attributed to some anionic effect like, for instance, some tendency of the sulfate ion to compete with water molecules for the occupancy of hydration shell positions.

2. *Complex Formation*

a. With Chloride Ions

Contrary to the usual, the study of halogen-complexes of Cu(II) meets with great difficulty.[106,126-128] In fact, just because of the Jahn-Teller distortion, if two different chemical species (H_2O and X^-) form the nearest environment of Cu(II), four different pair distances are possible, $r_{Cu^{2+}-H_2O_{eq}}$, $r_{Cu^{2+}-H_2O_{ax}}$, $r_{Cu^{2+}-X_{eq}}$, and $r_{Cu^{2+}-X_{ax}}$. The longest among them (in particular the $Cu^{2+}-X_{ax}$ one) are superimposed to the many H_2O-H_2O distances and are therefore difficult to isolate and quantify.

Table 26
HYDRATION OF Cu^{2+} IONS

A. First Shell Parameters

Sample no.	Sample	Ref.	Cu^{2+} conc. (*M*)	$r_{Cu^{2+}\text{-}H_2O_{I,eq}}$	$\sigma_{Cu^{2+}\text{-}H_2O_{I,eq}}$	$N_{Cu^{2+}\text{-}H_2O_{I,eq}}$	$r_{Cu^{2+}\text{-}H_2O_{I,ax}}$	$\sigma_{Cu^{2+}\text{-}H_2O_{I,ax}}$	$N_{Cu^{2+}\text{-}H_2O_{I,ax}}$	Geometry	Method	Anion
1	$Cu(ClO_4)_2$ 25.1 H_2O	124	1.9	1.98	0.05	4	2.34	0.10	2	See the text	TSM1	Not hydrated
2	$Cu(ClO_4)_2$ 15.4 H_2O	124	2.9	1.98	0.07	4	2.39	0.12	2	See the text	TSM1	Not hydrated
3	$CuSO_4$ 39.8 H_2O	125	1.4	2.00	0.09	4	2.33	0.23	2	See the text	TSM1	Hydrated

B. Second Shell Parameters

Sample no.	Ref.	Cu^{2+} conc. (*M*)	$r_{Cu^{2+}\text{-}H_2O_{II}}$	$\sigma_{Cu^{2+}\text{-}H_2O_{II}}$	$N_{Cu^{2+}\text{-}H_2O_{II}}$	$r_{H_2O_I\text{-}H_2O_{II}}$	$\sigma_{H_2O_I\text{-}H_2O_{II}}$
1	124	1.9	3.95	0.33	11.6	2.73	0.07
2	124	2.9	3.96	0.39	11.1	2.80	0.05
3	125	1.4	4.20	0.20	7.6	2.79	0.03

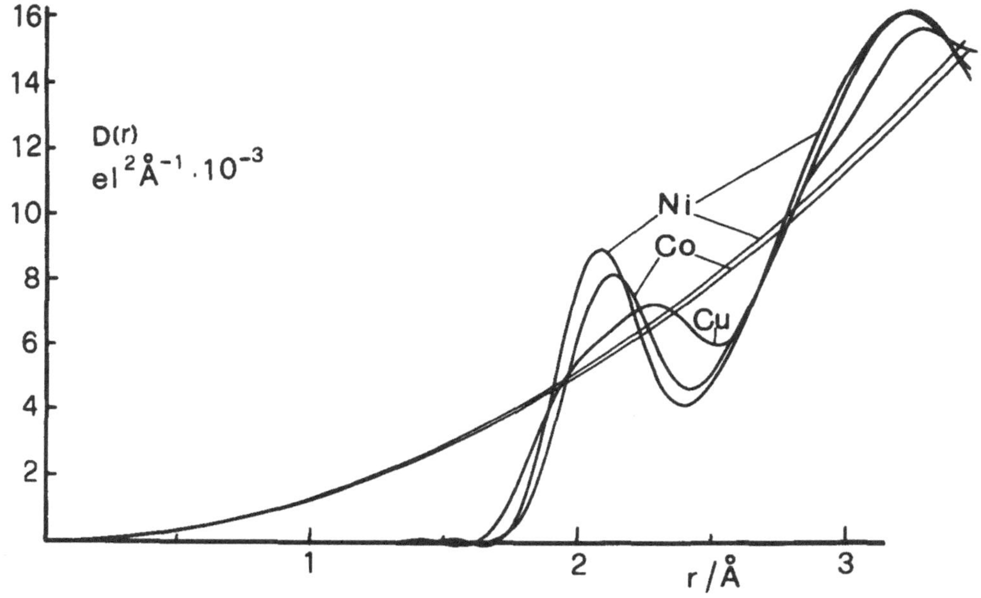

FIGURE 43. Radial distribution curves, D(r) (solid lines), and $4\pi r^2\rho_0$ functions (weak solid lines) for Ni, Co, and Cu chloride solutions. The $4\pi r^2\rho_0$ function of $CuCl_2$ lies between those of $CoCl_2$ and $NiCl_2$ and is not shown for clarity. (From Magini, M., *J. Chem. Phys.*, 74, 2523, 1981. With permission.)

The complexity of the Cu(II) environment is quite evident in the radial curve D(r) of a $CuCl_2$ solution, and even more if this D(r) is compared with those of equally concentrated solutions of $NiCl_2$ and $CoCl_2$ (Figure 43).[106] The quantitative analysis proved that on the average about 2.8 water molecules and 1.2 chloride ions occupy the nearest (equatorial) positions at 1.95 and 2.25 Å respectively. It can be seen in Figure 44 that no other possibility turned out consistent with the experimental data. However, neither the distribution curve analysis nor the direct fitting of intensity data gave information about the occupancy of the longer (axial) positions of Cu^{2+} coordination polyhedron.

b. With Bromide Ions

In order to overcome the mentioned difficulties, the use of a heavier anion turned out convenient, as in the case of Ni^{2+} solutions. Ichihashi et al.[128] performed a systematic investigation of $CuBr_2$ solutions, both stoichiometric ($Br^-/Cu^{2+} = 2$) and with $Br^-/Cu^{2+} > 2$ (up to 6.7). The radial curves (Figures 45 and 46) show that with increasing concentration and/or Br^-/Cu^{2+} ratio a peak at about 2.0 Å (probable short $Cu^{2+}-H_2O$ distances), lowers progressively and then disappears, while a peak at about 2.4 Å (probable short $Cu^{2+}-Br^-$ distances) arises, increases, and becomes the only one present at low r. At this point, the peak at 3.3 to 3.4 Å, due to $Br^- - H_2O$ distances, is replaced by a large peak at ~3.6 Å, which probably originates from Br − Br distances within higher bromo-complexes of Cu(II). Noteworthy is the lack of any peak at ~3.2 Å, where the longer $Cu^{2+}-Br^-$ distances are expected on the basis of crystal structure information, implying that no axial $Cu^{2+}-Br^-$ coordination takes place in solution.

The average complexation degree was deduced from the fittings of high-angle structure functions and the results, given in Table 27, confirm the qualitative indications reported above. The coordination numbers refer to the occupancy of the nearest sites and they are in all cases very close to 4. For the solution with the highest Br^-/Cu^{2+} ratio, the complexation with solely Br^- ions and the values of $Br^- - Br^-$ distance inferred from the D(r), indicate a $CuBr_4^{2-}$ complex with flattened tetrahedral structure. In the four stoichiometric solutions

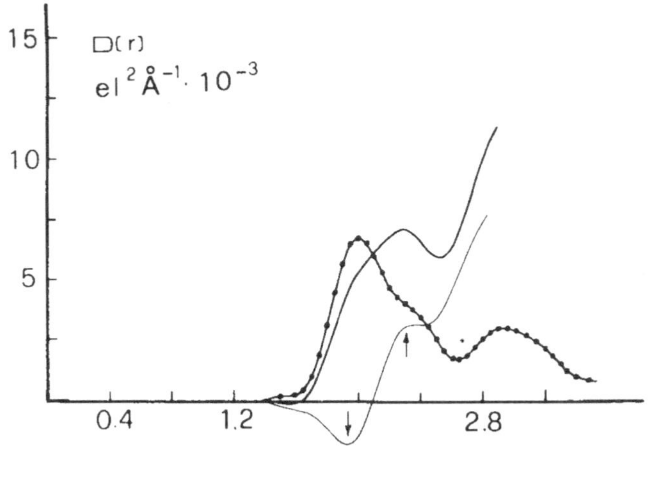

A

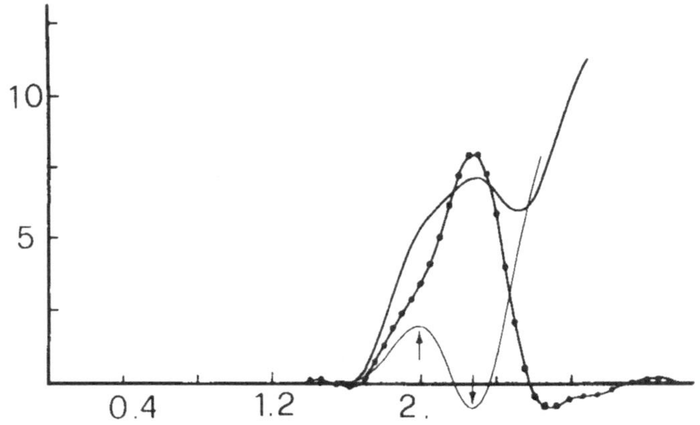

B

FIGURE 44. Comparison of the experimental distribution curve (solid line) of $CuCl_2$ solution with theoretical peaks (solid lines and dots) calculated for the following species: (A) $Cu(H_2O)_{4,eq}(H_2O)_{2,ax}$; (B) $Cu(H_2O)_{2,eq}Cl_{2,eq}$; (C) $Cu(H_2O)_{2.8,eq}Cl_{1.2,eq}$. The weak solid lines are the differences between experimental and calculated distribution curves. Only in (C), no residual peaks are practically left unresolved in the difference curve. (From Magini, M., *J. Chem. Phys.*, 74, 2523, 1981. With permission.)

the complexes with either H_2O or Br^- in the equatorial positions are probably completed by two H_2O molecules in axial positions; the sample 5 is likely to be a complicated mixture of different complexes.

It is interesting to observe that for the stoichiometric solution ~3.3 *M* the average number of $Cu^{2+}-Br^-$ contacts, $N_{Cu^{2+}-Br^-} = 1.1$, is very similar to the value of $N_{Cu^{2+}-Cl^-}$ proposed in $CuCl_2$ solution of similar concentration discussed above ($N_{Cu^{2+}-Cl^-} = 1.2$). This confirms that, as in Ni complexation, Cl^- and Br^- ions surround the metal atom almost to the same extent. These observations concordantly indicate that some previous studies, which suggested a very high complexation degree also in stoichiometric solutions, therefore involving polymeric associated structures, are not reliable.[126,127]

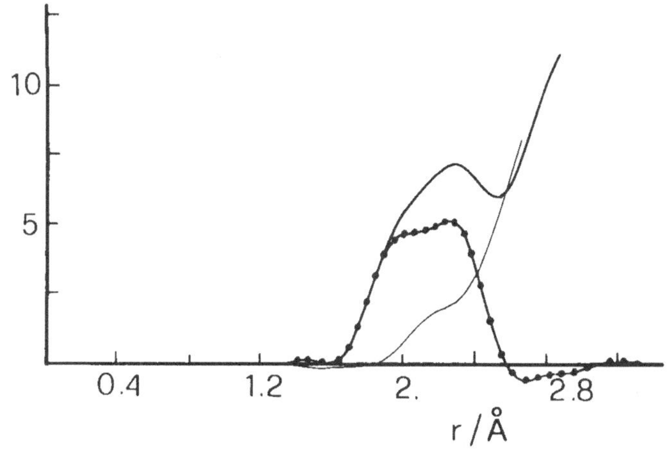

FIGURE 44C

c. With Nitrate Ions

The correlation functions from a study of two $Cu(NO_3)_2$ solutions are shown in Figure 47.[129] As in the case of $CuSO_4$, there is no evident proof in the G(r)s of distortion in the nearest coordination of Cu^{2+} (that is, some peak or shoulder in the r region between 2.0 and 2.5 Å). However, an indirect, still convincing, proof may be obtained also in this case by comparing the present results with a proper reference system, here represented by solutions of $Cr(NO_3)_3$. This system, in fact, has the same anion and the cation-water distances are very close to the shortest $Cu^{2+} - O_I$ distances. In the G(r) of $Cr(NO_3)_3$ solutions (Figure 19) the first peak at 2.0 Å is perfectly resolved on its right side from longer distance peaks, the G(r) being almost null at $r = 2.4$ Å. The peak includes the contribution of $O_N - O_N$ distances entirely, while no discrete distances fall between the peak of 2.0 and the one at 2.85 Å. On the contrary, in the G(r) of $Cu(NO_3)_2$ solutions, the minimum between the first two peaks has a value different from zero, thus suggesting that some correlation distance falls in this distance range. This may likely be the longer $Cu^{2+} - O_I$ distance, forming the cation-distorted coordination polyhedron.

Besides the mentioned peaks, a less conspicuous but very meaningful detail is apparent (especially in the 3*M* solution) as a shoulder at about 3.15 Å. Shoulders as prominent as this have not been observed in other solutions of completely dissociated nitrates (see, for example, the case of $Cr[NO_3]_3$, or that of $Mg[NO_3]_2$).[38,79] Since the N (in NO_3)$-H_2O$ distances are generally longer (see Section XV) and the $H_2O_{eq} - H_2O_{ax}$ distances alone do not account for this detail, this shoulder may be interpreted in terms of Cu-N distances, giving evidence of a considerable amount of inner-sphere complex formation.

The fitting procedure of the experimental structure functions confirmed that omitting Cu-N interactions produces unsatisfactory agreements. The coordination parameters of Cu^{2+} in nitrate solutions are given in Table 28. $O_{I,ax}$ and $O_{I,eq}$ indicate O atoms which, in the presence of anion-cation complexes, may belong to water molecules or nitrate ions. The presence of NO_3^- ions in the first coordination shell gives rise to not surprising differences between the values of the parameters describing the environment of the Cu^{2+} in nitrate solutions and those in perchlorate and sulfate solutions. The shortening of the distance $r_{Cu^{2+} - O_{I,ax}}$ should be noticed, which perhaps means that the anionic complexation takes place in the axial positions of Cu(II) polyhedron.

We wish to conclude this section by recalling that the diffraction experiments on the solutions discussed have been repeated in two different laboratories, resulting in an excellent reproducibility of experimental data (see Figure 21, Chapter 1).

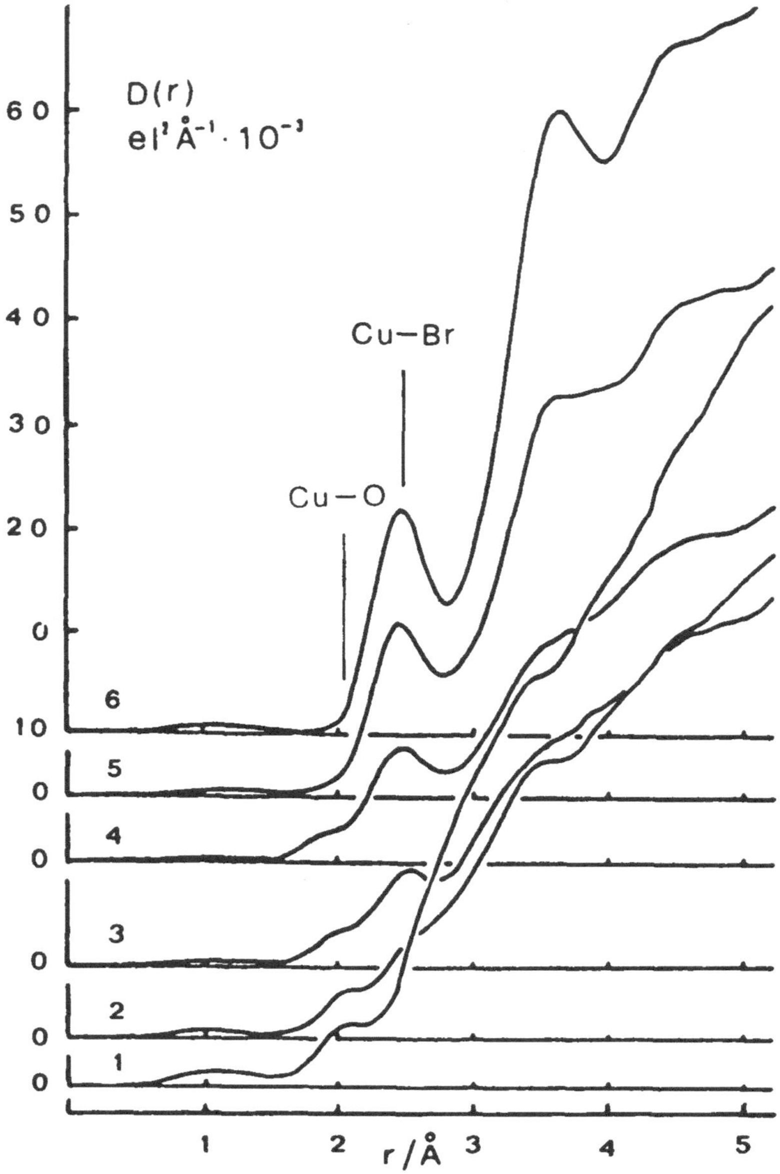

FIGURE 45. Radial distribution curves, D(r), for $CuBr_2$ solutions. (From Ichihashi, M., Wakita, H., Mibuchi, T., and Masuda, I., *Bull. Chem. Soc. Jpn.*, 55, 3160, 1982. With permission.)

B. Silver

The hydration of Ag^+ ions has been studied in solutions of $AgClO_4$ and $AgNO_3$.[130,131] The task was not completely easy, as in the experimental radial curves the peak $Ag^+ - H_2O$ came out at about 2.40 Å and was mixed with O − O interactions within the oxyanions and, still worse, it was not sufficiently separated from $H_2O - H_2O$ distances at ~2.85 Å. Thus, in a first study,[130] a low (~2) coordination number was proposed by high-angle fitting, while in a following and more complete investigation, the tetracoordination of Ag^+ was suggested.[131] This latter result was in agreement with the observation that Ag − O distances of ~2.40 Å are characteristic of fourfold Ag coordination in crystalline structures; further-

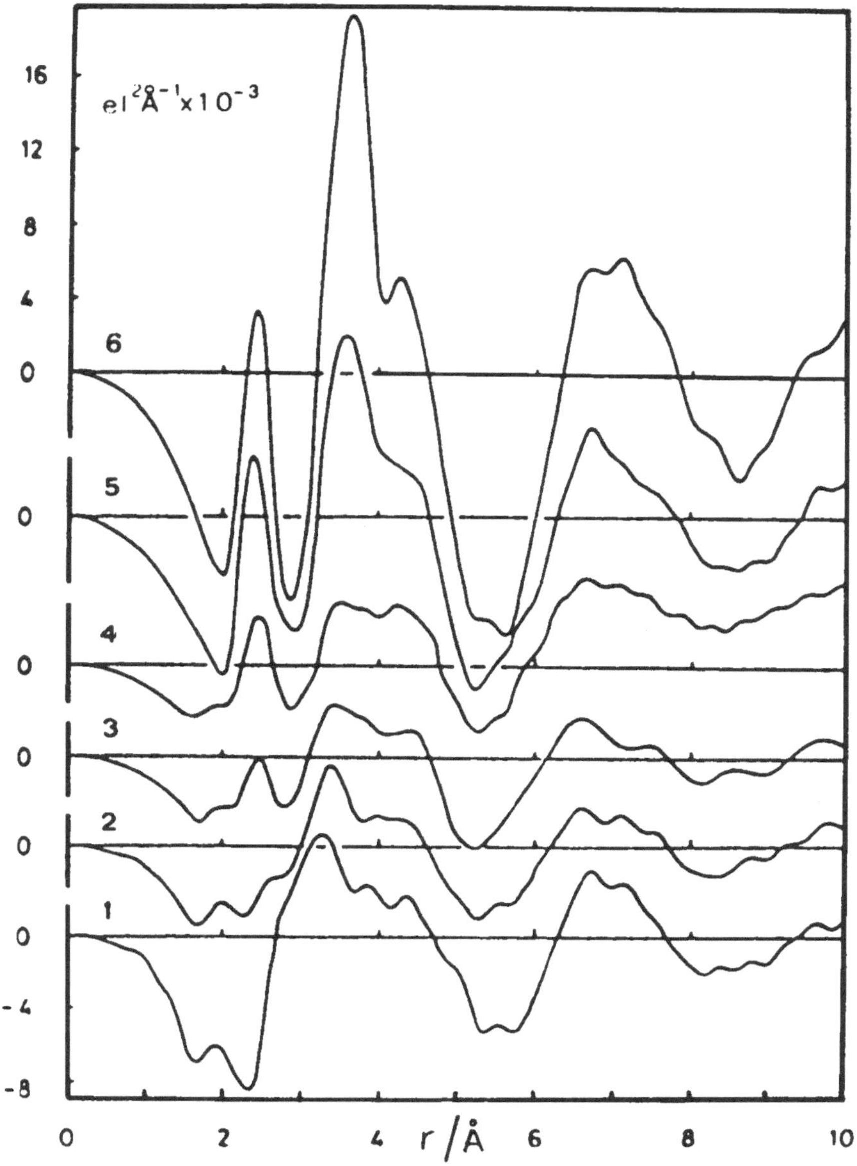

FIGURE 46. D(r) − $4\pi r^2\rho_0$ curves for $CuBr_2$ solutions. (From Ichihashi, M., Wakita, H., Mibuchi, T., and Masuda, I., *Bull. Chem. Soc. Jpn.*, 55, 3160, 1982. With permission.)

more, it was also confirmed by an analysis of a "depurated" Ag^+-O peak, obtained through subtraction procedures using properly chosen and normalized radial curves. The same procedures allowed to evidence the existence of $Ag^+-NO_3^-$ and $Ag^+-ClO_4^-$ (only in very concentrated solutions) direct interactions.

C. Gold

Solutions of tetrachloro- and tetrabromo-auric acids ($HAuCl_4$ and $HAuBr_4$) are the only examples of systems containing Au(III) ions investigated by X-ray diffraction.[132] Here the situation is very similar to that described for the Pt(IV). A first big peak appears in the radial curves at r values coinciding with the Au − Cl and Au − Br distances determined in

Table 27
COORDINATION OF Cu^{2+} IONS IN BROMIDE SOLUTIONS

Sample no.	Sample	Ref.	Cu^{2+} conc. (*M*)	Br/Cu ratio	$r_{Cu^{2+}\text{-}H_2O_{eq}}$	$N_{Cu^{2+}\text{-}H_2O_{eq}}$	$r_{Cu^{2+}\text{-}Br^-{}_{eq}}$	$N_{Cu^{2+}\text{-}Br^-{}_{eq}}$	Method
1	$CuBr_2$ 53.1 H_2O	128	1.0	2.0	1.96	3.7	2.42	0.3	HAA
2	$CuBr_2$ 25.6 H_2O	128	2.0	2.0	1.99	3.3	2.43	0.6	HAA
3	$CuBr_2$ 14.9 H_2O	128	3.3	2.0	1.96	2.7	2.44	1.1	HAA
4	$CuBr_2$ 10.8 H_2O	128	4.4	2.0	1.97	2.5	2.46	1.3	HAA
5	$CuBr_2$ 2.0 HBr 21.2 H_2O	128	2.1	4.0	1.93	1.1	2.42	2.9	HAA
6	$CuBr_2$ 4.7 HBr 26.4 H_2O	128	1.5	6.7			2.41	3.8_5	HAA

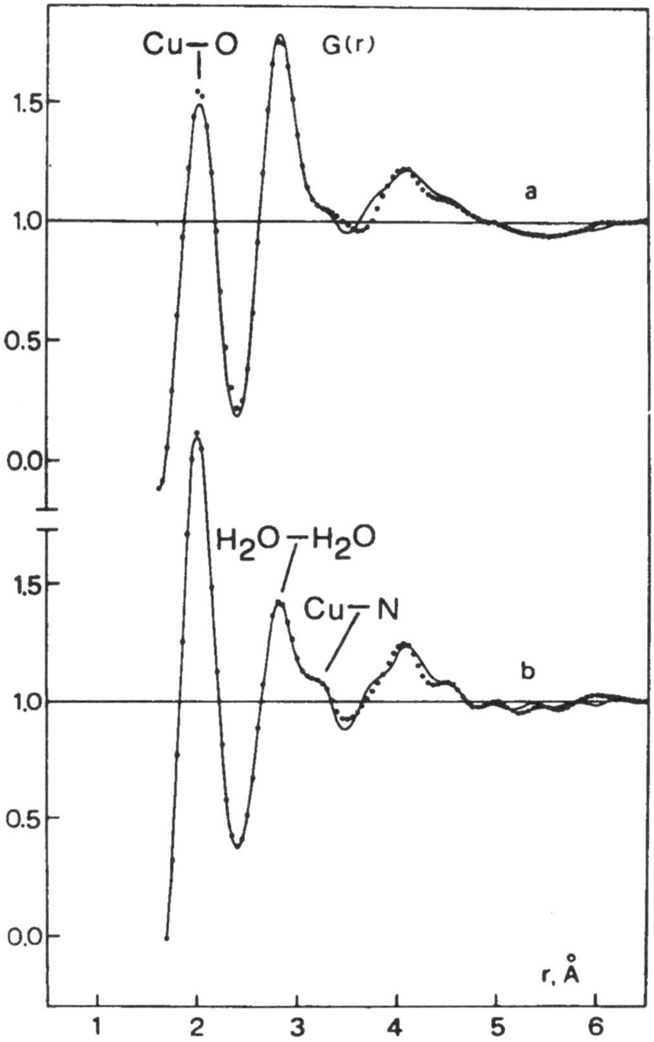

FIGURE 47. Experimental (dots) and model (solid line) correlation functions for $Cu(NO_3)_2$ solutions: (a) 1.5 *M*; (b) 3.0 *M*. (From Licheri, G., Musinu, A., Paschina, G., Piccaluga, G., Pinna, G., and Sedda, A. F., *J. Chem. Phys.*, 80, 5308, 1984. With permission.)

crystalline compounds containing the complexes $AuCl_4^-$ and $AuBr_4^-$ (r_{Au-Cl} = 2.30 Å and r_{Au-Br} = 2.45 Å). From the evaluation of the peak areas and from the analysis of the high-angle structure functions, the tetrahalo-coordination of Au(III) was determined unambiguously. On the other hand, the analysis of the two following peaks gave clear indications about the geometry of AuX_4 units. In fact, the maxima of the D(r)s fall at distances coincident with direct Au–X distance after multiplication by $2^{1/2}$ and 2, respectively; the peaks can be therefore ascribed to *cis* and *trans* X–X distances in square planar complexes, as already observed in solid state. Obviously, thanks to the higher scattering power of Br, this indication is clearer for the tetra-bromo complex.

Table 28
COORDINATION OF Cu^{2+} IONS IN NITRATE SOLUTIONS

A. First Shell Parameters

Sample no.	Sample	Ref.	Cu^{2+} conc. (*M*)	$r_{Cu^{2+}-O_{I,eq}}$	$\sigma_{Cu^{2+}-O_{I,eq}}$	$N_{Cu^{2+}-O_{I,eq}}$	$r_{Cu^{2+}-O_{I,ax}}$	$\sigma_{Cu^{2+}-O_{I,ax}}$	$N_{Cu^{2+}-O_{I,ax}}$	$r_{Cu^{2+}-N}$	$\sigma_{Cu^{2+}-N}$	$N_{Cu^{2+}-N}$	Method	Free anion
1	$Cu(NO_3)_2$ 34.7 H_2O	129	1.5	2.00	0.13	4	2.12	0.18	2	3.03	0.14	1.2	TSM1 + CF	Not hydrated
2	$Cu(NO_3)_2$ 15.7 H_2O	129	3.1	2.00	0.11	4	2.22	0.25	2	3.07	0.11	1.7	TSM1 + CF	Not hydrated

B. Second Shell Parameters

Sample no.	Ref.	Cu^{2+} conc. (*M*)	$r_{Cu^{2+}-H_2O_{II}}$	$\sigma_{Cu^{2+}-H_2O_{II}}$	$N_{Cu^{2+}-H_2O_{II}}$	$r_{H_2O_I-H_2O_{II}}$	$\sigma_{H_2O_I-H_2O_{II}}$
1	129	1.5	4.08	0.41	9.1	2.72	0.03
2	129	3.1	4.09	0.37	8.3	2.74	0.03

XIII. GROUP IIB

A. Zinc

1. Hydration

The studies of Zn^{2+} ions hydration performed on oxyanions solutions are rather concordant in proposing hexacoordination.[35,85,86,133-135] Among these studies, the ones by Bol et al.[35] and by Ohtaki et al.[86] deserve particular mention and have been repeatedly referred to previously. In the work by Bol et al., also the existence of a second coordination shell around Zn^{2+} ions was demonstrated.

Recently, five solutions of $ZnSO_4$, of concentration ranging from 0.6 to 3.1 *M*, have been systematically examined, in order to determine the cationic hydration parameters and to follow their dependence on the concentration.[134] The structure functions and the corresponding correlation functions are shown in Figures 48 and 49.

In the G(r)s, three sharp peaks, centered at 1.45 to 1.50, 2.10 to 2.15, and 2.85 Å, are clearly observable. Since the intensity data were normalized to $(ZnSO_4)_x\,(H_2O)_{1-x}$ unit (with variable Zn content), the heights of these peaks show the regular expected variations with the concentration. Thus, the first two peaks, due to $S-O$ (in the sulfate groups) pairs and to $Zn^{2+}-H_2O$ distances, are determined by the solute species and grow accordingly at increasing concentration; on the contrary, the peak at 2.85 Å, coming from H_2O-H_2O interactions, lowers when the bulk water decreases. The complex peak at 4.10 Å is likely to receive one of the most important contributions from second shell distances.

Least square refinement of model parameters showed that, in spite of the great concentration range explored, the same model (TSM1) is consistent with the experimental data from all the solutions. The parameters from the best fits are given in Table 29. They are very similar in the different solutions; this again suggests that in the concentration range examined, meaningful structural modifications do not take place. The only quantity changing appreciably is the second coordination number, which comprehensibly lowers with increasing concentration as the number of available H_2O molecules lowers correspondingly. The regularity of the parameters induced us to try a calculation of the tilt angle between $Zn^{2+}-O_I$ bond and the water molecule H_2O_I plane. Under the same assumptions made in the investigation about Mg^{2+}, this angle turned out randomly variable in a narrow range (37 to 42°).

Another study of a ~3 *M* solution of $ZnSO_4$,[135] though giving a description of the nearest environment of Zn^{2+} very similar to that reported in Table 29, still suggested that, together with the dominant complexes $Zn(H_2O)_6{}^{2+}$, also the complexes $Zn(H_2O)_5SO_4$ are present in amounts of about 40%; the $Zn-S$ distance proposed was 3.13 Å, so that this contribution is hidden under the envelope of H_2O-H_2O peak. Since the hypothesis of formation of sulfate complexes is consistent with some thermodynamic indications, it was also tested in Reference 134. It came out that even models including complex formation are consistent with the experimental data, but the improvement in the fits does not allow to prefer these models to those involving hydration only. As we shall see better in the Discussion, it is very difficult to assert the existence of cation-oxyanion complexes, the more so when the phenomenon involves minor percentages of the metal ions.

2. Halide Complex Formation

The tendency of Zn^{2+} ions to form halo-complexes is well known. Evidence of their existence has been given in several studies by XRD.[136-139] Quantitative structural determinations are complicated, since, as stability constants suggest, the existence range of the species overlap. In these cases, thermodynamic data valid for the solution examined by the diffraction method are essential to extend the structural investigation up to the determination of the structures of the various possible complexes.

This strategy was adopted in an investigation on $ZnBr_2$ solutions, in which Raman and

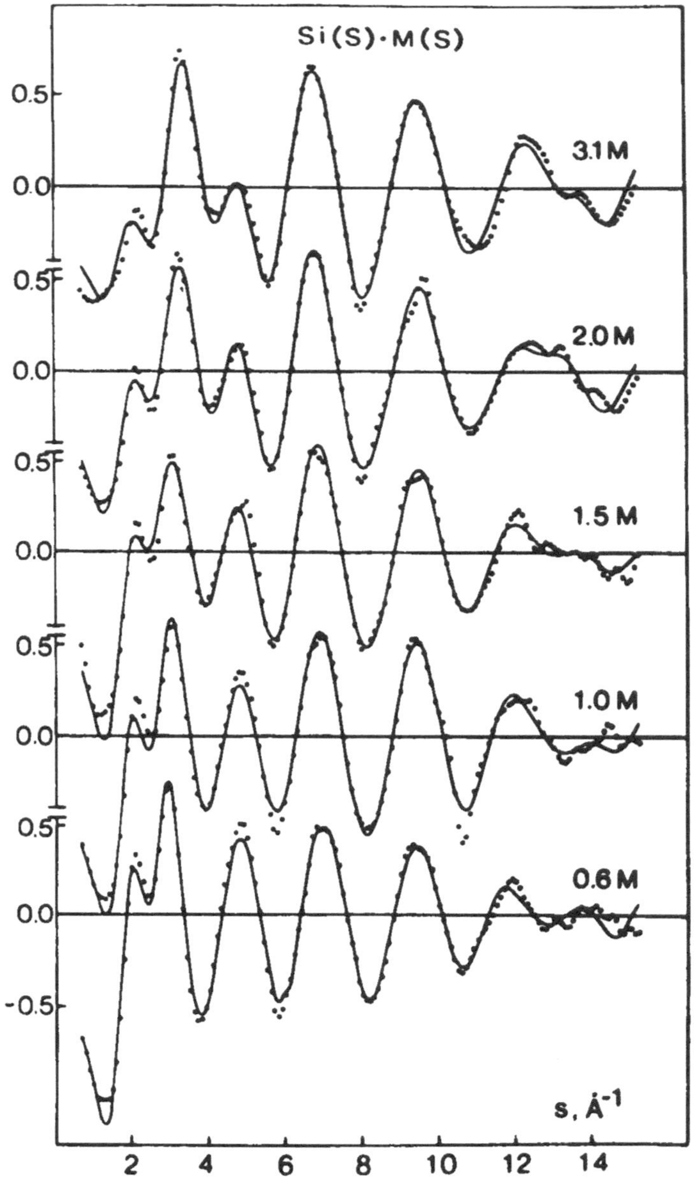

FIGURE 48. Experimental (dots) and model (solid lines) structure functions for $ZnSO_4$ solutions. (From Licheri, G., Paschina, G., Piccaluga, G., and Pinna, G., *Z. Naturforsch.*, 37a, 1205, 1982. With permission.)

IR spectroscopies were applied together with XRD to estimate concentrations of the complexes.[140,141] In particular, Goggin et al.[141] carried out a systematic research on nine solutions, divided into three groups, in which the values of the atomic ratio Br^-/Zn^{2+} and the concentrations of the ions were properly chosen in view of the analysis procedures of the experimental data (Table 30). Also a solution of only Zn perchlorate was included in one of the series. Bromides rather than chlorides were chosen to increase the possibility of identifying ligand-ligand interactions within the complexes.

In the radial curves, given in Figure 50, peaks occur at expected distances: $Zn^{2+} - H_2O$ at 2.1 Å, $Zn^{2+} - Br^-$ at 2.4 Å, $Br^- - Br^-$ at 4.0 Å and, for perchlorate solutions, Cl – O

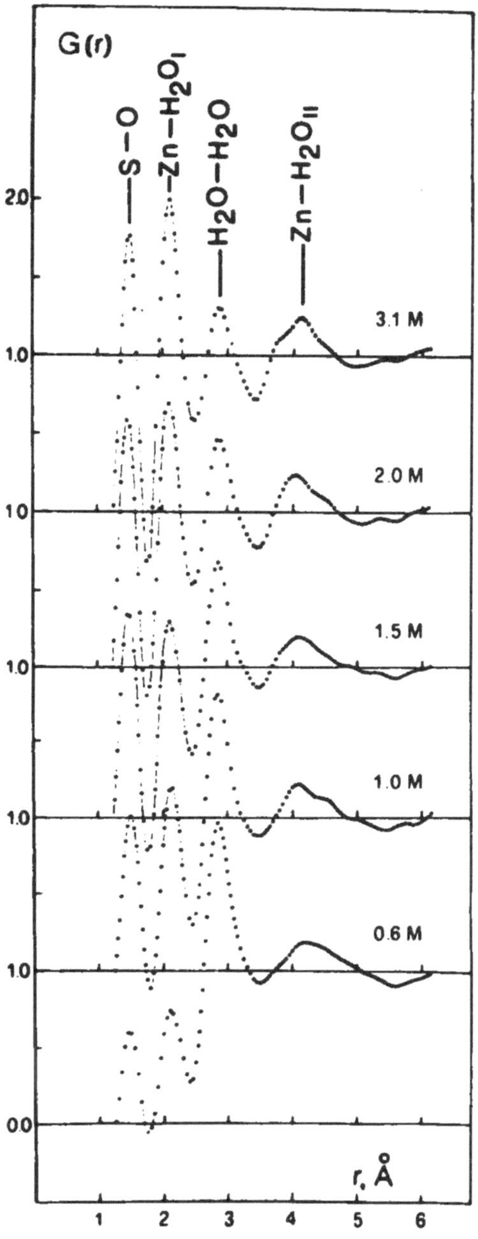

FIGURE 49. Experimental correlation functions for $ZnSO_4$ solutions (a continuous line has been drawn through experimental points for clarity). (From Licheri, G., Paschina, G., Piccaluga, G., and Pinna, G., *Z. Naturforsch.*, 37a, 1205, 1982. With permission.)

at 1.4 Å. The ratios between Br – Br and Zn – Br distances came out close to $(8/3)^{1/2}$ = 1.63, which is the expected value for a tetrahedral arrangement. Values for the parameters characterizing the Zn – Br and Br – Br interactions were obtained by least square refinements using the high-angle part of the intensity curves of the samples at Br^- concentration of ~15 and ~5 *M* (Table 30). N_{Br-Br} represents the number of Br^- – Br^- distances per metal atom. Zn^{2+} hydration parameters were determined interpreting the results from the $Zn(ClO_4)_2$

Table 29
HYDRATION OF Zn^{2+} IONS

Sample no.	Sample	Ref.	Zn^{2+} conc. (*M*)	$r_{Zn^{2+}-H_2O_I}$	$\sigma_{Zn^{2+}-H_2O_I}$	$N_{Zn^{2+}-H_2O_I}$	Geometry	$r_{Zn^{2+}-H_2O_{II}}$	$\sigma_{Zn^{2+}-H_2O_{II}}$	$N_{Zn^{2+}-H_2O_{II}}$	$r_{H_2O_I-H_2O_{II}}$	$\sigma_{H_2O_I-H_2O_{II}}$	Method	Anion
1	$ZnSO_4$ 100.0 H_2O	134	0.6	2.13	0.11	6	Octahedral	4.26	0.34	12.5	2.79	0.04	TSM1	Hydrated
2	$ZnSO_4$ 52.5 H_2O	134	1.0	2.14	0.09	6	Octahedral	4.22	0.40	13.2	2.78	0.04	TSM1	Hydrated
3	$ZnSO_4$ 36.0 H_2O	134	1.5	2.12	0.11	6	Octahedral	4.25	0.33	10.3	2.79	0.02	TSM1	Hydrated
4	$ZnSO_4$ 27.3 H_2O	134	2.0	2.12	0.10	6	Octahedral	4.21	0.26	9.9	2.80	0.03	TSM1	Hydrated
5	$ZnSO_4$ 16.9 H_2O	134	3.1	2.13	0.10	6	Octahedral	4.22	0.24	7.6	2.76	0.02	TSM1	Hydrated

Table 30
COORDINATION OF ZN^{2+} IONS IN BROMIDE AND PERCHLORATE SOLUTIONS

A. Structural Parameters

Sample no.	Sample	Ref.	Zn^{2+} conc. (*M*)	Br/Zn ratio	Br^- conc. (*M*)	$r_{Zn^{2+}-Br^-}$	$\sigma_{Zn^{2+}-Br^-}$	$N_{Zn^{2+}-Br^-}$	$r_{Br^--Br^-}$	$\sigma_{Br^--Br^-}$	$N_{Br^--Br^-}$	$r_{Zn^{2+}-H_2O}$	$N_{Zn^{2+}-H_2O}$	$\frac{r_{Br^--Br^-}}{r_{Zn^{2+}-Br^-}}$
A1	$ZnBr_2$ 4.5 H_2O	141	7.6	2.0	15.1	2.39	0.09	2.0	3.98	0.27	3.0			1.67
A2	$ZnBr_2$ 1.0 LiBr 6.1 H_2O	141	5.2	3.0	15.5	2.40	0.09	2.9	3.97	0.19	3.3			1.66
A3	$ZnBr_2$ 3.0 LiBr 11.0 H_2O	141	3.0	5.0	14.9	2.40	0.08	4.0	3.92	0.24	9.5			1.63
B1	$ZnBr_2$ 19.8 H_2O	141	2.5	2.0	5.0	2.37	0.11	1.3	4.00	0.14	1.0			1.69
B2	$ZnBr_2$ 1.0 LiBr 29.5 H_2O	141	1.6	3.0	5.1	2.38	0.08	1.8	3.95	0.24	3.1			1.66
B3	$ZnBr_2$ 3.0 LiBr 48.0 H_2O	141	1.0	5.0	5.0	2.38	0.09	2.2	3.95	0.24	3.5			1.66
C1	$Zn(ClO_4)_2$ 22.3 H_2O	141	3.0	0.0	0.0							2.10	6	
C2	$ZnBr_2$ 3$Zn(ClO_4)_2$ 20.8 H_2O	141	3.0	0.5	1.5									
C3	$ZnBr_2$ $Zn(ClO_4)_2$ 18.9 H_2O	141	3.0	1.0	3.0									

B. Percentages of the Complexes Used

Sample no.	Ref.	Zn^{2+} conc. (*M*)	Zn^{2+}	$ZnBr^+$	$ZnBr_2$	$ZnBr_3^-$	$ZnBr_4^{2-}$	Br-Zn-Br	Method
A1	141	7.6	28.2	0.0	26.2	30.1	15.5	113°	HAA
A2	141	5.2	8.5	0.0	16.8	38.5	36.2	112°	HAA
A3	141	3.0	0.0	0.0	0.0	0.0	100	109°	HAA
B1	141	2.5	56.6	0.0	10.0	19.1	14.3	115°	HAA
B2	141	1.6	39.7	0.0	7.9	25.0	27.4	112°	HAA
B3	141	1.0	27.0	0.0	4.0	25.0	44.0	112°	HAA
C1	141	3.0	100	0.0	0.0	0.0	0.0		HAA
C2	141	3.0	79.4	2.3	9.0	7.3	2.0		
C3	141	3.0	62.3	2.3	14.9	14.2	6.3		

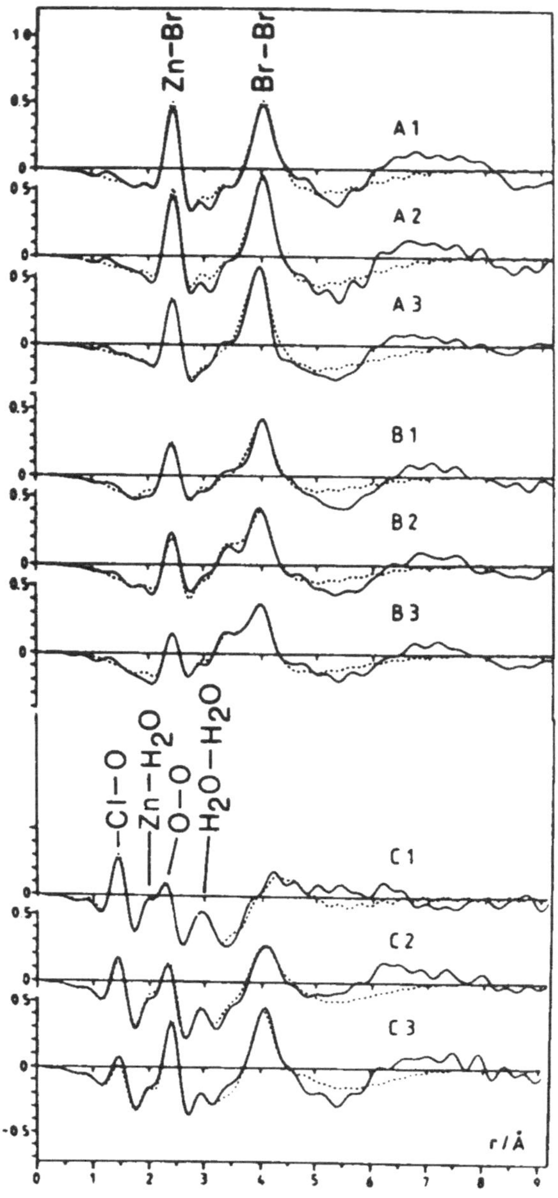

FIGURE 50. Radial distribution functions, D(r) – $4\pi r^2\rho_0$ (solid lines), compared with theoretical curves (dashed lines) for zinc bromide and perchlorate solutions (see Table 30 for symbols explanation). (From Goggin, P. L., Johansson, G., Maeda, M., and Wakita, H., *Acta Chem. Scand.*, A38, 625, 1984. With permission.)

solution, while the data from the two last solutions were only used in the final part of the analysis.

The results given in Table 30 show that, for the solution A3, the number of $Zn^{2+} - Br^-$ interactions per Zn atom is 4.0, that is the expected value if all Zn^{2+} ions occur as $ZnBr_4^{2-}$ complexes. Subtraction of $ZnBr_4^{2-}$ contributions from the experimental distribution curve leaves a smooth background curve, without features related to the presence of polynuclear complexes. The occurrence of all Zn^{2+} as tetrahedral $ZnBr_4$ units in the solution A3 is consistent with its Raman spectrum, which shows only one polarized peak at 171 cm^{-1}, which is the position of the symmetric stretching vibration, ν_1, for $ZnBr_4$. The parameters of solution A3 therefore just describe the structure of $ZnBr_4^{2-}$ ions.

Lowering either the Br^-/Zn^{2+} ratio or Zn^{2+} concentration, each Zn binds fewer Br^- ions, and "lower" Zn complexes form. The relative concentrations of the complexes can be estimated from Raman spectra by resolving the composite bands into their component peaks. By combining these values with Zn – Br contact numbers from least square refinements of the scattering curves, concentrations of the complexes could be estimated (Table 30B).

With the structural parameters of the complexes $ZnBr_4^{2-}$ and $Zn(H_2O)_6^{2+}$, coming from the analyses of the A3 and C1 solutions, and with the concentrations of all the species present in solution, the determination of the structure of the complexes $ZnBr_3^-$ and $ZnBr_2$ was attempted (the amount of $ZnBr^+$ is always too small to attempt any description of this complex). New refinements of synthetic structure functions against the data from the samples A1, A2, B1, B2, and B3 suggested only slight shortenings of the average $Zn^{2+} - Br^-$ distances in the lower complexes, accompanied by increase of the $Br - \hat{Zn} - Br$ angle, with respect to the values in Table 30A. The following indications were then derived: (1) pyramidal structure for the $ZnBr_3^-$ species; (2) bent structure for the $ZnBr_2$ complex; and (3) strict similarity between the two complexes. Since for $ZnBr_3^-$ the structural parameters proposed are very similar to those found for the species $ZnBr_3H_2O^-$ in crystalline α- and β-$KZnBr_3{\cdot}2H_2O$, pseudo-tetrahedral structures were proposed for both complexes, according to the compositions $ZnBr_2(H_2O)_2$ and $ZnBr_3H_2O^-$. However, it must not be forgotten that an unambiguous determination of the bound water was not possible, just because of the predominant contribution of Br^- ions. In any case, the tetracoordination of $ZnBr_3^-$ and $ZnBr_2$ complexes is in agreement with thermodynamic results, which suggest change in coordination at the second coordination step, that is when the second halide ligand adds up.[142]

Finally, with the structural parameters (and percentages) of the species $ZnBr_4^{2-}$, $ZnBr_3H_2O^-$, $ZnBr_2(H_2O)_2$, and $Zn(H_2O)_6^{2+}$, also the intensities and radial curves of the samples C2 and C3 could be reproduced (Figure 50).

B. Cadmium

1. Hydration

The early information about Cd^{2+} hydration was obtained by Bol et al.[35] who examined, by isomorphic substitution method, one ~1-*M* solution of $Cd(NO_3)_2$ and an equally concentrated solution of $Ca(NO_3)_2$. Sixfold coordination, with $Cd^{2+} - H_2O$ distance of 2.29 Å, was proposed and a second coordination shell was also suggested.

Full hydration of Cd^{2+} has been observed in two other cases reported in literature, a ~2.9-*M* solution of $Cd(ClO_4)_2$, slightly acidified with $HClO_4$,[26] and a ~1-*M* solution of $Cd(ClO_4)_2$ in ~1 *M* $HClO_4$.[143] In the first system, sixfold coordination and $r_{Cd^{2+}-H_2O}$ = 2.31 Å were deduced from an analysis combining high-angle simulation of the structure function and interpretation of the $Cd^{2+} - H_2O$ peak in the radial curve. In the second sample, the nearest neighbor coordination parameters were confirmed ($r_{Cd^{2+}-H_2O}$ = 2.29 Å, $N_{Cd^{2+}-H_2O}$ = 6) and, in addition, the whole structure function was reproduced by the introduction in the calculation of second-shell interactions.

2. Complex Formation

a. With Halide Ions

Because of the strong tendency of Cd^{2+} ions to form halo-complexes,it is not difficult to observe in XRD results the contributions of Cd^{2+}-halogen interactions. Thus, if we observe the correlation functions obtained for two $CdCl_2$ solutions (Figure 51),[114] we notice that the dominant peaks are centered at about 2.55 to 2.60 Å, that is, at the distance value frequently

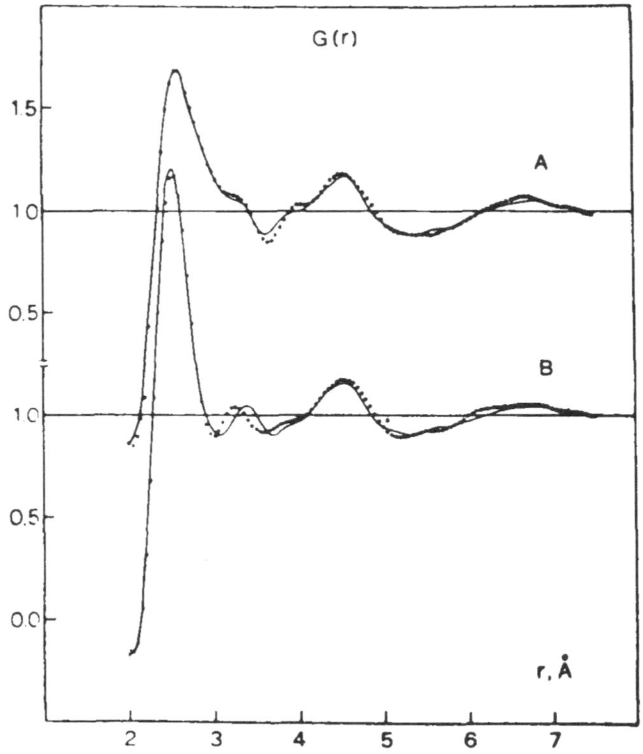

FIGURE 51. Experimental (dots) and model (solid lines) correlation functions for $CdCl_2$ solutions: (A) 0.85 *M*; (B) 1.25 *M*. (From Caminiti, R., Licheri, G., Paschina, G., Piccaluga, G., and Pinna, G., *Z. Naturforsch.*, 35a, 1361, 1980. With permission.)

observed for $Cd^{2+}-Cl^-$ pairs in hydrated crystals. Actually, these peaks are broad, and other interactions are likely to contribute to them; however, the most important contribution is supplied by $Cd^{2+}-Cl^-$ distances.

High-angle analysis of the structure functions allowed the determination of the total average Cd^{2+} coordination (~6) and of the partial contributions (~4 H_2O and ~2 Cl^-), so that from a model based on $Cd(H_2O)_4Cl_2$ complex the whole structure functions were satisfactorily reproduced. In the absence of information about amounts of different complexes, no deeper analysis (like that carried out for $ZnBr_2$ solutions) was possible.

The simple evaluation of the average Me − Cl coordinations permitted the study of the competitive behavior of Cd^{2+} and Zn^{2+} towards Cl^- complexation in a solution 1 *M* both in $ZnCl_2$ and $CdCl_2$.[144] Here the conditions are right for the simultaneous study of the environment of the two cations, since both scattering factors and ionic radii of the species are greatly different. The results of the mixed solution were compared with those obtained for two solutions in which the same salts were separately dissolved. Zn^{2+} ions turned out to be hexacoordinated, both in the single and the mixed solution, the coordination shell being composed on the average of about five water molecules and one Cl^- ion. On the contrary, the coordination of Cd^{2+} ions seems to have undergone important variations in the mixing. The number of $Cd^{2+}-H_2O$ contacts decreases drastically going from 4.4 to 1.9; the number of $Cd^{2+}-Cl^-$ contacts increases from 1.75 to 2.2, and this change, although less striking, is even more reliable, since the contribution of $Cd^{2+}-Cl^-$ pair to the i(s) is the most important one. Thus, the results of the mixing seem to be a significant increase of the $Cd^{2+}-Cl^-$ complexation degree and formation of a great amount of tetracoordinated complexes.

So far, no XRD investigation on Cd^{2+} bromo-complexes formation has been carried out, while Ohtaki et al.[26] have reported a study of the structural configuration of the CdI_4^{2-} complex in a 2.9 *M* solution of CdI_2, in which the addition of NaI brought the atomic ratio I^-/Cd^{2+} to a value of 5. Thus, the cation iodo-complexation equilibria were shifted towards the almost complete formation of tetraiodocadmate ions. Two large peaks dominated the radial distribution curve, centered at about 2.8 and 4.6 Å. The first one was ascribed to the $Cd^{2+}-I^-$ pair and Cd^{2+} tetracoordination was clearly indicated from the area evaluation; the second peak is due to $I^- - I^-$ distances. On the other hand, the ratio of the two distances, $\sim(8/3)^{1/2}$, was that expected for tetrahedral geometry of the CdI_4^{2-} ion.

b. With Sulfate Ions

The tendency of Cd^{2+} ions to form inner-sphere complexes with anions has been evidenced by XRD also in oxyanion solutions, thus confirming the expectations of thermodynamic and spectroscopic studies.

Sulfate complexation has been examined by Caminiti and Johansson[143] in a 3-*M* solution of $CdSO_4$ and in two other solutions with the same SO_4^{2-} concentration, but where 2 mol of Cd^{2+} had been replaced by NH_4^+ or H^+ ions, so that SO_4^{2-}/Cd^{2+} ratios >2 were reached. The solution of $Cd(ClO_4)_2$ mentioned above was used as reference for analyses of the cation structure in sulfate solutions; to this end, equal concentrations of perchlorate and sulfate ions were considered. The ClO_4^- ion has the same number of electrons and the same structure as the SO_4^{2-} ion, but it does not form inner-sphere complexes with Cd^{2+} ions, so that $Cd^{2+}-Cl^-$ distances do not contribute to the radial curves in the 3- to 4-Å region. The radial functions of the sulfate solutions, given in Figure 53, look very similar to that of the perchlorate solution, given in Figure 52, but display a small additional peak at about 3.5 Å. It was ascribed to Cd – S interactions on the ground of the average distance (3.44 Å) found for Cd – S pairs in crystalline $CdSO_4(H_2O)_{8/3}$, with Cd – Ô – S angle of 131°.

To evaluate the number of sulfate groups bound to each Cd^{2+} ion, $N_{Cd^{2+}-S}$, difference radial curves were calculated in the form $[D(r) - 4\pi r^2\rho_o]_{sulf} - [D(r) - 4\pi r^2\rho_o]_{perchl}$. Adding to these functions the "intramolecular" contributions determined for the $Cd(ClO_4)_2$ solution, a residual curve was obtained containing the contributions to the radial curves from the species SO_4^{2-} and $Cd(H_2O)_y(SO_4)_x$. Simulation of the latter showed that the inner coordination sphere of Cd^{2+} in the sulfate solution forms a regular octahedron, as it does in the perchlorate solution. The average coordination number $N_{Cd^{2+}-S}$ turned out increasing with the concentration of SO_4^{2-} actually available in each solution, going from 0.8 for the stoichiometric solution to 1.2 for the acid solution and finally, to 1.6 for the solution containing NH_4^+ ions (part of the sulfate ions are in the form HSO_4^- in the acid solution). The average complexation number for the stoichiometric solution was in rough agreement with the value calculated from stability constants, assuming these to be valid in concentrated solutions. In the other two cases, $N_{Cd^{2+}-S} > 1$ would indicate that negative complexes also form.

According to these indications, structural models were proposed, and short- and medium-range radial curves calculated. Only monodentate coordination of sulfate ions was considered, with Cd – Ô – S angle variable. The results are shown in Figure 54; synthetic curves from models not including complex formation are also drawn for comparison. In Table 31, the final parameter values are reported; it is interesting to observe that, apart from $N_{Cd^{2+}-S}$, the parameters were put identical in the three solutions. Since TSM2 was used, two water-water terms, namely the $H_2O_I - H_2O_{II}$ term and the $H_2O - H_2O$ one of the free water, had to be considered and they were described by a single term. X in the table represents the central atom of the various oxyanions.

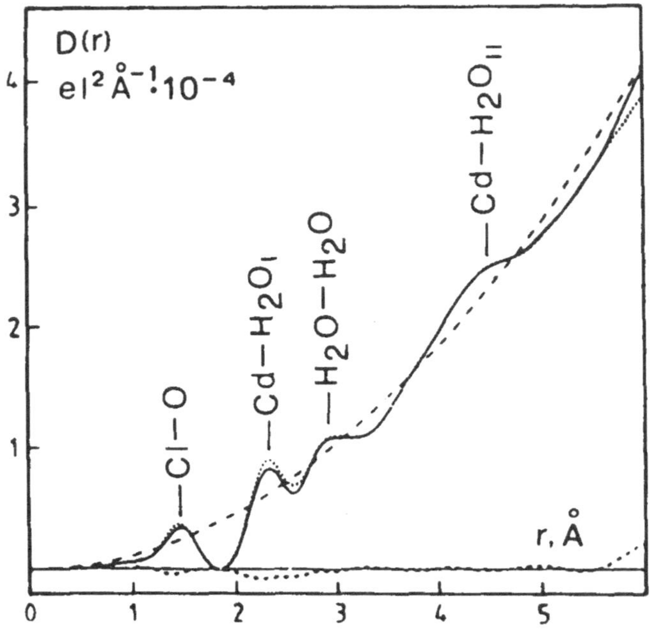

FIGURE 52. The observed (solid line) and the calculated (dotted line) D(r) function for a 1.0-*M* cadmium perchlorate solution. The difference between them is given by the dashed line. Long dashes indicate the $4\pi r^2\rho_0$ function. (From Caminiti, R., and Johansson, G., *Acta Chem. Scand.*, A35, 373, 1981. With permission.)

Sulfate complex formation was also demonstrated in a less concentrated (~2 *M*) solution of $CdSO_4$, with $N_{Cd^{2+}-S}$ smaller than that found in the 3 *M* solution.[145] In this study, the solution was examined at two different temperatures (9 and 62°C) and increasing complexation with increasing T was found.

c. With Nitrate Ions

Five solutions of $Cd(NO_3)_2$, in the concentration range from 1.0 to 4.5 *M*, have been investigated by Raman spectroscopy, and inner-sphere nitrate-Cd complexes were indicated.[146] In the same study, the most concentrated solution was also examined by XRD; the correlation function confirmed the existence of nitrate complex formation, thanks to a small peak at about 3.15 Å, due to the pair $Cd^{2+}-N$ (Figure 55).

The quantitative analysis was performed by the usual refinement of a synthetic structure function; it was calculated from a structural model containing the complexes $Cd(H_2O)_{6-N}(ONO_2)_N$, surrounded by an external shell of water molecules. The best fit parameter values are reported in Table 31. $N_{Cd^{2+}-N}$ turned out ~1; obviously, equilibria involving higher complexes or fully hydrated ions are likely to be present, but only the "average" complex $Cd(H_2O)_5ONO_2$ was introduced in the final calculation.

d. With Phosphate Ions

A situation similar to that just described has been found in a solution of Cd phosphate.[120] Clear indication of complex formation is provided by the correlation function, where the peak at 3.55 Å is close to the $Cd^{2+}-P$ distance value found in some crystalline structures. The same analysis as in the nitrate solution was carried out; the final parameters are again listed in Table 31.

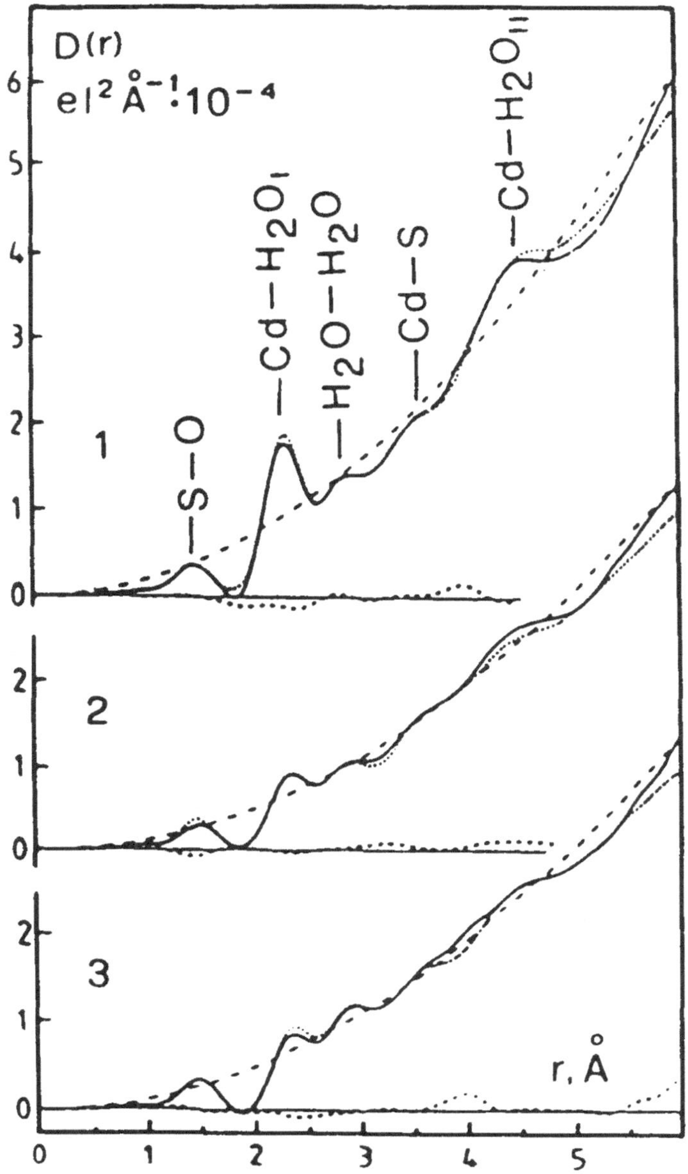

FIGURE 53. The experimental distribution functions D(r) for cadmium sulfate solutions (solid lines) compared with calculated functions (dotted lines). The difference between experimental and calculated functions are given by dashed lines. Long dashes indicate the $4\pi r^2\rho_0$ functions. (From Caminiti, R. and Johansson, G., *Acta Chem. Scand.*, A35, 373, 1981. With permission.)

C. Mercury (Hg_2^{2+})

The radial curves of four Hg(I) perchlorate solutions, obtained by Johansson,[147] are presented in Figure 56. They are characterized by a peak at ~2.5 Å, in the region where Hg – Hg distances are expected in the Hg_2^{2+} ion. The area under this peak is larger than that corresponding to a single Hg – Hg interaction, as water molecules close to the mercury atoms should give substantial contributions to the same peak. The distance value was obtained by comparing theoretical scattering curves for the Hg_2^{2+} ion with the high-angle portion of

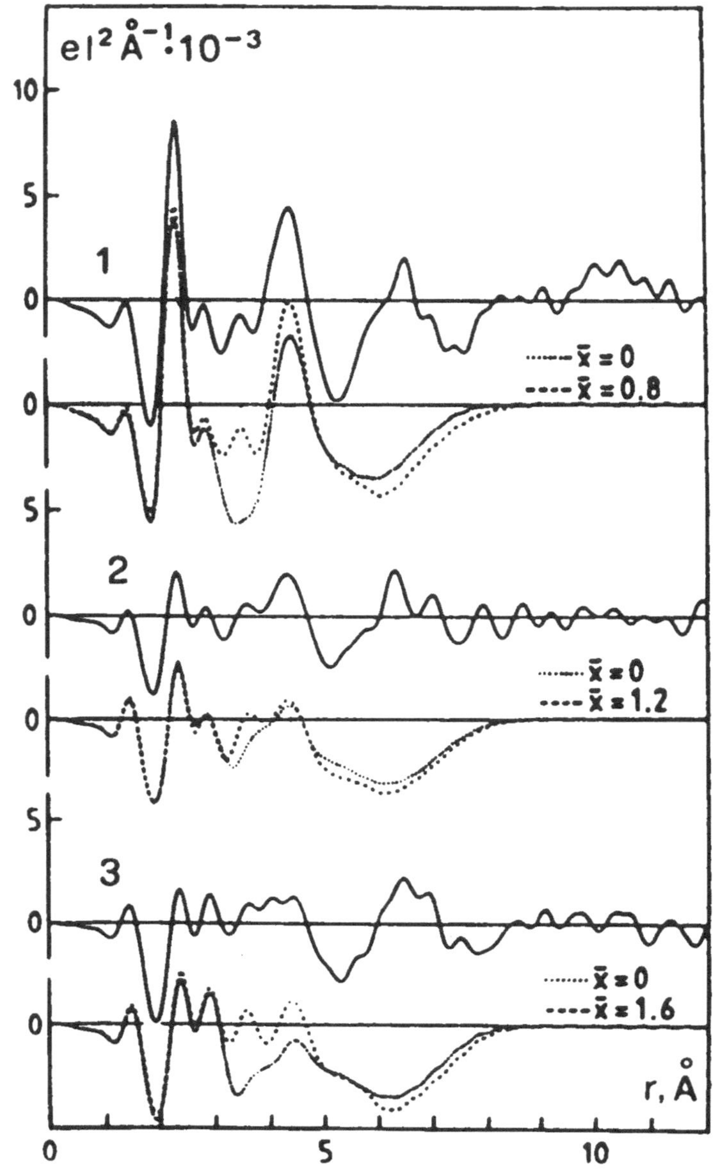

FIGURE 54. The experimental $D(r) - 4\pi r^2\rho_0$ functions (solid lines) for the Cd sulfate solutions compared with functions calculated with (dashed lines) and without (dotted lines) an assumed complex formation with the anions. (From Caminiti, R. and Johansson, G., *Acta Chem. Scand.*, A35, 373, 1981. With permission.)

the si(s) curves, to which Hg – O bonds are expected to contribute very little as compared to Hg – Hg bonds. Very small variations in the r_{Hg-Hg} value around 2.52 Å were observed either by changing the s ranges used in the refinements, or going from one solution to another.

The refined Hg – Hg peak was then subtracted from the radial curves and the residual curves were used to estimate the Hg_2^{2+} hydration. The results indicated that each mercury atom has approximately one oxygen at a distance of about 2.0 Å; about three more oxygens are placed at a distance of 2.5 to 2.6 Å, and additional oxygen seems to be placed at about

Table 31
COORDINATION OF Cd^{2+} IONS IN OXYANIONS SALT SOLUTIONS

Sample no.	Sample	Ref.	Cd^{2+} conc. (*M*)	Anion /metal ratio	$r_{Cd^{2+}\text{-}H_2O_I}$	$\sigma_{Cd^{2+}\text{-}H_2O_I}$	$N_{Cd^{2+}\text{-}H_2O_I}$	$r_{Cd^{2+}\text{-}H_2O_{II}}$	$N_{Cd^{2+}\text{-}H_2O_{II}}$	$r_{H_2O_I-H_2O_{II}}$	$r_{Cd^{2+}\text{-}X}$	$N_{Cd^{2+}\text{-}X}$	Method	Free anion
1	$CdSO_4$ 17.6 H_2O	143	3.1	1.0	2.29	0.10	6—0.8	4.33	12		3.48	0.8	TSM2 + CF	Hydrated
2	$CdSO_4$ 2.1 H_2SO_4 48.5 H_2O	143	1.0	3.1	2.29	0.10	6—1.2	4.33	12		3.48	1.2	TSM2 + CF	Hydrated
3	$CdSO_4$ 2.1 $(NH_4)_2SO_4$ 46.7 H_2O	143	1.0	3.1	2.29	0.10	6—1.6	4.33	12		3.48	1.6	TSM2 + CF	Hydrated
4	$Cd(NO_3)_2$ 8.9 H_2O	146	4.5	2.0	2.28	0.04	6—1.0	4.31	11.9	2.69	3.10	1.0	TSM1 + CF	Hydrated
5	$Cd(H_2PO_4)_2$ 3.0 H_3PO_4 41.7 H_2O	120	1.0	5.0	2.30	0.14	6—0.9_5	4.33	9.3	2.76	3.52	0.9_5	TSM1 + CF	Hydrated

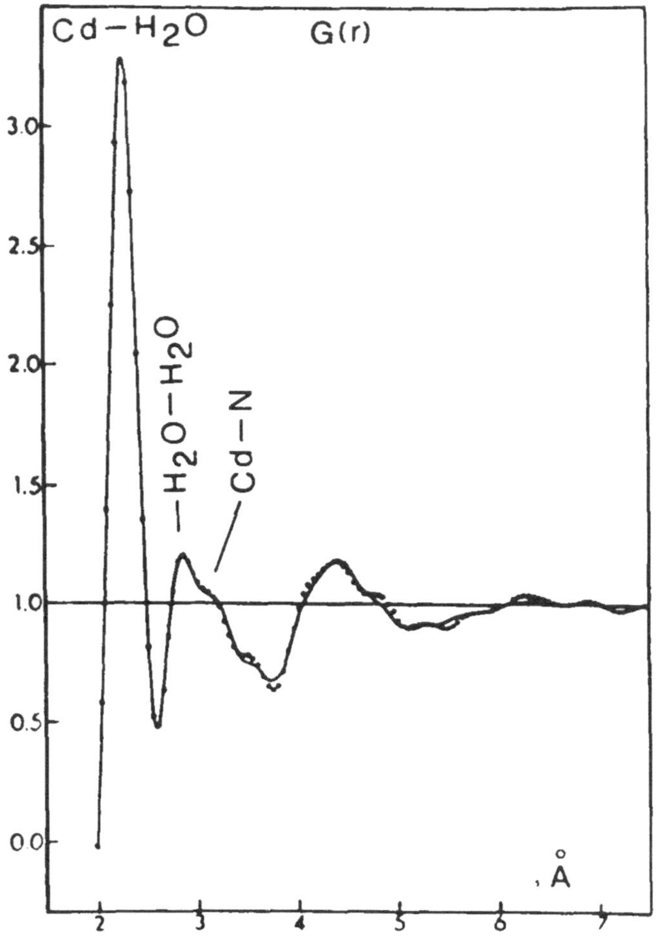

FIGURE 55. Experimental (dots) and model (solid line) correlation functions for a 4.5-*M* $Cd(NO_3)_2$ solution. (From Caminiti, R., Cucca, P., and Radnai, T., *J. Phys. Chem.*, 88, 2382, 1984. With permission.)

3.10 Å. The complete model proposed was very similar to that of Hg(I) in the crystalline hydrated Hg(I) perchlorate as regards either pair distances or frequency factors of the Hg − Hg and Hg − O interactions.

D. Mercury (Hg^{2+})

1. Hydration

The studies of Hg^{2+} hydration are very few.[148-150] A systematic investigation has been performed on perchlorate solutions,[148] but the interest was mainly devoted to the characterization of hydrolysis products. In one sample, the hydrolysis was inhibited by slight addition of $HClO_4$. In the radial curve of this sample, two peaks are evident, the first at about 2.4 Å, which corresponds to the expected Hg(II) − O distance for water molecules in the first coordination sphere of the Hg^{2+} ion, and the second, very broad, at about 4.2 Å, which is probably an indication of a second coordination sphere. High-angle si(s) analysis gave mean $Hg-O_I$ distance (2.42 Å) and hydration number (5.7), without any assumption on the geometry of the hydrated complex. The same data were examined again,[149] and they turned out consistent with a model in which Hg^{2+} ions are surrounded by six water molecules set at the vertices of a regular octahedron.

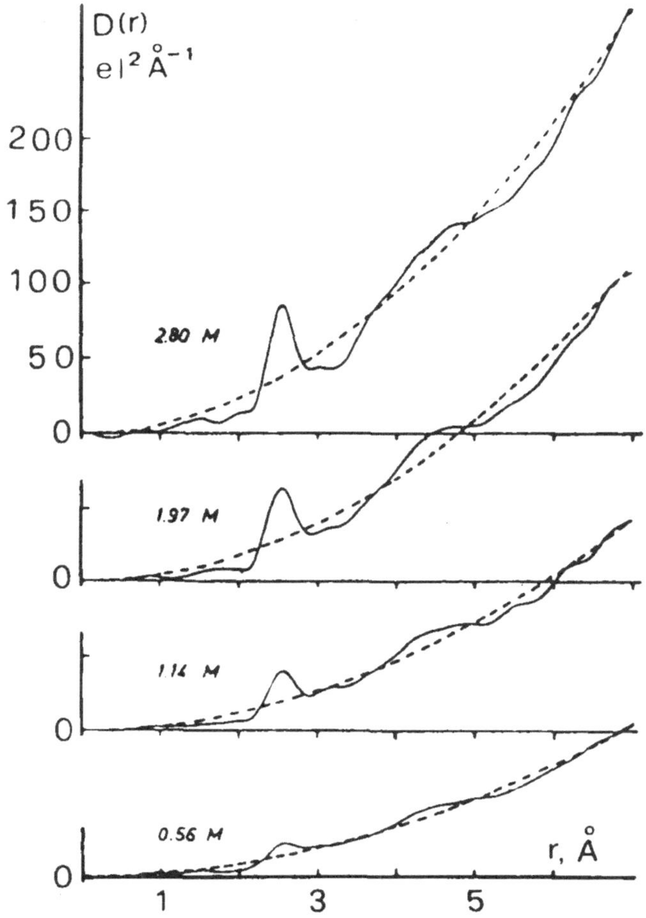

FIGURE 56. Experimental radial distribution functions for Hg(I) perchlorate solutions; dashed lines represent $4\pi r^2\rho_0$. (From Johansson, G., *Acta Chem. Scand.*, 20, 553, 1966. With permission.)

2. Complex Formation

From thermodynamic information, mercury(II) is known to form complexes HgX^+, HgX_2, HgX_3^-, and HgX_4^{2-} with halogens (and, under proper conditions, polynuclear complexes as well). The distribution of halo-complexes, calculated by potentiometric formation constants, shows that the stability ranges of the different complexes are well separated (Figure 57) and, assuming these constants valid also at high concentrations, the curves of Figure 57 may be used to choose solutions in which a single complex dominates or is the only species present. The solutions indicated in Figure 57 were then prepared and examined by XRD.[151-152] Their compositions are reported in Table 32; of the chloride solutions, only those containing mononuclear complexes will be accounted for here. In all cases, accurate structural determinations of lower halide complexes were not possible because of the low solubility of HgX_2 salts.

All the distribution curves in Figure 58, calculated assuming composition units containing one Hg atom, show two prominent peaks which lower remarkably with decreasing the electron number of the halogen. The first peaks correspond to the Hg – X distances, the second ones to X – X interactions within the complex ions. These interactions form almost the total contribution to the structure functions, with the exception of the low-angle region. Therefore,

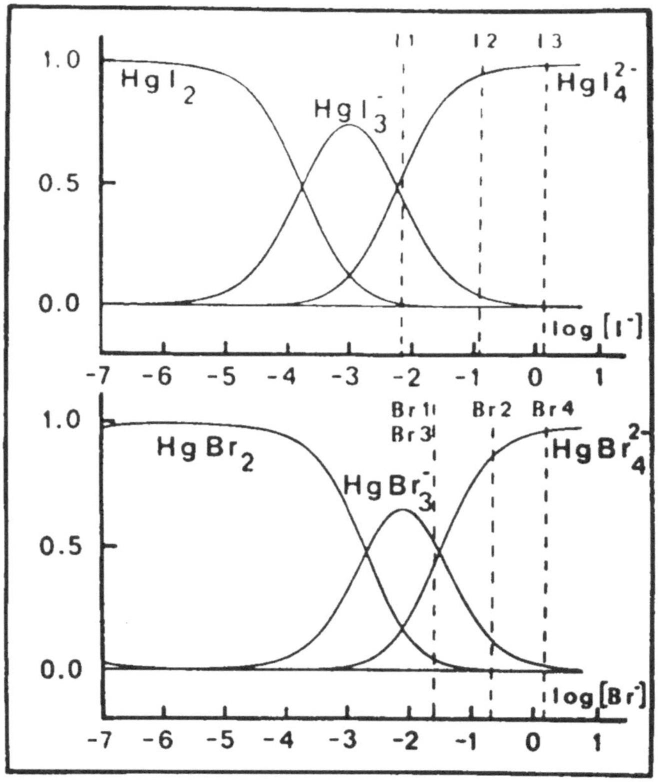

A

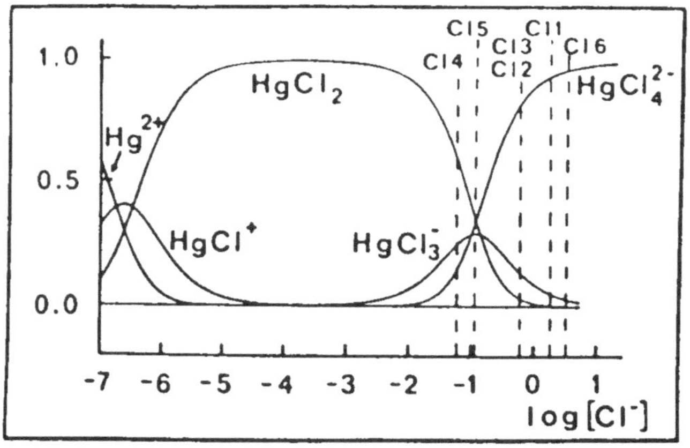

B

FIGURE 57. Fraction of Hg(II) bonded in the different complexes as a function of the free halide concentration: (A) iodide and bromide solutions; (B) chloride solutions. The calculated complex distributions of the solutions investigated, assuming the equilibrium constants to be valid, are indicated. (Fig. A from Sandström, M. and Johansson, G., *Acta Chem. Scand.*, A31, 132, 1977. With permission. Fig. B from Sandström, M., *Acta Chem. Scand.*, A31, 141, 1977. With permission.)

Table 32
COORDINATION OF Hg^{2+} IONS IN CHLORIDE, BROMIDE, AND IODIDE SOLUTIONS

Sample no.	Sample	Ref.	Hg^{2+} conc.	Halide/Hg^{2+} ratio	Halide ion conc.(*M*)	$r_{Hg^{2+}-X^-}$	$\sigma_{Hg^{2+}-X^-}$	$N_{Hg^{2+}-X^-}$	$r_{X^--X^-}$	$\sigma_{X^--X^-}$	$N_{X^--X^-}$	$r_{X^--X^-}/r_{Hg^{2+}-X^-}$	Method
Cl1	$HgCl_2$ 3.8 LiCl 48.1 H_2O	152	1.0 *M*	5.8	5.8	2.47	0.11	4.3					HAA
Cl2	$HgCl_2$ 2.5 LiCl 49.7 H_2O	152	1.0 *M*	4.5	4.5	2.45	0.12	4.0					HAA
Cl3	$HgCl_2$ 2.5 NaCl 52.7 H_2O	152	1.0 *M*	4.5	4.5	2.44	0.09	3.3					HAA
Cl4	$HgCl_2$ 3.0 LiCl 9.94 H_2O	152	3.4 *M*	5.0	17.0	2.47	0.13	3.7					HAA
Br1	$HgBr_2$ 1.4 NaBr 23.2 H_2O	151	1.9 *M*	3.4	6.5	2.60	0.11	3.2	4.26	0.32	9	1.63	HAA
Br2	$HgBr_2$ 2.0 NaBr 27.4 H_2O	151	1.6 *M*	4.0	6.5	2.61	0.10	3.9	4.26	0.28	7	1.63	HAA
Br3	$HgBr_2$ 1.4 NaBr 9.6 H_2O	151	3.6 *M*	3.4	12.2	2.59	0.11	3.2	4.26	0.32	9	1.64	HAA
Br4	$HgBr_2$ 2.5 NaBr 12.9 H_2O	151	2.7 *M*	4.5	12.2	2.61	0.10	3.9	4.25	0.28	8	1.63	HAA
I1	HgI_2 1.5 NaI 8.1 H_2O	151	3.5 *M*	3.5	12.2	2.77	0.10	3.5	4.55	0.24	6	1.64	HAA
I2	HgI_2 2.0 NaI 9.2 H_2O	151	3.0_5 *M*	4.0	12.2	2.79	0.11	3.9	4.55	0.28	8	1.63	HAA
I3	HgI_2 2.5 NaI 10.6 H_2O	151	2.7 *M*	4.5	12.2	2.79	0.10	3.9	4.55	0.24	7	1.63	HAA

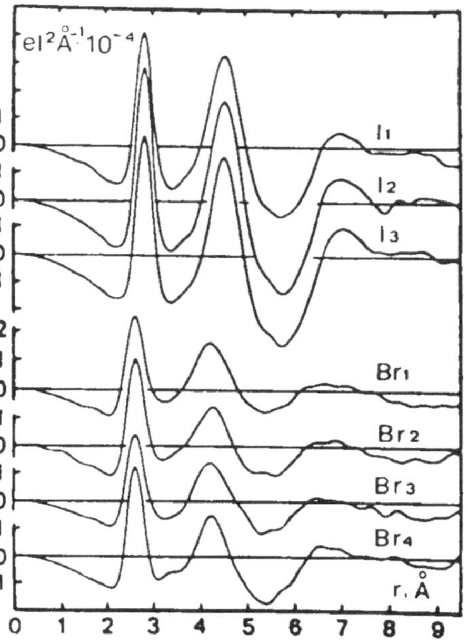

A

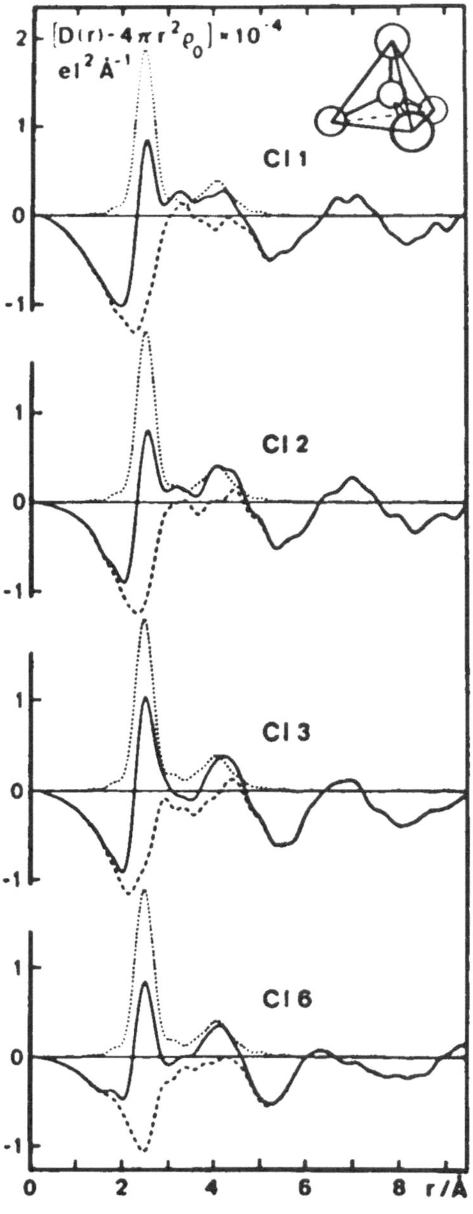

B

FIGURE 58. (A) $D(r) - 4\pi r^2\rho_0$ functions for the solutions of Hg(II) iodide and Hg(II) bromide; (B) experimental $D(r) - 4\pi r^2\rho_0$ functions for the solutions of Hg(II) chloride (solid lines), compared with the theoretical peaks calculated for a regular $HgCl_4^{2-}$ tetrahedron (dotted lines). The difference is shown by the dashed lines. (Figure A. from Sandström, M. and Johansson, G., *Acta Chem. Scand.*, A31, 132, 1977. With permission. Figure B from Sandström, M., *Acta Chem. Scand.*, A31, 141, 1977. With permission.)

by the use of high-angle analysis, structural parameters could be determined, which have the greatest reliability for the heaviest halogens. In this way, the experimental structure functions of bromide and iodide solutions were satisfactorily reproduced from $s = 3$ Å^{-1} onwards, with no meaningful differences in parameter values depending on refinement strategy, data range used, and so on. On the contrary, the experimental data at $s < 8$ Å^{-1}

could not be introduced in refinements of chloride solutions, and Cl – Cl interaction parameters were not obtained accurately.

The parameter values from least square refinements are given in Table 32. For bromide and iodide samples with X^-/Hg^{2+} ratios >4 the number of halogen atoms bound to each Hg is nearly equal to 4; furthermore, the ratios of X – X and Hg – X distances do not differ significantly from the value expected for a regular tetrahedral coordination. The difficulties in refinements of medium-range contributions led in some cases to N_{X-X} values which somehow deviate from the value expected for HgX_4^{2-} units. However, the assumption of regular tetrahedral structure for HgX_4^{2-} units allowed the calculation of synthetic structure functions in good agreement with the experimental data. Also for chloride solutions (all of them with Cl^-/Hg^{2+} ratio >4) the parameters indicate the almost sole occurrence of tetracoordinated complexes, and only the solution C13, prepared with NaCl instead of LiCl, yielded to a number of contacts meaningfully smaller than 4. In this case, however, the parameter determination was more difficult for the interference of $Na^+ - H_2O$ and $Hg^{2+} - Cl^-$ distances, both falling between 2.40 and 2.50 Å.

For the samples with X^-/Hg^{2+} <4, N_{Hg-X} turned out <4 and almost equal to the halide/mercury ratio. This suggested the formation of some amount of HgX_3^- complexes, at the same time excluding the formation of polynuclear species sharing halide atoms, which would result in significantly higher N_{Hg-X} values. The absence of polynuclear complexes was confirmed by calculating difference radial curves between samples with ratio X^-/Hg^{2+} >4 and <4, respectively. These functions were just those expected if HgX_3 complexes were present, and did not show any peak which might indicate the presence of polynuclear complexes.

In Table 32, a slight shortening of the average Hg – X distance can be observed when X^-/Hg^{2+} becomes smaller than 4. This may depend on different Hg – X distances in the complexes HgX_3^- and HgX_4^{2-}. Then, using the structural parameters for HgX_4^{2-} determined in the solutions where only this complex is present, and assuming the relative amounts of HgX_3^- and HgX_4^{2-} to be given by the X^-/Hg^{2+} ratio, estimates of the Hg – X bond lengths in HgX_3 could be obtained, again through least square refinements. The shortening of the bond turned out small (about 0.03 Å). This suggests that HgX_3^- and HgX_4^{2-} have similar shape; pseudo-tetrahedral coordination should occur in HgX_3^-, with the fourth position occupied by a water molecule.

XIV. HALIDE IONS

A. Fluorine

Solutions of NH_4F are one of the few systems in which F^- ions have been investigated.[153] The experimental radial curves almost exactly superimpose on the one of pure water; this could follow from the fact that cation, anion, and water molecules have very similar size and the same electron number. The conclusion was drawn that F^- ions (like NH_4^+) can replace water molecules without apparent breaking off of the solvent structure. Similar description of the F ion was given by Bertagnolli et al.[32] for CsF solutions.

B. Chlorine

Cl^- is the species for which the greatest amount of results exists, since it has been used as counterion in the studies of almost all the cations mentioned (References 5 to 11, 14, 17 to 19, 21, 24, 28, 31, 33, 36, 37, 39, 42, 44, 45, 47, 48, 50, 51, 56, 58, 63 to 68, 70, 78, 84, 90, 95 to 102, 105 to 108, 113, 114, 116, 127, 136, 138, 139, 144, 152, and 153). As soon as the Fourier inversion was used in the interpretation of diffraction data of solutions, the existence of preferential $Cl^- - H_2O$ distances, calling for a definite hydration shell, was deduced from radial functions inspection. In fact, the peak at 2.85 Å, characteristic of pure water and dilute solutions, shifts its maximum to about 3.2 Å on addition of chloride salts and sometimes clearly appears doubly composed. Since 3.2 Å is close to the sum of Cl^- and H_2O radii, the phenomenon has been interpreted straightforwardly as due to the existence of discrete $Cl^- - H_2O$ distances.

This observation gave rise to much interest, since the until-then accepted view of feeble anionic hydration did not imply the existence of definite anionic hydration shells. Therefore, several attempts towards more quantitative characterization of $Cl^- - H_2O$ interactions have been carried out. Early efforts were not too fruitful; the $Cl^- - H_2O$ peak is little or not at all resolved in the radial curves (as shown in the Figures previously reported) and its analysis is very difficult. The procedures adopted aimed at some isolation of the peak at issue from the adjacent ones, through more or less arbitrary assumptions; in particular, it is worth mentioning the well known study by Lawrence and Kruh,[17] in which, using extreme hypotheses about solvent structure, a wide range of possible values of Cl^- coordination number was proposed.

An important turning point in the studies of Cl^- hydration (and, in general, of structural studies of solutions) is represented by the paper by Narten et al.[19] on LiCl solutions, in which the hydrated Cl^- was described as $Cl(H_2O)_6^-$, using the first neighbor model; the small cationic contributions to XRD intensities made this determination easier. Furthermore, refinements of the $O - O$ distances implied by this coordination number yielded to a good fit for octahedral coordination about Cl^-.

This result has been confirmed by many following studies, with few exceptions that will be mentioned. The values of the Cl^- coordination parameters are given in Table 33. These values have been obtained in solutions in which the hydrated Cl^- is the only one or by far the prevalent form containing the chloride ion. If this condition is not fulfilled, different $Cl^- - H_2O$ interaction parameters must be defined for free and bound (in chloride-metal complexes) ions and the choice of meaningful Cl^- hydration parameters becomes fuzzy.

The comprehensive view of the results clearly shows that the Cl^- hydration is described similarly in many different systems, without apparent dependence on counterions and on concentration. Though the accuracy of parameters is strongly dependent on the weight of the terms in which they appear, the spreading of the values is very moderate. All the values of $r_{Cl^- - H_2O}$ stay within a range of 0.12 Å, but, indeed about 90% of the results are between 3.10 and 3.18 Å. The reproducibility of the coordination numbers seems even greater, but this is illusory. In fact, this parameter has been seldom refined freely; more frequently, it has been chosen among a few options (4,6,8) connected with definite geometries and, finally, in the light of the previous uniform results, it has been simply put equal to 6 and associated to an octahedral geometry in the most recent studies.

With due caution, the results can be seen as pointing to a discrete coordination shell around the Cl^- ion; any different hypothesis should contrast with the convergence of so many refinements towards the same solution and the same parameter values. A further argument supports this conclusion. Some refinements including only $Cl^- - H_2O$ direct distances (thus omitting $H_2O - H_2O$ interactions within $Cl\,[H_2O]_6^-$ polyhedra) have been carried out, giving rise, in general, to definitely worse fits. It implies that a definite geometry must be ascribed to the mentioned units (which, however, is not necessarily a regular octahedron). In any case, the order around the Cl^- ions is less than that usually present in the environment of the cations (in this connection, see the values of $\sigma_{Cl^- - H_2O}$) and it does not go beyond the first coordination shell.

We have anticipated the existence of some results indicating structural arrangements for the Cl^- ion different from the one previously outlined. In this class, we do not include evaluations obtained in conditions objectively too unfavorable, like those found in a 1-*M* solution of $CaCl_2$.[45,47] In fact, the correlation function of this sample reported in Figure 8 exhibits only one broad peak under which envelope $Cl^- - H_2O$, $H_2O - H_2O$ and even Ca^{2+} –

Table 33
COORDINATION OF Cl^- IONS

Sample no.	Sample	Ref.	Salt conc.	$r_{Cl^- \text{-} H_2O}$	$\sigma_{Cl^- \text{-} H_2O}$	$N_{Cl^- \text{-} H_2O}$	Geometry	Method
1	LiCl 136.0 H_2O	19	0.4 *m*					
2	LiCl 68.0 H_2O	19	0.8 *m*					
3	LiCl 33.3 H_2O	19	1.7 *m*	3.10	0.20	6	Octahedral	FNM1
4	LiCl 16.2 H_2O	19	3.4 *m*					
5	LiCl 8.15 H_2O	19	6.8 *m*					
6	LiCl 4.0 H_2O	19	13.9 *m*	3.18	0.19	6	Octahedral	FNM1
7	LiCl 3.0 H_2O	19	18.5 *m*	3.19	0.20	6	Octahedral	FNM1
8	LiCl 27.8 H_2O	21	2.0 *m*	3.15	0.19	6	Octahedral	FNM2
9	LiCl 13.9 H_2O	21	4.0 *m*	3.08	0.19	6	Octahedral	FNM2
10	NaCl 27.8 H_2O	21	2.0 *m*	3.14	0.16	6	Octahedral	FNM2
11	NaCl 13.9 H_2O	21	4.0 *m*	3.14	0.17	6	Octahedral	FNM2
12	NaCl 10.2 H_2O	28	5.0 *M*	3.16	0.24	6	Octahedral	FNM1
13	KCl 27.8 H_2O	21	2.0 *m*	3.14	0.17	6	Octahedral	FNM2
14	KCl 13.9 H_2O	21	4.0 *m*	3.14	0.17	6	Octahedral	FNM2
15	CsCl 27.8 H_2O	21	2.0 *m*	3.09	0.14	6	Octahedral	FNM2
16	CsCl 13.9 H_2O	21	4.0 *m*	3.16	0.14	6	Octahedral	FNM2
17	$MgCl_2$ 55.5 H_2O	37	1.0 *M*	3.13	0.20	6	Octahedral	FNM1
18	$MgCl_2$ 27.1 H_2O	37	1.9_5 *M*	3.14	0.21	6	Octahedral	FNM1
19	$MgCl_2$ 11.5 H_2O	37	4.3 *M*	3.16	0.21	6	Octahedral	FNM1
20	$MgCl_2$ 50.5 H_2O	39	1.1 *m*	3.16	0.15	6	Octahedral	FNM2
21	$CaCl_2$ 26.6 H_2O	45	2.0 *M*	3.14	0.22	6	Octahedral	FNM1
22	$CaCl_2$ 12.3 H_2O	45	4.0 *M*	3.15	0.23	6	Octahedral	FNM1
23	$SrCl_2$ 34.6 H_2O	50	1.5 *M*	3.19	0.13	6	Octahedral	FNM1
24	$SrCl_2$ 26.6 H_2O	50	2.0 *M*	3.18	0.19	6	Octahedral	FNM1
25	$AlCl_3$ 53.9 H_2O	51	1.0 *M*	3.14	0.20	6	Octahedral	FNM1
26	$AlCl_3$ 23.8 H_2O	51	2.0 *M*	3.17	0.22	6	Octahedral	FNM1
27	$CrCl_3$ 0.6 HCl 64.8H_2O	80	0.8_5 *m*	3.08	0.25	6	Octahedral	FNM1

Table 33 (continued)
COORDINATION OF Cl^- IONS

Sample no.	Sample	Ref.	Salt conc.	$r_{Cl^- \cdot H_2O}$	$\sigma_{Cl^- \cdot H_2O}$	$N_{Cl^- \cdot H_2O}$	Geometry	Method
28	$CrCl_3$ 26.5 H_2O	84	2.0 *M*	3.17	0.23	6	Octahedral	FNM1
29	$CrCl_3$ 17.9 H_2O	84	2.8 *M*	3.16	0.25	6	Octahedral	FNM1
30	$CoCl_2$ 17.4 H_2O	106	3.0 *M*	3.11	0.21	6	Octahedral	FNM1
31	$CoCl_2$ 2.2 LiCl 26.0 H_2O	107	2.0 *M*	3.14	0.21	6	Octahedral	FNM1
32	$CoCl_2$ 2.1 LiCl 55.5 H_2O	108	0.9 *M*	3.14	0.18	6	Octahedral	FNM1
33	$CoCl_2$ 4.1 LiCl 50.4 H_2O	108	1.0 *M*	3.19	0.21	6	Octahedral	FNM1
34	$CoCl_2$ 5.2 LiCl 49.9 H_2O	108	1.0 *M*	3.16	0.22	6	Octahedral	FNM1
35	$CoCl_2$ 1.4 LiCl 23.2 H_2O	11	2.1 *M*	3.19		6.0		PS + HAA
36	$CoCl_2$ 20.1 LiCl 64.3 H_2O	11	0.6 *M*	3.19		6.2		PS + HAA
37	$NiCl_2$ 27.1 H_2O	114	2.0 *M*	3.12	0.18	6	Octahedral	FNM1
38	$NiCl_2$ 17.8 H_2O	106	2.9_5 *M*	3.06	0.20	6	Octahedral	FNM1
39	$NiCl_2$ 13.3 H_2O	115	3.9 *M*	3.17	0.20	6	Octahedral	FNM1
40	$NiCl_2$ 2.0 LiCl 24.5 H_2O	116	2.0 *M*	3.15	0.21	6	Octahedral	FNM1
41	$NiCl_2$ 2.0 HCl 24.8 H_2O	116	2.0 *M*	3.12	0.23	6	Octahedral	FNM1
42	$NiCl_2$ 3.0 LiCl 23.2 H_2O	116	2.0 *M*	3.12	0.22	6	Octahedral	FNM1

H_2O distance distributions are mixed. In this condition, it is clearly hopeless to find unequivocal structural parameters.

Instead, a reliable result may be considered the one obtained in solutions strongly acidified with HCl by Triolo and Narten[10] and Lee et al.,[11] who proposed coordination numbers <6. The formation of $H_3O^+ - Cl^-$ ion pairs has been put forward as possible explanation for this unusual coordination; in fact, the ion-dipole repulsion between the oxonium ion and the water molecules nearest neighbors to the chloride ion becomes so great that some water molecules may be expelled from the solvation shell of the chloride ion.

A different situation arises in very concentrated solutions, which have been interpreted by lattice-like models or by building blocks somehow inspired by crystalline structures. It can happen here that the Cl^- ion, rather than have its own hydration shell, "sees" water molecules that are around as free water, or because of packing of hydrated cations. This is

Table 34
COORDINATION OF Br^- IONS

Sample no.	Sample	Ref.	Salt Conc. (*M*)	$r_{Br^- \cdot H_2O}$	$\sigma_{Br^- \cdot H_2O}$	$N_{Br^- \cdot H_2O}$	Geometry	Method
1	LiBr 25.0 H_2O	20	2.1	3.29	0.23	6	Octahedral	FNM1
2	LiBr 10.8 H_2O	20	4.5	3.29	0.25	6	Octahedral	FNM1
3	LiBr 8.4 H_2O	20	5.6	3.29	0.26	6	Octahedral	FNM1
4	$CaBr_2$ 44.0 H_2O	46	1.2	3.32	0.24	6	Octahedral	FNM1
5	$CaBr_2$ 26.0 H_2O	46	2.0	3.34	0.26	6	Octahedral	FNM1
6	$NiBr_2$ 26.5 H_2O	118	2.0	3.4		6	Octahedral	PS + HAA
7	$NiBr_2$ 11.8 H_2O	118	4.1	3.4		6	Octahedral	PS + HAA
8	$NiBr_2$ 9.6 H_2O	118	4.8	3.4		6	Octahedral	PS + HAA
9	$NiBr_2$ 24.7 H_2O	119	2.1	3.35	0.21	6	Octahedral	FNM1
10	$NiBr_2$ 11.9 H_2O	119	4.1	3.33	0.20	6	Octahedral	FNM1

well illustrated by a study of melted $CaCl_2 \cdot 6H_2O$.[48] Another example could be represented by a NH_4Cl solution, for which Narten adopted the same Ice-I-like model used for pure water;[153] the Cl^- ion is inserted in the structural holes of this lattice, and, as a consequence, though keeping the sixfold coordination, it has not the octahedral geometry. The model, however, was based on the hypothesis of isostructurality of NH_4^+ ions and water molecules, which has been questioned in a recent study by molecular dynamics and XRD,[154] as we shall discuss thoroughly in the section on oxyanions. Therefore, the result by Narten cannot be considered conclusive, but it is an interesting example of Cl^- coordination in lattice-like models.

In conclusion, the existence of $Cl(H_2O)_6^-$ units with roughly octahedral geometry has been demonstrated in many solutions. These units, however, are not as stable as many hydrated cations are and the possibility of different structures must always be taken into account, particularly in special situations.

C. Bromine

In the radial curves obtained by XRD from many bromide solutions, the qualitative evidence of discrete $Br^- - H_2O$ distances is as strong as (or stronger than) the one obtained for Cl^-; the higher scattering factor makes the observation of the interaction, falling at about 3.4 Å, easier (Figures 35, 38, and 45 and References 17, 18, 20, 46, 69, 91, 109, 117 to 119, 128, 137, 140 and 141). Quantitative results about Br^- hydration are, on the contrary, much less numerous. In fact, the studies on this subject are few, and, moreover, many bromide solutions have been studied in which extended bromide complexation takes place. These cases do not provide reliable information on $Br^- - H_2O$ interactions, either because more than one type of Br^- (free and bound ions) is present, or because the main interest in the investigation is given by the complex formation, and data on $Br^- - H_2O$ interactions are not reported explicitly. The available results are summarized in Table 34.

The reproducibility of Br^- hydration parameters is apparent from the table, and, again, it supports the existence of a discrete solvation shell around the anions. As for Cl^-, the coordination numbers have not always been chosen among alternatives. However, in the cases in which different options were tested, the choice of the best model was easier thanks to the strong contribution of Br^- to the total scattering. A single coordination shell was sufficient to explain the experimental data; as far as the hydration geometry is concerned, the same considerations made about the Cl^- are valid. For a study on a NH_4Br solution, the same remarks made in the previous section with reference to NH_4Cl hold.[153]

D. Iodine

Apart from the pioneer work by Lawrence and Kruh,[17] the hydration of I^- ions has been studied only in two solutions, one ~7 *M* of NaI,[27] and one ~2.2 *M* of LiI.[22] No general conclusion will be thus drawn here. $I^- - H_2O$ interaction gives rise to a clear peak in the radial curves, centered at ~3.60 to 3.65 Å. The coordination numbers proposed (6.0 and 6.9) coincide with those evaluated for Cl^- and Br^- ions, within the error limits ± 1.0 estimated by several authors. Contrary to the cases of Cl^- and Br^-, a good simulation of the experimental si(s) does not require a definite geometry for the anionic hydration shell. This behavior indicates more loosely bound water molecules in the hydration shell of I^-, as expected in virtue of the different sizes of the ions.

XV. OXYANIONS

Ascertaining the formation of discrete hydration shells around of oxyanions has been much more difficult than for the halide ions; similarly, greater difficulties have also been met within the modeling of oxyanions-water interactions. This might depend on the characteristics of the interactions themselves; or, more likely, it may arise because of resolution problems, typical of single scattering experiments. In fact, in the hypothetical units $XO_n - H_2O_m$, the shortest intermolecular pair distances, due to the $O_n....H - O_m$ interactions, superimpose on and cannot be distinguished from $H_2O - H_2O$ distances always present in aqueous solutions. The new $X - O_m$ distances are longer and hardly give rise to peaks which can be considered proof of anionic hydration shells. This can be realized looking at the figures where radial curves from oxyanions solutions are reported. The $X - O_m$ interactions are not visible or, at most, they appear as broad details superimposed to more important peaks. Even worse is the situation if the oxyanions give rise to inner-sphere complexes, since in this case also $Me - X$ distances may overlap to the $X - O_m$ ones.

When the evaluation of a given pair interaction is related with such weak traces, the precision of the experimental data obviously becomes a critical point. First, termination errors in the Fourier transforms give rise to spurious ripples beside the real peaks, mainly emerging in the flat region beyond the $H_2O - H_2O$ peak. Secondly, possible differences among different experiments are dramatically reflected on minor structural details. The occurrence of spurious termination details was evident when some peaks present in the experimental radial curves appeared even in the synthetic functions, although the models adopted did not imply any direct interaction at those distances. An example is given by the G(r) of a $Cr(NO_3)_3$ solution (Figure 19) in which the peak at 3.3 Å in both experimental and theoretical curves does not correspond to any pair distance in the model. An example of systematic errors is given by a comparative study of two ~2-*M* $ZnSO_4$ solutions, run independently in two different laboratories.[155] The two structure functions (Figure 59) are in good agreement in the whole angular range explored, except in the region between $s = 6$ Å^{-1} and $s = 9$ Å^{-1}. The two correlation functions are in very good agreement as far as the peaks at 2.10 and 2.85 Å are concerned; however, they show some differences in the region 3.5 to 4.5 Å, that is, just where oxyanions-water (and/or Me-oxyanions) interactions should give their characteristic contributions (Figure 60).

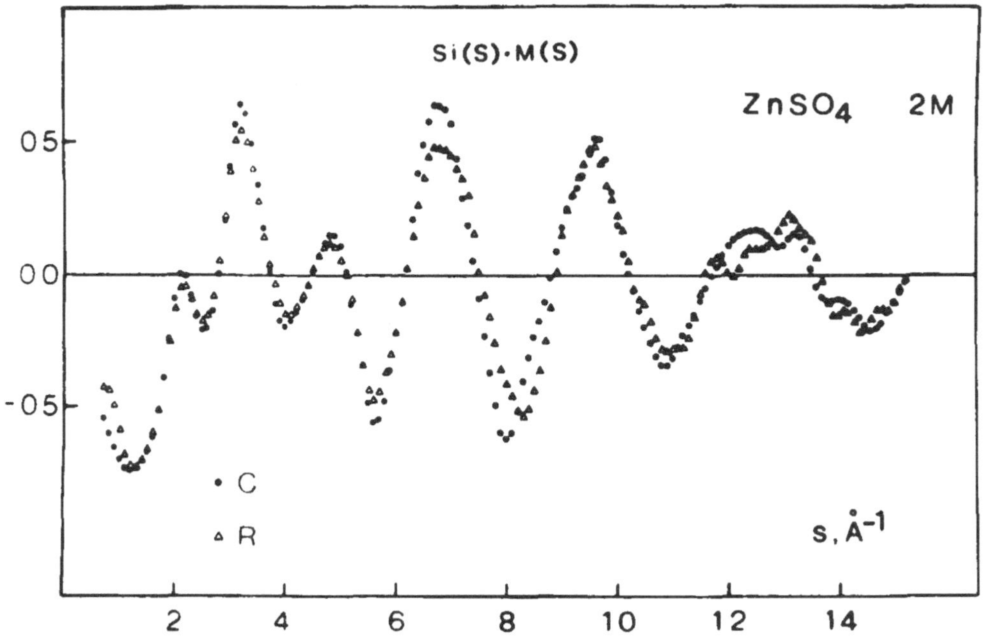

FIGURE 59. Comparison of the structure functions calculated from two sets of experimental data collected in two different laboratories (C = Cagliari data, R = Rome Data). (From Musinu, A., Paschina, G., Piccaluga, G., and Magini, M., *J. Appl. Cryst.*, 15, 621, 1982. With permission.)

Summing up, we can say that the structural evidence of oxyanions-water interactions is poor in general, their contributions to the radial curves may mingle with spurious details, and their detectability is strongly influenced by the accuracy of the experiments. On the other hand, the small peculiar contributions of oxyanions-water interactions to the diffracted intensities involve the advantage that the modeling of cation coordination is practically not conditioned by the modeling of anionic coordination.

In this perspective, the treatment of oxyanions solutions was carried on for a long time adopting the drastic choice of neglecting anion-water interactions completely, so that the only contributions from the anions to diffracted intensities and radial curves were considered to be those coming from intramolecular $X-O$ and $O-O$ distances. This procedure was followed when the HAA or PSA methods were used for quantitative analysis, but it was also used in connection with FNM or TSM interpretation techniques (in these cases, the expression "not hydrated" has been put in the tables in the column, "Anion").

The need for a more detailed description of oxyanions hydration arose when the study of ammonium salt solutions demonstrated unequivocally the existence of discrete $XO_n-H_2O_m$ interactions. These solutions offer favorable conditions for the examination of the anionic environment. In fact, thanks to the remarkable structural similarity of NH_4^+ ions and water molecules, the major structural effects are here due to anion-water interactions; it has been said,[153] maybe overemphasizing, that ammonium solutions may be considered, to a good approximation, as solutions of anions in water. Anyway, $X-O_m$ interactions become manifest in the radial curves through important peaks which are proof of anionic hydration shells. As examples, the correlation functions of one NH_4NO_3 and one $(NH_4)_2SO_4$ solution are shown in Figures 61 and 62.[156,157] The peaks at ~3.40 Å and at ~3.80 Å in the two solutions, respectively, can be explained only through discrete $N-H_2O$ and $S-H_2O$ distances. This strong evidence, however, has not been followed by a clear quantitative description. This ensues essentially from the fact that, as suggested in a study by XRD and molecular dy-

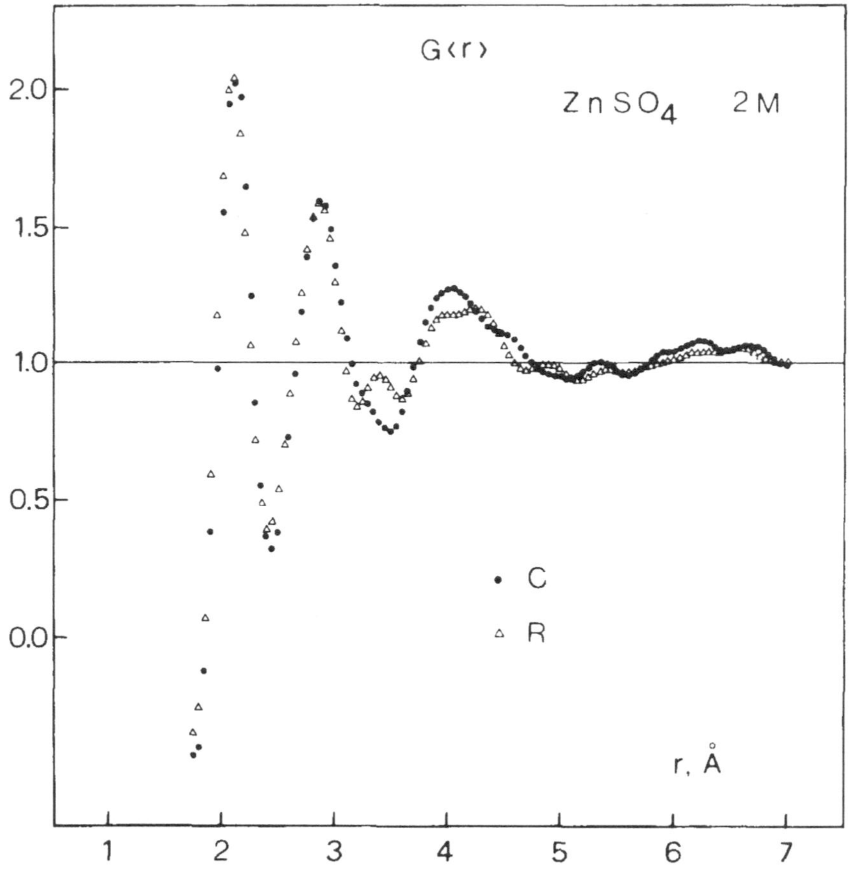

FIGURE 60. Comparison of the experimental correlation functions, G(r), evaluated from the structure functions shown in Figure 59. (From Musinu, A., Paschina, G., Piccaluga, G., and Magini, M., *J. Appl. Cryst.*, 15, 621, 1982. With permission.)

namics,[154] the hypothesis of isostructurality of NH_4^+ and H_2O is probably wrong, and alternative descriptions of the NH_4^+ ions in water yield to different values of oxyanion hydration parameters.[158] Anyway, these results are sufficient to recommend the inclusion of oxyanions-water interactions in oxyanion solutions models; their description, though not univocal, should give a better approximation than their complete omission. A similar conclusion was achieved in studying the same interactions in oxyacid solutions.[159] So, all the recent studies consider explicitly oxyanion hydration and an increasing amount of results is piling up, particularly about sulfate ions.[40]

Undoubtedly, part of the difficulties met in the study of oxyanion-water interactions comes from the weakness of these interactions and from the ensuing softness of the hydration shell. Therefore, the hydration models of oxyanions only consist of one shell of water molecules, with no geometrical arrangement.

Obviously, the secondary role of the oxyanions-water interactions in the fitting work has not allowed a description of oxyanion hydration as clear as that of the Cl^- ion. In particular, the coordination parameters are very sensitive to the treatment of the transition to a continuous distribution of distances. When the continuum terms are not considered at all, the calculated parameters do not have a strict physical meaning and must better be regarded as indicative values.[93,103] But, even when these terms are introduced, the precision of the final parameters may feel the effect of different situations in which the determinations are made. As a favorable

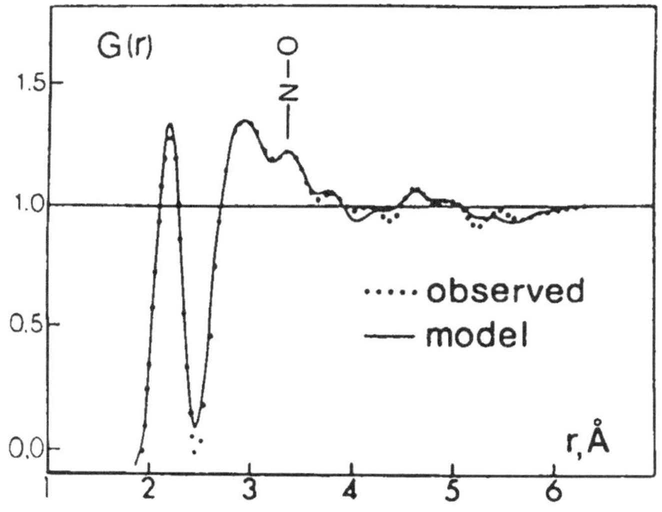

FIGURE 61. Observed (dots) and model (solid line) correlation functions for a 8.0-*M* NH_4NO_3 solution. (From Caminiti, R., Licheri, G., Piccaluga, G., and Pinna, G., *J. Chem. Phys.*, 68, 1967, 1978. With permission.)

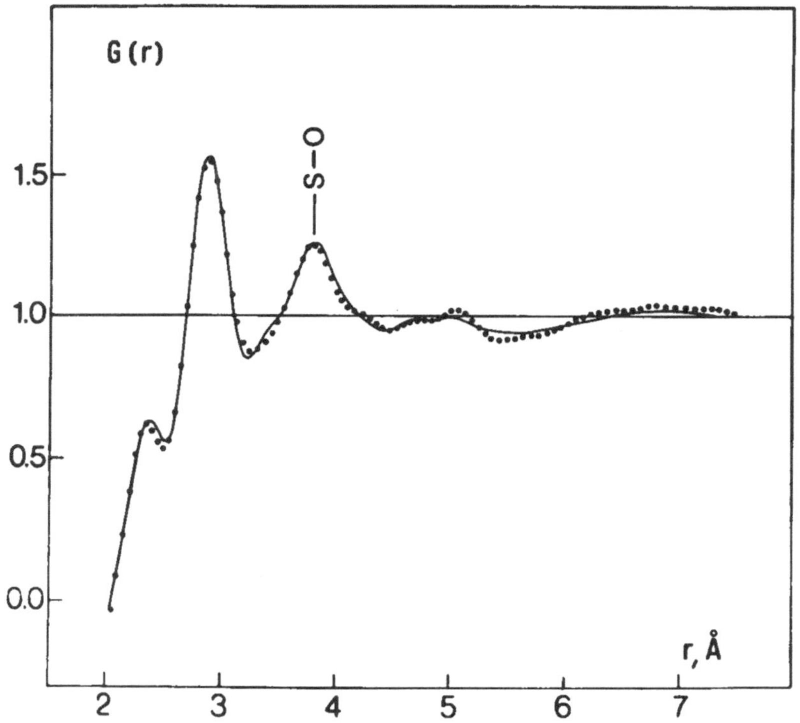

FIGURE 62. Observed (dots) and model (solid line) correlation functions for a 3.55-*M* $(NH_4)_2SO_4$ solution. (From Caminiti, R., Paschina, G., Pinna, G., and Magini, M., *Chem. Phys. Lett.*, 64, 391, 1979. With permission.)

case we may mention a $MgSO_4$ solution, in which the strongly hydrated but light Mg^{2+} ion does not eclipse the S − O peak (ion-pairs are absent).[40] As an unfavorable case we may recall a $Ce(NO_3)_3$ solution, in which the proposed Ce − N and N − H_2O distances (ion pairs are present) are nearly identical and it is difficult to independently refine the two terms.[71] Obviously, the greatest uncertainty regards the determination of the coordination number, always difficult to define; also X − O_m distances display not negligible variations (±0.1 Å) around the following values: r_{N-H_2O} = 3.45 Å, r_{Cl-H_2O} = 3.8 Å, R_{S-H_2O} = 3.8 Å, and r_{P-H_2O} = 3.75 Å. Further careful and systematic experimental work is necessary for progressing in this field.

XVI. GENERAL DISCUSSION

Before commenting on the results presented, some general remarks are in order. The great number of cases investigated and the variety of problems faced demonstrate that the XRD studies surveyed have acquired the great merit of showing which kind of problems may be confronted, in which way, and in which reliability range the results are to be expected. We emphasize that, when the investigations have been carefully and extensively performed, i.e., exploring the behavior of a given ion in a wide range of chemical conditions, a complete description of both hydration and complex formation has been obtained. However, in spite of the considerable research work in this field, mainly done in the last 10 years by high quality instrumentation, the overall results cannot be regarded as a systematic description of the state of ions in solution, a picture from which to draw without hesitation data on ionic coordination. The matter reported is fragmentary; most of the studies reviewed here, with some relevant exceptions, may be regarded as investigations defining the potentialities of XRD technique, rather than providing systematic answers. To this end, actually, were some investigations explicitly aimed in borderline cases, as, for example, the studies on melted hydrated salts or on mixed solutions. But, at the same time, the situation has been caused also by the misuse of "one sample" investigations which, although justified in an explorative stage, are not acceptable in modern research.

Strictly connected with the above considerations, is the problem of the reliability of the information provided by XRD. We will comment in the following on the confidence degree which may be assigned to the various results reported in the tables. Here, we wish to confirm again that, while to ascertain the absolute values of some quantities can be difficult, studying differences among cases investigated in rigorously standardized conditions may come easier. The study of analogies in the behavior, trends as function of particular variables, comparisons among analogous situations, and so on, yield straightforwardly to convincing information even about subtle problems. In this regard, it is worth recalling the evidence of the coordination evolution of the lanthanides,[63-65] obtained thanks to the systematic investigation of the entire group, the experimental proof of the Jahn-Teller distortion of the Cu(II), obtained by comparison of solutions of Cu(II) salts with those of Zn and Co,[106,125] the structural evolution of $FeCl_3$ and $CoCl_2$ solutions obtained by studying small variations of the radial curves with the composition,[101,108] and so on. This kind of procedure is to be highly recommended, as the increasing amount of information directly obtained on a system is of help in the choice of the most plausible models, thus reducing, at least partially, the consequences of the nonuniqueness of the models used in the analysis of XRD data of liquids.

Though the nonuniqueness of the model is a well-known problem for every student in the field, its importance has been sometimes underevaluated. Cases in which only one model has been tried are not rare. Although this procedure may be justified when strong arguments are advanced in favor of one particular interpretation, it is nevertheless unacceptable when, on the basis of XRD data, one model is claimed as more reliable than others which only differ in minor details. Clearly, this has been the cause of major difficulties for the evaluation and classification of the results that we have undertaken previously.

Finally, it has to be recalled that, for obvious reasons, the enormous majority of the solutions investigated are of concentrations >1 *M;* only at these high values, in fact, the contributions of the interactions involving the ions become important and measurable. Undoubtedly, this is a limitation, to which, on the other hand, a nonnegligible advantage, often not valued enough, corresponds. At the ionic strengths considered, it is meaningless to interpret XRD data in terms of modifications of the solvent structure, but the structures around the ions, within their strong electric fields, are to be described. So, contrary to ones opinion, it is unnecessary to have good models available, describing the structure of the water, which is still an open problem in the physical chemistry of solutions. However, in solutions about 1 *M,* some amount of water not involved in hydration is present. In FNM1 the hypothesis is advanced that this water keeps the structure of pure water, and its contribution to diffraction is evaluated by using the structure function of pure water, opportunely "weighted". This is a point open to criticism from the physical point of view, but the matter is in our opinion unimportant in practice as, when the hypothesis is less credible (increasing concentrations), the relative term vanishes. In FNM2, the weighted water is substituted by an explicit term H_2O-H_2O, which parameters are used to understand if the water-water bonds in the bulk are stronger or weaker than in pure water.[21] It is not clear whether this information can be obtained coherently, as the weight of the water-water term is small in many cases and, if two hydration shells are present, the H_2O-H_2O term of bulk water and the $H_2O_I-H_2O_{II}$ one overlap inextricably. On the other hand, it is sure that the distinction between FNM1 and FNM2 disappears when there is no free water in the system, that is, in the greatest part of the cases investigated.

Let us comment on the parameter values reported in the tables and on the relevant information connected to them.

A. First Coordination Shell

1. Bond Lengths

Direct ion-ligand distances are the structural parameters which XRD studies provide with the greatest realibility. It is apparent from the tables that determinations of a certain ion-ligand distance, obtained by different researchers on the same system, result in good agreement, with differences in the order of a few hundredths of an Angstrom (in most papers, the standard error of the parameters, obtained from fitting procedures based on the goodness of fit in a nonlinear regression, falls in the third decimal figure; however, it has been specified in various papers, and the comparison of the cases clarifies and confirms, that these standard errors are substantially smaller than the real uncertainties of the parameters). Obviously, the more weighing is the term to which the parameter refers and the more structured is the complex in which the distance has been determined, the truer this observation becomes. So, in the less halogen-rich halo-complexes, metal ion-water distances come out better defined than the metal ion-halogen one depending on their relative weight, while metal ion-first neighbor distances can be evaluated with greater precision than halogen-first neighbor ones because of their different structuring.

In the great majority of cases, ion-ligand distances are very close to the average of those found in crystalline compounds of analogous composition. Many papers, in fact, report tables of crystallographic data, either to support the findings of fitting procedures, or to validate the interpretations of certain peaks (especially minor ones) as due to a specific pair. More details on this topic will be given after commenting on coordination number results.

2. Coordination Numbers

The coordination number represents the time- and space-averaged number of atoms or ions, which is present in the immediate neighborhood of a certain species. The term hydration number has been used instead of coordination number, when the atoms bound to the central

Table 35
BOND VALENCES, $V_{Me\text{-}O}$, AND ROOT MEAN SQUARE DEVIATIONS, $\sigma_{Me\text{-}O}$, FOR SOME ION-WATER PAIR DISTANCES

Ion	$V_{Me\text{-}O}$	$\sigma_{Me\text{-}O}$
Li^+	0.25	0.28
Sr^{2+}	0.25	0.13
Mg^{2+}	0.33	0.09
Ca^{2+}	0.33	0.13
Mn^{2+}	0.33	0.11
Co^{2+}	0.33	0.11
Ni^{2+}	0.33	0.10
Al^{3+}	0.50	0.10
In^{3+}	0.50	0.11
Cr^{3+}	0.50	0.08
Fe^{3+}	0.50	0.08
Rh^{3+}	0.50	0.10
Zn^{2+}	0.33	0.10
Cd^{2+}	0.33	0.10

species are all oxygen atoms from water molecules; in this case, coordination number or hydration number can be used indifferently. No distinction between the two terms is possible, when dealing with XRD results, based on the lifetime of ion-solvent aggregates, as the correlation functions by XRD are time-independent.

If we look at the values of hydration numbers given in the tables, we get the idea that they are obtained with great precision, as in many cases a constant integer number is reported. Actually, this follows from the use of FNM or TSM, where coordination numbers are fixed at integer values, chosen among different options after rough preliminary estimates. In practice, the coordination number is used as adjustable parameter only when analyses like HAA or PSA are carried out, by which the accuracy of this quantity can be evaluated. Coordination numbers are undoubtedly more critical than interatomic distances. The latter, in fact, depend on the frequency of the structure function oscillations, while coordination numbers depend on their amplitude which may be more easily influenced by systematic errors due to the experimental measurements or introduced by data treatments. Besides, coordination number values are correlated with σs, and this implies additional uncertainties when the fitting procedure is carried out in limited ranges of the experimental structure function. However, when the ion-water peaks are well resolved, determinations of coordination numbers lead to values not too far (less than 10%) from integer numbers (as expected), which are generally the ones suggested by NMR or other spectroscopic investigations. Obviously, coordination number determinations become more difficult when ion-water peaks are less evident or less resolved (see alkali ions).

3. Root Mean Square Deviations

The σ values represent the root mean square variations of distances under the hypothesis that the distance distribution around the mean value is gaussian-type. The good reproduction of peaks in the radial curves in a great number of cases shows that the use of gaussian distributions represents a good approximation of the real system. If this is true, σ values might be regarded as a measure of the disorder existing around the ions and of the strength of nearest neighbor ion-solvent interactions. In Table 35 we report values of σ_{Me-O} (averaged over different determinations) vs. bond valence values, v (v = ionic charge/coordination number), for cases of fully hydrated ions and ions partially complexed with oxyanions.

It seems that not much information may be inferred from the table. Leaving apart Li^+, because of the uncertainty connected with XRD studies on its solutions, we notice a remarkable leveling of values, with only a slight trend towards smaller σs as v increases. Clearly, σ is affected by imprecisions deriving from having kept coordination numbers fixed at integer values. In the parameter triplet (r, N, σ) describing a pair interaction, σ is surely the least meaningful. It is also worth noticing that σ values (hence the disorder around the ions) do not seem to depend on stability of aggregates, as expressed by exchange rates of hydration water molecules. For example, Cr^{3+} or Rh^{3+}, with low exchange rates, do not look dissimilar from Al^{3+} or Fe^{3+}.

4. Geometry

The geometry of hydration polyhedra, as reported in the tables, is in most cases just a recall of the assumptions made in the calculations, and does not represent the result of a choice among different options actually tested. It would be quite arduous, in fact, to discriminate among several geometries at fixed coordination numbers. Indications in this sense should come from distance values (and from frequency factors of distances as well) within the ligands, that is, from O−O distances within coordination polyhedra of hydrated ions. Such distances, however, intertwine with various water-water distances which are present in solution; in fact, in an octahedral complex with Me−O distances around 2.0 Å, O−O distances fall at about 2.8 Å. It is obviously impossible to refine independently many different O−O terms meaningfully. Thus, with rare exceptions, which have been mentioned, the adopted geometry is the one thought as the most probable on the basis of crystallographic and/or spectroscopic information.

More detailed descriptions may be achieved when the ligands are exclusively or prevalently halogen ions. In fact, the distances among ligands have here values different from the others in the system. Besides, the relative terms become "heavy", mainly in the cases in which the ligand is Br^- or I^-. Then, from X−X distances and from their frequency, indications on polyhedra geometries have been obtained quite convincingly (see cases of Tl[III], Au[III], Pt[IV], and Zn[II]). In any case, whichever be the ligands (even O atoms), it is indispensable to fix a coordination geometry to obtain good fits, which implies strong order of first coordination shells of ions. We want to remark that, surprisingly, this statement has proved to be valid in many cases even for Cl^- ions.

5. The "Bond Valence" Point of View

We wish to examine the results reported in the tables in the light of the bond valence model, as suggested by Brown for cation-oxygen interactions. Following Pauling's concepts of bond strength and bond valence, Brown and Shannon[160] proposed an empirical expression which permits the bond valence v to be calculated from observed bond length (and vice versa):

$$v = (r/R)^{-M} \tag{1}$$

where R and M are parameters fitted empirically for a given atom pair. The bond valences are such that their sum around a cation is equal to the cationic charge. Specifically, Brown and Kang Kun Wu[161] reported a set of parameters R and M which cover most of the cation-oxygen bonds for which a sufficient number of suitable structures is available. In a following paper, Brown[81] suggested that the same parameters could be used to discuss the coordination of ions in aqueous solutions. From the coordination numbers proposed, we evaluated the cation-oxygen bond valence (v = ionic charge/coordination number, if all the bonds are

Table 36
ION-WATER DISTANCES AND COORDINATION NUMBERS FROM XRD EXPERIMENTS (r_{XRD} and N_{XRD}), BOND VALENCES ($v_{Me\text{-}O}$ = IONIC CHARGE/CN_{XRD}), AND ION-WATER DISTANCES FROM THE BOND VALENCE MODEL (r_{calc})

Ion	r_{XRD}	N_{XRD}	$v_{Me\text{-}O}$	r_{calc}
Li^+	1.95—2.25	4	0.25	1.94
Mg^{2+}	2.09—2.12	6	0.33	2.09
Ca^{2+}	2.40—2.44	6	0.33	2.34
Sr^{2+}	2.64	8	0.25	2.61
Al^{3+}	1.87—1.90	6	0.50	1.91
In^{3+}	2.16	6	0.50	2.16
La^{3+}	2.58	9	0.33	2.57
Pr^{3+}	2.54	9	0.33	2.55
Nd^{3+}	2.51	9	0.33	2.53
Eu^{3+}	2.45	8.5	0.35	2.45
Tb^{3+}	2.41	8	0.375	2.40
Dy^{3+}	2.40	8	0.375	2.39
Er^{3+}	2.37	8	0.375	2.36
Tm^{3+}	2.36	8	0.375	2.35
Lu^{3+}	2.34	8	0.375	2.33
Cr^{3+}	1.99—2.00	6	0.50	1.98
Mn^{2+}	2.19—2.22	6	0.33	2.19
Fe^{2+}	2.12	6	0.33	2.14
Fe^{3+}	2.01—2.05	6	0.50	2.01
Co^{2+}	2.08—2.10	6	0.33	2.10
Ni^{2+}	2.03—2.07	6	0.33	2.06
Zn^{2+}	2.12—2.14	6	0.33	2.09
Cd^{2+}	2.28—2.30	6	0.33	2.31

equivalent), then, from Equation 1, the bond lengths. The results are reported in Table 36. Bond lengths from XRD analysis are also given in the same table; the extreme results have been indicated, when more than one determination exists.

The agreement between calculated distances values and XRD values is quite good in almost all the cases; as the calculated value implies the knowledge of the coordination number (in the definition of v), this finding supports both distances and coordination numbers on the basis of crystallographic data (as so have to be interpreted the r_{calc} values, after all). Leaving apart Li^+, for which the great uncertainty of results prevents any judgement, the only result in poor agreement is the one regarding Ca^{2+}; this is again a confirmation of the already discussed difficulty in defining the coordination of this ion.

The overall agreement suggests that the bond valence model can be used to get a rationalized picture of the experimental results, and to make reasonable predictions in dubious cases. An example is given by the Na^+ ion for which different coordination number values have been proposed. With $N_{Na^+-H_2O}$ = 4, the expected distance is 2.24 Å, with $N_{Na^+-H_2O}$ = 6, the distance becomes 2.46 Å. The latter is closer to the experimental one (2.40 to 2.42 Å), which supports the sixfold coordination as the most probable. The results of the lanthanide series look quite interesting, where the transition of the coordination number from 9 to 8 occurs, according to the findings by Habenschuss and Spedding.[65] It is worth noticing that assuming coordination number 8 for the lighter ions, or 9 for the heavier ones, implies worsening of the agreement between experimental and calculated r values. For example, r_{calc} of La^{3+} would become 2.52 Å for coordination number 8, while r_{calc} of Lu^{3+}

would go to 2.37 Å for coordination number 9, both of them in worse agreement with r_{XRD} values reported in Table 36.

The case of the Cu^{2+} ion must be considered apart. In diffraction studies of aqueous $CuSO_4$ and $Cu(ClO_4)_2$, hydration number 6 was assigned to the Cu^{2+} ion, but, because of the Jahn-Teller distortion, the water molecules are not equidistant from the central ion. Since in this case it is impossible to assign "*a priori*" the bond valence to the different interactions, we evaluated these valences from the experimental ion-water distances using Equation 1; the valences of the single bonds were then summed over the entire hydration shell and this value compared with the ionic formal charge. On the whole, a fairly good match was obtained; for example, a valence sum of 1.93 was calculated in $CuSO_4$,[125] against the value of 2 of copper ion charge.

B. Second Coordination Shell

1. Me-(X)$_{2nd}$

The existence of order beyond the first shell has been unequivocally demonstrated for the most part of cations, but the parameters describing second coordination shells are comprehensibly uncertain and not always given in the papers. However, the various r_{Me-H_2O} distances proposed for a given species show variations generally not too spread. This is important if they are to be used to attempt evaluations of conformation of water molecules in the first coordination shells of the cation, as done in cases like Mg^{2+}, Al^{3+}, and Cr^{3+}.

More uncertain is the evaluation of second coordination numbers, for the very reasons already discussed for coordination numbers in general; moreover, many procedures have been used to get to coordination number values, from the extreme case of a fixed value coupled with a rigid model, to the other extreme of numbers freely refined without any geometrical assumption. A reasonable pattern, however, seems to emerge, that is, the number of ordered molecules tends to decrease at increasing concentrations (hence, decreasing of water/salt ratio).

Whichever the interpretative model used in the fitting procedure be, hydrated cations and anions have been considered as independent, noncorrelated units. Similarly, when cations and anions give rise to inner-sphere complexes, the various complex ions have been again treated separately the one from the others. The assumption of independence of the structural units appears to be something more than a mere hypothesis. In fact, if we look at the parameters describing the first coordination shell of a given ion, we realize that in most cases they do not depend on concentration or counterion. Differences between different determinations are small, look uncorrelated with variations in experimental conditions, and only seem to reflect the intrinsic uncertainty of the method. We call attention to the fact that, when the environment of a cation is formed by only oxygen atoms, the structural parameters turn out identical, independently from the belonging of O atoms to water molecules or oxyanions. The same behavior has been noticed even for oxygen atoms pertaining to organic ligands. For instance, an investigation on solutions of acetates of Mg, Cd, Co, Zn, and Mn may be quoted,[41] in which the environment of the cations, all of them bound to acetate ions, was described with parameters in strict agreement with those reported for fully hydrated cations.

Obviously, the hypothesis of independence of hydrated ions must be correctly interpreted in its structural meaning: the ion-ion interactions, which are largely responsible in determining the thermodynamic properties of the electrolyte solutions, appear not to give rise to order phenomena which can be analyzed by XRD, at present. The hypothesis, however, is difficult to be accepted in the case of very concentrated solutions where the water molecules available are not enough for cation and anion hydration without some sharing of molecules by coordination shells of near ions, that is, some sort of outer-sphere ionic association. This picture becomes obviously necessary even at not too high concentrations when cations are surrounded by a second hydration shell.

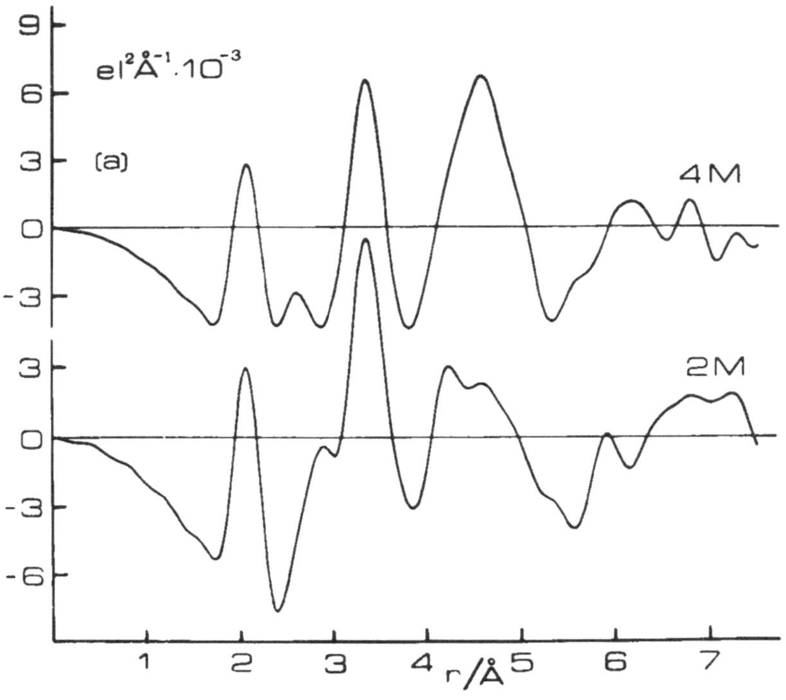

A

FIGURE 63. (A) Difference radial distribution curves, $D(r) - 4\pi r^2\rho_0$, for two $NiBr_2$ solutions; (B) Difference radial curve of the 4-*M* solution (solid line), radial contributions coming from $Br(H_2O)_6^-$ complexes (dotted solid line), and radial difference curve after subtraction of the dotted curve (dashed line); (C) Difference radial curves of the two solutions after subtraction of the contributions coming from $Br(H_2O)_6^-$ complexes. (From Magini, M., de Moraes, M., Licheri, G., and Piccaluga, G., *J. Chem. Phys.*, 83, 5797, 1985. With permission.)

In these cases, the lack of experimental evidence of cation-anion interactions cannot be anything but a matter of weight, as it has been recently proven. If we take a chloride solution (for example, of $MgCl_2$ or $NiCl_2$ ~4 *M*), a model accounting for outer sphere cation-anion interactions comes out consistent with the experimental data, but it does not yield to any meaningful improvement with respect to the model in which these interactions were neglected.[114] A quite different situation is offered by concentrated bromide solutions. Taking the example of $NiBr_2$ solutions, interesting features appear in the radial distribution functions (Figure 63).[119] Besides the peaks due to the shorter pair distances, an important multicomponent peak appears in the 4- to 5-Å range. Important contributions to peaks in this distance region are usually provided by interactions among cations and water molecules set in second hydration spheres. The comparison of curves of the two solutions shows that the peaks at issue are different both in shape and in height; oddly, the peak is higher in the more concentrated solution where less water is available for second ordered hydration shells. This suggests that, in the 4-*M* solution, some Br^- ions, Br_{II}, set themselves in the second cationic coordination shells. To check this hypothesis, the contributions coming from $Br^- - H_2O$ and $H_2O - H_2O$ interactions in the anionic complexes $Br(H_2O)_6^-$ were subtracted from the difference radial curves; these contributions are sketched as a dotted line in Figure 63B. The results of the subtraction are shown in Figure 63C. The residual peaks at 4 to 5 Å, while retaining their complex nature, shift their maximum from 4.1 in the 2-*M* sample

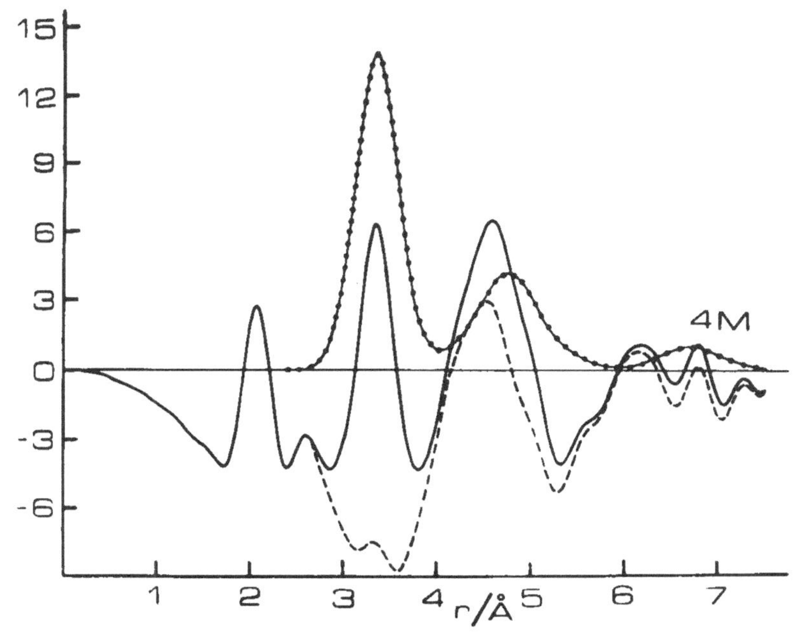

FIGURE 63B

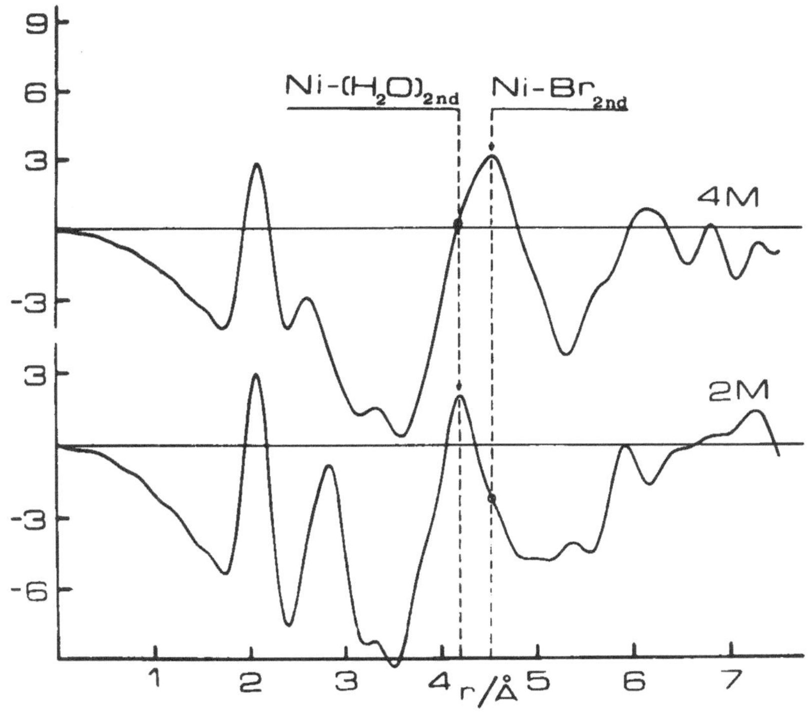

FIGURE 63C

Table 37
BOND VALENCES FOR SOME ION-WATER PAIR DISTANCES. $v_{Me\text{-}O}$, AND DISTANCES BETWEEN HYDRATION AND EXTERNAL WATER MOLECULES AROUND THE VARIOUS IONS, $r_{H_2O_I-H_2O_{II}}$

Ion	$v_{Me\text{-}O}$	$r_{H_2O_I-H_2O_{II}}$
Mg^{2+}	0.33	2.76
Sr^{2+}	0.25	2.84
Al^{3+}	0.50	2.70
In^{3+}	0.50	2.67
Cr^{3+}	0.50	2.70
Mn^{2+}	0.33	2.75
Fe^{3+}	0.50	2.71
Co^{2+}	0.33	2.76
Ni^{2+}	0.33	2.73
Rh^{3+}	0.50	2.69
Zn^{2+}	0.33	2.78
Cd^{2+}	0.33	2.73

(possible $Ni^{2+}-H_2O_{II}$ distance) to 4.6 Å in the 4-*M* sample (possible $Ni^{2+}-Br_{II}$ distance), in agreement with the explanation proposed.

2. The $H_2O_I-H_2O_{II}$ Problem

Let us consider the distance values between water molecules pertaining to two subsequent hydration shells (Table 37). We have already said that when two hydration shells are set around a small ion with +2 or +3 charge, the distances $r_{H_2O-H_2O}$ determined are rather short, surely shorter than water-water distances in pure water. We may attempt an explanation of the characteristics of the interactions between hydrated ions and external water molecules, resorting again to the bond valence model. These interactions are hydrogen bonds, that is, they consist of a system of two bonds, one donor, O – H, and one acceptor, H...O, and the O – O distance is essentially controlled by the weaker acceptor bond.

Going through the results of many crystal structures determined by neutron diffraction, Brown was able to show that O – O distance shortens when the valence used by the acceptor bond increases. Besides, Brown reached the conclusion that strong and weak hydrogen bonds are different in kind, the stronger ones (O – O distance shorter than about 2.73 Å, O...H bond valence >0.2) involve strain and are linear, while the weaker ones (O – O distance longer than about 2.73 Å, O...H bond valence <0.2) have an extra degree of freedom and are generally bent.

Following this idea, the linear kind of bond should form around cations of charge +3 and coordination number 6. In fact, the only possible distribution of the total valence is the one sketched below. If more than two H-bonds formed per molecule, the O – H valence should be less than 0.75 and the H...O one greater than 0.25, thus giving rise to O – O bonds even shorter.

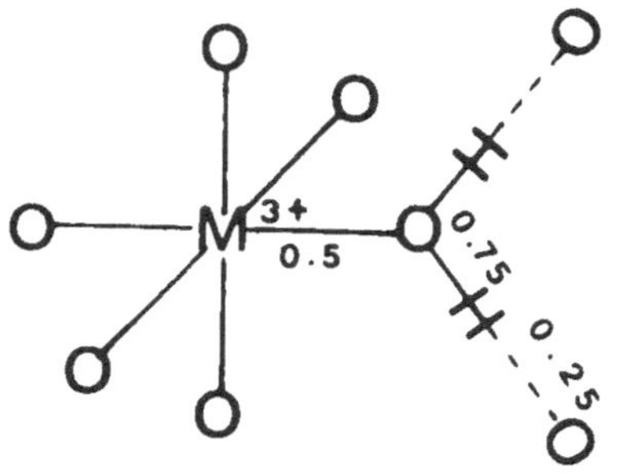

For an ion with charge +2 and coordination number 6, we may have a situation which implies longer H-bonds (below). Other distributions are possible by forming more than two H-bonds per water molecule and strengthening the acceptor bonds at least up to the limit of 0.2 valence units. In any case, the situation should not be too far from that typical of the transition point between the two different bond types, so that even these bonds, though longer than in the previous case, should be almost linear. Let us see, then, if the results reported in Table 37 confirm these expectations.

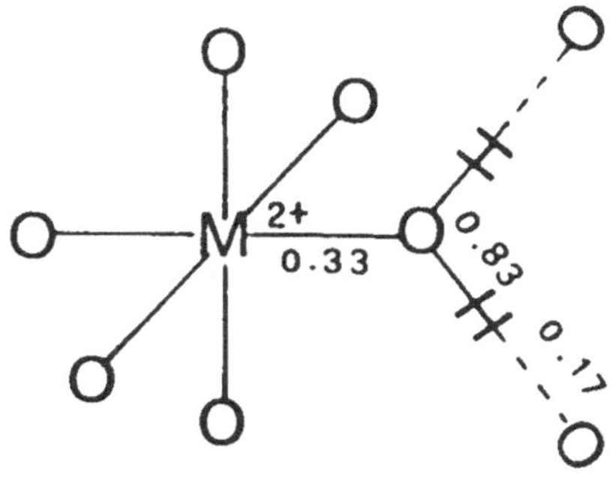

The picture seems on the whole quite encouraging. The shortest distances (~2.70 Å) are found for V_{Me-O} = 0.5, the intermediate (~2.75 Å) for v_{Me-O} = 0.33 and the longest is related to the only one case with V_{Me-O} = 0.25.

The interpretation discussed above has been criticized by some authors.[135] Starting from the consideration that short distances $H_2O_I-H_2O_{II}$ were also found by themselves around sixfold coordinated Li^+,[22] they suggested that the parameter in point is essentially representative of what happens to O−O bonds in bulk water. In other words, the distance values we ascribe to $H_2O_I-H_2O_{II}$ should occur in bulk water rather than in hydration shells; short or long values should then reflect the structure making or structure breaking power of cations. In our opinion, even leaving apart the questionable case of LiI, this interpretation cannot be maintained. In fact, neither short nor long bulk H_2O-H_2O distances may form in these solutions in which, because of the high concentration and/or on account of the demonstrated formation of second hydration shells, no bulk water is left, while a term H_2O-H_2O, which provide the parameter values reported in the table, is necessary even in these cases.

Less conclusive explanations can be given about values related to this type of distance. In fact, the calculations often provide very small σs even with not too short $r_{H_2O_I-H_2O_{II}}$. Undoubtedly, some obscure point still remains, which may be reflection of inadequacies of the models, or, simply, the result of too many O−O interactions overlapping in the same distance range.

C. Information on Complex Formation

It is beyond the aim of this book to provide a systematic review on all complex formation phenomena occurring in solution which have been investigated by XRD. In the previous paragraphs of this chapter, indeed, we limited ourselves to review the mononuclear complex formation between cations and the simplest anions in inorganic aqueous solutions, i.e., halides and oxyanions. A complete review of this field should account for the considerable work performed by the Swedish School on polynuclear hydrolysis complexes[60,61,148,152] as well as on large polymolibdate and polyarsenate solution complexes.[73-76] The work by Ohtaki[2] on the structure of ammine complexes should be also recalled.

We will limit ourselves here to some comments related to the cases discussed previously and, specially, to a further discussion about the modeling technique in presence of complex formation.

1. Inner Complex Formation

From the previous paragraphs of this chapter it is clear that different levels of difficulties are encountered in the studies of inner – complex formation either with halide ions or with oxyanions. The first kind of complex is much more easily observed, due to relatively short metal-halide distances which often result in well-defined peaks in the low r part of the radial functions. If the degree of complex formation is high (e.g., Fe^{3+} or Cd^{2+} with Cl^-), the position of the first peak is centered on Me – X rather than Me – H_2O distance.

In the case of complex formation with oxyanions, XO_n, the immediate surrounding of the cation is always constituted by oxygen atoms and the qualitative evidence of the complex formation is bound to the appearing of Me – X (X = S, N, P, etc.) peaks in the 3- to 4-Å region of the radial curves. In this region, however, the overlapping with other interactions is unavoidable and, in addition, nonphysical spurious details may occur. Furthermore, a high degree of complex formation is not usual with oxyanions, so that, often, even the qualitative evidence of this phenomenon is not straightforward; difference methods are frequently used to have clear indications when complex formation is not eye-evident in the experimental radial curves.

All this is reflected in the different realibility of the quantitative information. While for the halo-complexes concordant results about the average degree of complexation have been presented by now (e.g., $NiCl_2$, $CoCl_2$ solutions), in the case of oxyanions solutions, debates exist about the same existence of complex formation.

It is clear that to go more thoroughly in the structural analysis is very problematic. In fact, when complex formation takes place, it is generally true that various step complexes are present in solution and unless suitable chemical conditions are purposely chosen so as to favor the predominancy of only one species, all of them should be considered in order to fully describe the structure of the system. Problems arise then about the various complexes that can be formed, about their geometries, and finally, about the amounts of each complex.

Clearly, very useful indications could be provided to diffraction workers by thermodynamic studies of chemical equilibria in solution. However, in this respect, one must consider that thermodynamic data:

1. Are normally obtained for very low metal concentration (usually in the range from 10^{-3} to 10^{-1} *M*) at a given constant ionic strength (obtained by addition of a suitable electrolyte) so that they are available for solutions quite different from those examined by XRD.
2. Do not distinguish between inner or outer complex formation.
3. Do not give the total ion coordination number, as only ion-ligand bond formation is considered, and nothing can be said on the water molecules present around the cation.

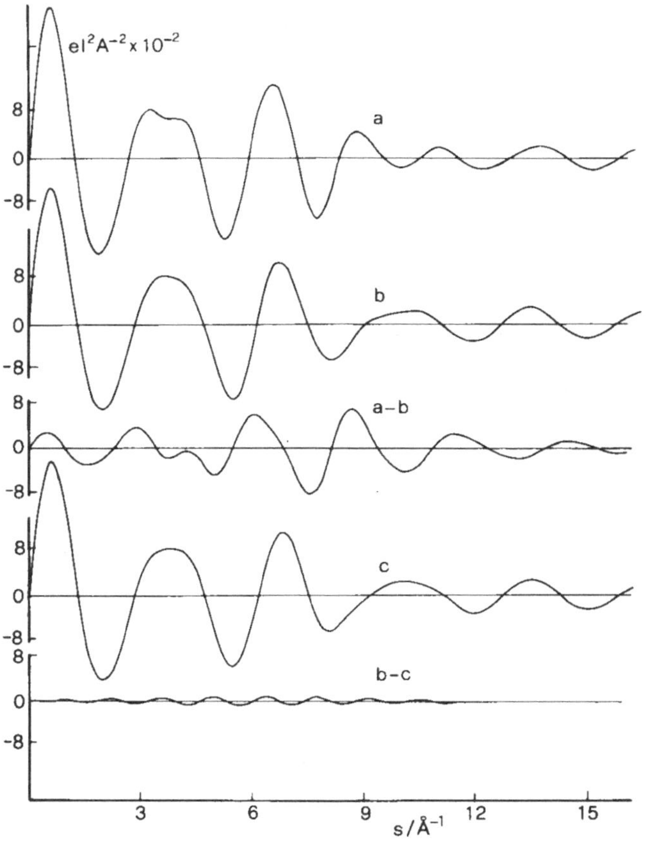

FIGURE 64. Calculated intensity values, in the si(s)M(s) form, from: a, $Cr(H_2O)_4Cl_2^+$ ($N_{Cr^{3+}-Cl^-}$ = 2); b, $Cr(H_2O)_5Cl^{2+}$ ($N_{Cr^{3+}-Cl^-}$ = 1); and c, 50% of $Cr(H_2O)_6^{3+}$ and 50% of $Cr(H_2O)_4Cl_2^+$ ($N_{Cr^{3+}-Cl^-}$ = 1). a-b and b-c are self-explaining difference curves. (From Magini, M., *J. Chem. Phys.*, 73, 2499, 1980. With permission.)

Other help is then welcome and can come from:

1. Knowledge of solid structure from which total coordination numbers and geometries of the various species can be inferred.
2. Spectroscopic determinations of the amount of the different inner complexes existing (e.g., by NMR or Raman) in the same solutions investigated by XRD (see the studies on $ZnBr_2$ or $TlBr_3$ solutions).

In any case, when thermodynamic equilibrium constants were used to tentatively calculate the complexation degree expected in the solutions investigated by XRD, the expectations were often matched by XRD data analysis (see $CrCl_3$ or $Fe_2(SO_4)_3$ solutions). Under these circumstances the same K_{eq}s may be reasonably used for calculating probable percentages of the various complexes.

In the absence of this preliminary information, the XRD analysis cannot go further than the determination of mean numbers of bound anions per metal atom, N_{Me-X} . In fact, if several complexes are taken into account in the modeling procedure, very similar synthetic structure functions are produced by different percentages of the various complexes corresponding to a prefixed N_{Me-X} . The following example is enlightening. Figure 64 reports contributions from: (a) the complex $Cr(H_2O)_4Cl_2^+$ ($N_{Cr^{3+}-Cl^-} = 2$), (b) the complex $Cr(H_2O)_5Cl^{2+}$ ($N_{Cr^{3+}-Cl^-} = 1$), (c) the mixture 50% to 50% of $Cr(H_2O)_6^{3+}$ and $Cr(H_2O)_4Cl_2^+$ ($N_{Cr^{3+}-Cl^-} = 1$). Despite the same pseudo-octahedral configurations, monochloro- and dichloro-complexes give calculated intensity contributions appreciably different, as better shown by the a-b difference curve. On the contrary, models (b) and (c), having the same average complexation number, give calculated intensities not appreciably different, as the difference curve b-c clearly confirms.

The final general conclusion is that when several species are coexisting in the solution under investigation, the least square fitting only gives the average degree of complex formation, but nothing can be said about amounts of single species and their structure. Therefore, structures of complexes have been only determined when solutions containing one or, at most, two complexes could be prepared (see the cases of Tl and Hg halide complexes).

2. Outer Complex Formation

As far as outer complex formation is concerned, it is seldom evidenced. As discussed in Section XVI.B, it is to be regarded as a general phenomenon at high concentrations, which extent is almost impossible to detect.

D. Continuous Distribution of Distances

Whenever the interpretation of XRD results has been achieved by modeling the entire structure function, the term accounting for the transition from discrete to continuous distance distribution (see Chapter 2) had to be considered in the calculations. This implies that for every pair ij (or for the major terms, at least) two extra parameters are to be added, that is, the starting distance of the "continuum", r_{0ij}, and its root mean square deviation, σ_{0ij}. The final values of these parameters have not been reported in the tables, as their physical meaning is vague and their numerical values roughly approximate. However, although strongly influenced by all the inadequacies and approximations of the method, their presence in the calculations is not marginal. In fact, if the continuum region is not properly simulated in the structure function, the medium-long-range interactions cannot be adequately described (see Sections X.B.2 and XV).

E. Final Comments

As this long discussion shows, the number of independent parameters that have to be used in calculations of simulated functions is not small, and this fact might cast doubts on the meaning of fitting procedures. In the simplest cases, that is, solutions with two ions, each of them surrounded by single, uncorrelated coordination shells, about a dozen parameters have to be used, and the number increases when two coordination shells must be accounted for, when cation-anion complex formation occurs, and so on. So, one may find obvious that, playing **freely** with so many variables, a structure (or radial) function can be always reproduced, and one may be surprised that a physical meaning can be attached to the final parameters, as shown here. However, there is a catch here, perhaps not explicitly mentioned so far. The point is that **the parameters**, at least not all of them, **are not really free to move**. In fact, the experimental radial functions provide many pieces of information straightforwardly, which are then fed into the fitting procedure. This is the case of interatomic distances which are already indicated, with good approximation, by the position of the major peaks in the radial curves. This is also the case of coordination numbers which may be estimated by the area of the peaks themselves, at least when they are resolved. So, the main parameters have a definite physical meaning and undergo in the fitting procedure only minor variations around mean values already roughly determined. The same considerations also apply when the "*a priori*" information comes from other studies, like spectroscopic or thermodynamic data.

Corollary of the above is, that the parameters describing direct interactions are almost completely uncorrelated with the ones describing medium-range order. Direct interaction terms alone reproduce most of the structure function (HAA method is based on it actually), so that very few parameters are needed to account for the experimental functions for s ~5 $Å^{-1}$. These are to be regarded with full confidence, while decreasing reliability must be attached to longer interaction descriptions, down to the almost meaningless parameters appearing in continuum terms.

Considerations on precision of experimental data go on accordingly. The experimental accuracy which is sufficient when the investigation is aimed at the study of the dominant interactions may prove to be quite unsuitable when the interest comes to structural details connected with minor contributions to the total scattering. A good example is offered in the comments about $ZnSO_4$ solutions, where two different measurements on the same sample provided concordant results on the main interactions $Zn-H_2O_I$, and different interpretations about minor details. Going through the results discussed in this chapter, it may be observed that the experimental data are in general accurate enough for non equivocal determinations of the major features of di- and trivalent ion coordination shells. However, different outcomes have been sometimes reported, which are clearly linked to serious discrepancies in the experimental data (see discussions on lanthanides or $FeCl_3$ solutions). A question comes to mind: is it possible to check the quality of the experiments, so as to prevent the introduction of unreliable data in the literature body? Our experience with repeated experiments on the same system shows that the answer is positive within the diffractometric work on condition that the researcher wants to take the trouble to check the calibration of his own experimental set up systematically. Would this be impracticable, or not sufficient (for example, for lack of known standards, or difficulties in comparing data differently processed and so on), the information obtained has to be tested anyway against others deriving from other studies. These precautions should have prevented unnecessary debates.

In only seeming contradiction with what was stated above, a major purpose of this review was to examine the applications of XRD to structural investigations on solutions and to present results obtained independently by this technique. In a few cases, where absolutely necessary, we resorted to thermodynamical or spectroscopic information to give support to otherwise uncertain conclusions. On the other hand, we agree with Pálinkás and Kálmán when they affirm that "it is difficult not to influence the interpretation of XRD data with assumptions reflecting general knowledge of the systems under study." More than that, it would not even be desirable! The full meaning of the results discussed here, their uselfulness, and limits will only come out from the comparison and integration with the information obtained by other routes on the same systems. This subject will be approached in the following chapter.

REFERENCES

1. **Hinton, J. F. and Amis, E. S.,** Solution numbers of ions, *Chem. Rev.*, 71, 627, 1971.
2. **Ohtaki, H.,** X-ray diffraction studies on the structure of metal complexes in solution, *Rev. Inorq. Chem.*, 4, 103, 1982.
3. **Burgess, J.,** *Metal Ions in Solution*, Ellis Horwood, Chichester, 1978, chap. 5.
4. **Enderby, J. E. and Neilson, G. W.,** X-ray and neutron scattering by aqueous solutions of electrolytes, in *Water: A Comprehensive Treatise*, Vol. 6, Franks, F., Ed., Plenum Press, New York, 1979, chap. 1.
5. **van Panthaleon van Eck, C. L., Mendel, H., and Booy, W.,** X-ray diffraction of aqueous electrolyte solutions, *Discuss. Faraday Soc.*, 24, 200, 1957.
6. **Terekhova, D. S.,** X-ray investigation of aqueous hydrochloric acid solutions, *Zh. Strukt. Khim.*, 11, 530, 1970.

7. **Lee, S. C. and Kaplow, R.,** Hydrogen bonding in hydrochloric acid solutions, *Science*, 169, 477, 1970.
8. **Licheri, G., Piccaluga, G., and Pinna, G.,** X-ray diffraction study on HCl aqueous solutions, *Chem. Phys. Lett.*, 12, 425, 1971.
9. **Wertz, D. L.,** Halide-oxygen interactions in solutions. Chloride-oxygen interactions in aqueous hydrogen chloride, *J. Solution. Chem.*, 1, 489, 1972.
10. **Triolo, R. and Narten, A. H.,** Diffraction pattern and structure of aqueous hydrochloric acid solutions at 20, *J. Chem. Phys.*, 63, 3624, 1975.
11. **Lee, H., Matsumoto, Y., Yamaguchi, T., and Ohtaki, H.,** X-ray diffraction studies on the structures of hydrated oxonium ion, and the chlorocobalt(II) and tetrachlorocobaltate(II) complexes in aqueous solutions, *Bull. Chem. Soc. Jpn.*, 56, 443, 1983.
12. **Caminiti, R.,** Sulphate-H_2O interactions in a concentrated aqueous H_2SO_4 solution, *Chem. Phys. Lett.*, 96, 390, 1983.
13. **Triolo, R. and Ruffo, I.,** Termination errors in Fourier Analysis of diffraction data of aqueous electrolyte solutions, *J. Chem. Soc. Faraday Trans. I*, 73, 186, 1977.
14. **Brady, G. W.,** Structure of ionic solution. II, *J. Chem. Phys.*, 28, 464, 1958.
15. **Shapovalov, I. M., Radchenko, I. V., and Lesovitskaya, M. K.,** X-ray investigation of aqueous sulfate solutions, *Zh. Strukt. Khim.*, 4, 10, 1963.
16. **Radchenko, I. V. and Ryss, A. I.,** X-ray diffraction study of aqueous solutions of ammonium and lithium tetrafluoroborates, *Zh. Strukt. Khim.*, 6, 182, 1965.
17. **Lawrence, R. M. and Kruh, R. F.,** X-ray diffraction studies of aqueous alkali-metal halide solutions, *J. Chem. Phys.*, 47, 4758, 1967.
18. **Licheri, G., Piccaluga, G., and Pinna, G.,** X-ray diffraction studies of alkali halide solutions, *J. Appl. Cryst.*, 6, 392, 1973.
19. **Narten, A. H., Vaslow, F., and Levy, H. A.,** Diffraction pattern and structure of aqueous lithium chloride solutions, *J. Chem. Phys.*, 58, 5017, 1973.
20. **Licheri, G., Piccaluga, G., and Pinna,G.,** X-ray diffraction study of LiBr aqueous solutions, *Chem. Phys. Lett.*, 35, 119, 1975.
21. **Pálinkás, G., Radnai, T., and Hajdu, F.,** Ion-solvent and solvent-solvent interactions. X-ray study of aqueous alkali chloride solutions, *Z. Naturforsch.*, 35a, 107, 1980.
22. **Radnai, T., Pálinkás, G., Szász, Gy. I., and Heinzinger, K.,** The second hydration shell of Li^+ in aqueous LiI from X-ray and M.D. studies, *Z. Naturforsch.*, 36a, 1076, 1981.
23. **Narten, A. H., Danford, M. D., and Levy, H. A.,** X-ray diffraction study of liquid water in the temperature range 4-200 °C, *Discuss. Faraday Soc.*, 43, 97, 1967.
24. **Licheri, G., Piccaluga, G., and Pinna, G.,** unpublished results.
25. **Ryss, A. I., and Radchenko, I. V.,** An X-ray diffraction study of aqueous sodium tetrafluoroborate solutions, *Zh. Strukt. Khim.*, 5, 530, 1964.
26. **Ohtaki, H., Maeda, M., and Ito, S.,** X-ray diffraction studies of aqueous solutions of cadmium perchlorate and sodium tetraiodocadmate, *Bull. Chem. Soc. Jpn.*, 47, 2217, 1974.
27. **Maeda, M. and Ohtaki, H.,** An x-ray diffraction study of a concentrated aqueous sodium iodide solution, *Bull. Chem. Soc. Jpn.*, 48, 3755, 1975.
28. **Caminiti, R., Licheri, G., Piccaluga, G., and Pinna, G.,** X-ray diffraction study of an aqueous NaCl solution, *Rend. Semin. Fac. Sci. Univ. Cagliari*, 47, 19, 1977.
29. **Bosi, P., Felici, R., Rongoni, E., and Sacchetti, F.,** On the structure of alkali halide solutions. II. Neutron and X-ray diffraction, *Nuovo Cimento D*, 3, 1029, 1984.
30. **Caminiti, R., Licheri, G., Paschina, G., Piccaluga, G., and Pinna, G.,** Interactions and structure in aqueous $NaNO_3$ solutions, *J. Chem. Phys.*, 72, 4522, 1980.
31. **Brady, G. W., and Krause, J. T.,** Structure in ionic solution. I., *J. Chem. Phys.*, 27, 304, 1957.
32. **Bertagnolli, von H., Weidner, J., and Zimmermann, H. W.,** Röntgenstrukturuntersuchung wäbriger Caesium fluorid-lösungen, *Ber. Bunsenges. Phys. Chem.*, 78, 2, 1974.
33. **Dorosh, A. K. and Skryshevskii, A. F.,** The structural characteristics of the immediate environment of cations in aqueous solutions, *Zh. Strukt. Khim.*, 5, 911, 1964.
34. **Ryss, A. I. and Radchenko, I. V.,** An x-ray diffraction study of aqueous solutions of magnesium tetrafluoroborate, *Zh. Strukt. Khim.*, 6, 449, 1965.
35. **Bol, W., Gerrits, G. J. A., and van Panthaleon van Eck, C. L.,** The hydration of divalent cations in aqueous solution. An X-ray investigation with isomorphous replacement, *J. Appl. Cryst.*, 3, 486, 1970.
36. **Alves Marques, M., and De Barros Marques, M. I.,** X-ray diffraction and structure of ionic solutions, *K. Ned. Akad. Wet. Versl. Gewone Vergad. Afd. Naturrkd.*, Proc. B 77, 286, 1974.
37. **Caminiti, R., Licheri, G., Piccaluga, G., and Pinna, G.,** X-ray diffraction study of $MgCl_2$ aqueous solutions, *J. Appl. Cryst.*, 12, 34, 1979.
38. **Caminiti, R., Licheri, G., Piccaluga, G., and Pinna, G.,** Diffraction of x-rays and hydration phenomena in aqueous solutions of $Mg(NO_3)_2$, *Chem. Phys. Lett.*, . 61, 45, 1979.

39. **Pálinkás, G., Radnai, T., Dietz, W., Szász, Gy. I., and Heinzinger, K.,** Hydration shell structure in an $MgCl_2$ solution from X-ray and MD studies, *Z. Naturforsch.*, 37a, 1049, 1982.
40. **Caminiti, R.,** An x-ray diffraction study on SO_4^{2-}-H_2O interactions in the presence of nickel and magnesium ions, *Chem. Phys. Lett.*, 88, 103, 1982.
41. **Caminiti, R., Cucca, P., Monduzzi, M., Saba, G., and Crisponi, G.,** Divalent metal-acetate complexes in concentrated aqueous solutions. An x-ray diffraction and N.M.R. spectroscopy study, *J. Chem. Phys.*, 81, 543, 1984.
42. **Albright, J. N.,** X-ray diffraction studies of aqueous alkaline-earth chloride solutions, *J. Chem. Phys.*, 56, 3783, 1972.
43. **Hamilton, W. C., and Ibers, J. A.,** *Hydrogen Bonding in Solids*, Benjamin, New York, 1968.
44. **Caminiti, R., Licheri, G., Piccaluga, G., and Pinna, G.,** X-ray diffraction study of a three-ion aqueous solution, *Chem. Phys. Lett.*, 47, 275, 1977.
45. **Licheri, G., Piccaluga, G., and Pinna, G.,** X-ray diffraction study of the average solute species in $CaCl_2$ aqueous solutions, *J. Chem. Phys.*, 64, 2437, 1976.
46. **Licheri, G., Piccaluga, G., and Pinna, G.,** X-ray diffraction study of $CaBr_2$ aqueous solutions, *J. Chem. Phys.*, 63, 4412, 1975.
47. **Probst, M. M., Radnai, T., Heinzinger, K., Bopp, P., and Rode, B. M.,** Molecular dynamics and x-ray investigation of an aqueous $CaCl_2$ solution, *J. Phys. Chem.*, 89, 753, 1985.
48. **Licheri, G., Piccaluga, G., and Pinna, G.,** Structural order in liquid $CaCl_2{\cdot}6H_2O$, *J. Am. Chem. Soc.*, 101, 5438, 1979.
49. **Caminiti, R., Musinu, A., Paschina, G., Piccaluga, G., and Pinna, G.,** On the structure in aqueous superconcentrated calcium nitrate, *Z. Naturforsch.*, 36a, 831, 1981.
50. **Caminiti, R., Musinu, A., Paschina, G., and Pinna, G.,** X-ray diffraction study of aqueous $SrCl_2$ solutions, *J. Appl. Cryst.*, 15, 482, 1982.
51. **Caminiti, R., Licheri, G., Piccaluga, G., Pinna, G., and Radnai, T.,** Order phenomena in aqueous $AlCl_3$ solutions, *J. Chem. Phys.*, 71, 2473, 1979.
52. **Caminiti, R., and Radnai, T.,** X-ray diffraction study of a concentrated $Al(NO_3)_3$ solution, *Z. Naturforsch.*, 35a, 1368, 1980.
53. **Bol, W. and Welzen, T.,** The interpretation of x-ray diffraction by aqueous solutions of aluminum(III) nitrate and chromium(III) nitrate, *Chem. Phys. Lett.*, 49, 189, 1977.
54. **Maeda, M. and Ohtaki, H.,** An x-ray diffraction study on the structure of the aqua indium(III) ion in the perchlorate solution, *Bull. Chem. Soc. Jpn.*, 50, 1893, 1977.
55. **Caminiti, R. and Paschina, G.,** An x-ray diffraction study of the structure of the aqua indium(III) ion in indium sulfate solution, *Chem. Phys. Lett.*, 82, 487, 1981.
56. **De Barros Marques, M. I., Cabaço, M. I., Sousa Oliveira, M. A., and Alves Marques, M.,** X-ray diffraction investigations of the structure of concentrated ionic solutions, *Chem. Phys. Lett.*, 91, 222, 1982.
57. **Glaser, J. and Johansson, G.,** On the structure of hydrated thallium(III) ion and its bromide complexes in aqueous solution, *Acta Chem. Scand.*, A36, 125, 1982.
58. **Glaser, J.,** On the structures of thallium(III) chloride complexes in solution. I. Evidence for tetrahedral $TlCl_4^-$ and octahedral $TlCl_6^{3-}$ species in aqueous solution, *Acta Chem. Scand.*, A36, 451, 1982.
59. **Glaser, J. and Henricksson, U.,** Thallium-205 nuclear magnetic resonance study of thallium(III) halide complexes in aqueous solutions, *J. Am. Chem. Soc.*, 103, 6642, 1981.
60. **Johansson, G. and Olin, A.,** On the structure of the dominating hydrolysis products of lead(II) in solution, *Acta Chem. Scand.*, 22, 3197, 1968.
61. **Johansson, G. and Ohtaki, H.,** An x-ray investigation of the hydrolysis products of tin(II) in solution, *Acta Chem. Scand.*, 27, 643, 1973.
62. **Yamaguchi, T., Lindquist, O., Claeson, T., and Boyce, J. B.,** EXAFS and X-ray diffraction studies of the hydration structure of stereochemically active Sn(II) ions in aqueous solution, *Chem. Phys. Lett.*, 93, 528, 1982.
63. **Habenschuss, A. and Spedding, F. H.,** The coordination (hydration) of rare earth ions in aqueous chloride solutions from x-ray diffraction. I. $TbCl_3$, $DyCl_3$, $ErCl_3$, $TmCl_3$, and $LuCl_3$, *J. Chem. Phys.*, 70, 2797, 1979.
64. **Habenschuss, A. and Spedding, F. H.,** The coordination (hydration) of rare earth ions in aqueous chloride solutions from x-ray diffraction. II. $LaCl_3$, $PrCl_3$, and $NdCl_3$, *J. Chem. Phys.*, 70, 3758, 1979.
65. **Habenschuss, A. and Spedding, F. H.,** The coordination (hydration) of rare earth ions in aqueous chloride solutions from x-ray diffraction. III. $SmCl_3$, $EuCl_3$, and series behavior, *J. Chem. Phys.*, 73, 442, 1980.
66. **Brady, G. W.,** Structure in ionic solutions. IV, *J. Chem. Phys.*, 33, 1079, 1960.
67. **Smith, L. S., Jr. and Wertz, D. L.,** Solute structuring in aqueous lanthanum(III) chloride solutions, *J. Am. Chem. Soc.*, 97, 2365, 1975.
68. **Steele, M. L. and Wertz, D. L.,** Solute complexes in aqueous gadolinium(III) chloride solutions, *J. Am. Chem. Soc.*, 98, 4424, 1976.

69. **Smith, L. S. and Wertz, D. L.**, On the coordination of La^{3+} in aqueous $LaBr_3$ solutions, *J. Inorg. Nucl. Chem.*, 39, 95, 1977.
70. **Steele, M. L. and Wertz, D. L.**, Solvent effects on the coordination of Nd^{3+} in concentrated $NdCl_3$ solutions, *Inorg. Chem.*, 16, 1225, 1977.
71. **Caminiti, R., Cucca, P., and D'Andrea, A.**, Hydration phenomena in a concentrated aqueous solution of $Ce(NO_3)_3$. X-ray diffraction and Raman spectroscopy, *Z. Naturforsch.*, 38a, 593, 1983.
72. **Johansson, G. and Wakita, H.**, X-ray investigation of the coordination and complex formation of lanthanoids ions in aqueous perchlorate and selenate solutions, *Inorg. Chem.*, 24, 3047, 1985.
73. **Johansson, G., Petterson, L., and Ingri, N.**, A large-angle x-ray scattering study of 9-molybdomonophosphate complexes in aqueous solution, *Acta Chem. Scand.*, A32, 407, 1978.
74. **Johansson, G., Petterson, L., and Ingri, N.**, A large-angle x-ray scattering investigation of some polynuclear molybdoarsenate complexes in aqueous solution, *Acta Chem. Scand.*, A32, 681, 1978.
75. **Johansson, G., Petterson, L., and Ingri, N.**, On the formation of hepta- and octamolybdates in aqueous solution. X-ray scattering and Raman measurements, *Acta Chem. Scand.*, A33, 305, 1979.
76. **Johansson, G., Petterson, L., and Ingri, N.**, On the structure of hexamolybdodiarsenate complexes in aqueous solution, *Acta Chem. Scand.*, A35, 181, 1981.
77. **Hunt, J. P.**, *Metal Ions in Aqueous Solution*, Benjamin, New York, 1963.
78. **Caminiti, R., Licheri, G., Piccaluga, G., and Pinna, G.**, Solute structuring in aqueous $[Cr(H_2O)_6]Cl_3$, *J. Chem. Phys.*, 65, 3134, 1976.
79. **Caminiti, R., Licheri, G., Piccaluga, G., and Pinna, G.**, X-ray diffraction study of $Cr(NO_3)_3$ aqueous solutions, *Chem. Phys.*, 19, 371, 1977.
80. **Caminiti, R., Licheri, G., Piccaluga, G., and Pinna, G.**, Hydration water-external water interactions around Cr^{3+} ions, *J. Chem. Phys.*, 69, 1, 1978.
81. **Brown, I. D.**, On the geometry of O-H···O hydrogen bonds, *Acta Cryst.*, A32, 24, 1976.
82. **Friedman, H. L. and Lewis, L.**, The coordination geometry of water in some salt hydrates, *J. Solution Chem.*, 5, 445, 1976.
83. **Caminiti, R.**, Complex formation and SO_4^{2-}-H_2O interactions in a concentrated aqueous $Cr_2(SO_4)_3$ solution, *Chem. Phys. Lett.*, 86, 214, 1982.
84. **Magini, M.**, X-ray diffraction study of concentrated chromium(III) chloride solutions. I. Complex formation analysis in equilibrium conditions, *J. Chem. Phys.*, 73, 2499, 1980.
85. **Shapovalov, I. M., and Radchenko, I. V.**, X-ray study of water and aqueous solutions of manganese, zinc and copper sulfates, *Zh. Strukt. Khim.*, 12, 769, 1971.
86. **Ohtaki, H., Yamaguchi, T., and Maeda, M.**, X-ray diffraction studies of the structures of hydrated divalent transition-metal ions in aqueous solution, *Bull. Chem. Soc. Jpn.*, 49, 701, 1976.
87. **Caminiti, R., Marongiu, G., and Paschina, G.**, A comparative x-ray diffraction study of aqueous $MnSO_4$ and crystals of $MnSO_4{\cdot}5H_2O$, *Z. Naturforsch.*, 37a, 581, 1982.
88. **Licheri, G., Paschina, G., Piccaluga, G., and Pinna, G.**, X-ray diffraction study of aqueous solutions of $NiSO_4$ and $MnSO_4$, *J. Chem. Phys.*, 81, 6059, 1984.
89. **Caminiti, R., Cucca, P., and Pintori, T.**, Hydration and ion-pairing in concentrated aqueous $Mn(NO_3)_2$ solutions. An x-ray and Raman spectrosopy study, *J. Chem. Phys.*, 88, 155, 1984.
90. **Magini, M. and Sandström, M.**, unpublished results.
91. **Ichihashi, M., Wakita, H., and Masuda, I.**, Structure of iron(II) and cobalt(II) bromide complexes in aqueous solution by x-ray diffraction analysis, *J. Solution Chem.*, 13, 505, 1984.
92. **Magini, M. and Caminiti, R.**, On the structure of highly concentrated iron(III) salt solutions, *J. Inorg. Nucl. Chem.*, 39, 91, 1977.
93. **Magini, M.**, An x-ray investigation on the structure of iron(III) perchlorate solutions, *J. Inorg. Nucl. Chem.*, 40, 43, 1978.
94. **Caminiti, R. and Magini, M.**, An x-ray diffraction study on the first and the second hydration shell of the Fe(III) ion in nitrate solutions, *Chem. Phys. Lett.*, 61, 40, 1979.
95. **Brady, G. W.**, Structure in ionic solutions. III., *J. Chem. Phys.*, 29, 1371, 1958.
96. **Standley, C. L. and Kruh, R. F.**, On the structure of ferric chloride solutions, *J. Chem. Phys.*, 34, 1450, 1961.
97. **Brady, G. W., Robin, M. B., and Varimbi, J.**, The structure of ferric chloride in neutral and acid solutions, *Inorg. Chem.*, 3, 1168, 1964.
98. **Wertz, D. L. and Steele, M. L.**, Coordination of Fe^{3+} in concentrated aqueous solutions with chloride ligand, *Inorg. Chem.*, 19, 1652, 1980.
99. **Wertz, D. L. and Luter, M. D.**, Evolving cation coordination in aqueous solutions prepared from iron(III) chloride hexahydrate, *Inorg. Chem.*, 20, 3118, 1981.
100. **Giubileo, G., Magini, M., Licheri, G., Paschina, G., Piccaluga, G., and Pinna, G.**, On the structure of iron(III) chloride solutions, *Inorg. Chem.*, 22, 1001, 1983.
101. **Magini, M. and Radnai, T.**, X-ray diffraction study of ferric chloride solutions and hydrated melt. Analysis of the iron(III)-chloride complexes formation, *J. Chem. Phys.*, 71, 4255, 1979.

102. **Magini, M.**, Solute structuring in iron(III) chloride solutions. II. Evidence of polynuclear complex formation in the $FeCl_3 \cdot 6H_2O$ melt by the "isoelectronic solutions" method, *J. Chem. Phys.*, 76, 1111, 1982.
103. **Magini, M.**, Solute structuring in aqueous iron(III) sulphate solutions. Evidence for the formation of iron(III)-sulphate complexes, *J. Chem. Phys.*, 70, 317, 1979.
104. **Magini, M. and Giubileo, G.**, Solute structuring in concentrated aqueous Co(II) and Ni(II) perchlorate solutions, *Gazz. Chim. Ital.*, 111, 449, 1981.
105. **Wertz, D. L. and Kruh, R. F.**, X-ray diffraction study of some concentrated cobalt(II) chloride solutions, *J. Chem. Phys.*, 50, 4313, 1969.
106. **Magini, M.**, Hydration and complex formation study on concentrated MCl_2 solutions [M = Co(II), Ni(II), Cu(II)] by x-ray diffraction technique, *J. Chem. Phys.*, 74, 2523, 1981.
107. **Paschina, G., Piccaluga, G., Pinna, G., and Magini, M.**, X-ray investigation of Co-Cl bonding in a concentrated aqueous solution of $CoCl_2$ and LiCl, *Chem. Phys. Lett.*, 98, 157, 1983.
108. **Musinu, A., Paschina, G., Piccaluga, G., and Magini, M.**, X-ray diffraction study of $CoCl_2$-LiCl aqueous solutions, *J. Chem. Phys.*, 80, 2772, 1984.
109. **Wertz, D. L. and Kruh, R. F.**, Solute-solvent interactions in some concentrated cobalt(II) bromide solutions, *Inorg. Chem.*, 9, 595, 1970.
110. **Ryss, A. I. and Radchenko, I. V.**, X-ray study of aqueous solutions of nickel tetrafluoroborate, *Zh. Strukt. Khim.*, 6, 507, 1965.
111. **Neilson, G. W., Howe, R. A., and Enderby, J. E.**, The quasi-lattice structure in concentrated aqueous solutions, *Chem. Phys. Lett.*, 33, 284, 1975.
112. **Enderby, J. E.**, Neutron and x-ray scattering from aqueous solutions, *Proc. R. Soc. London*, A 345, 107, 1975.
113. **Caminiti, R., Licheri, G., Piccaluga, G., and Pinna, G.**, X-ray diffraction and structure of $NiCl_2$ aqueous solutions, *Faraday Discuss. Chem. Soc.*, 64, 62, 1977.
114. **Caminiti, R., Licheri, G., Paschina, G., Piccaluga, G., and Pinna, G.**, X-ray diffraction and structural properties of aqueous solutions of divalent metal-chlorides, *Z. Naturforsch.*, 35a, 1361, 1980.
115. **Licheri, G., Paschina, G., Piccaluga, G., and Pinna, G.**, Unpublished results.
116. **Magini, M., Paschina, G., and Piccaluga, G.**, Ni-Cl bonding in concentrated Ni(II) aqueous solutions at high Cl^-/Ni^{2+} ratios. An x-ray diffraction investigation, *J. Chem. Phys.*, 76, 1116, 1982.
117. **Caminiti, R., and Cucca, P.**, X-ray diffraction study on a $NiBr_2$ aqueous solution. Experimental evidence of the Ni(II)Br contacts, *Chem. Phys. Lett.*, 89, 110, 1982.
118. **Wakita, H., Ichihashi, M., Mibuchi, T., and Masuda, I.**, The structure of nickel(II) bromide in highly concentrated aqueous solution by x-ray diffraction analysis, *Bull. Chem. Soc. Jpn.*, 55, 817, 1982.
119. **Magini, M., de Moraes, M., Licheri, G., and Piccaluga, G.**, Composition of the first coordination sphere of Ni^{2+} in concentrated aqueous $NiBr_2$ solutions by x-ray diffraction, *J. Chem. Phys.*, 83, 5797, 1985.
120. **Caminiti, R.**, Nickel and cadmium phosphates in aqueous solution. Cation-anion complex formation and phosphate-H_2O interactions, *J. Chem. Phys.*, 77, 5682, 1982.
121. **Caminiti, R. and Cucca, P.**, An x-ray diffraction study of Rh(III) coordination in a dilute aqueous solution of $Rh(ClO_4)_3$, *Chem. Phys. Lett.*, 108, 51, 1984.
122. **Maeda, M., Akaishi, T., and Ohtaki, H.**, An x-ray diffraction study of the structures of the hexachloro- and hexabromoplatinate(IV) ions in aqueous solution, *Bull. Chem. Soc. Jpn.*, 48, 3193, 1975.
123. **Ohtaki, H. and Maeda, M.**, An x-ray diffraction study of the structure of hydrated copper(II) ion in a copper(II) perchlorate solution, *Bull. Chem. Soc. Jpn.*, 47, 2197, 1974.
124. **Magini, M.**, Coordination of copper(II). Evidence of the Jahn-Teller effect in aqueous perchlorate solutions, *Inorg. Chem.*, 21, 1535, 1982.
125. **Musinu, A., Paschina, G., Piccaluga, G., and Magini, M.**, Coordination of copper(II) in aqueous $CuSO_4$ solution, *Inorg. Chem.*, 22, 1184, 1983.
126. **Bell, J. R., Tyvoll, J. L., and Wertz, D. L.**, Solute structuring in aqueous copper(II) chloride solutions, *J. Am. Chem. Soc.*, 95, 1456, 1973.
127. **Wertz, D. L. and Tyvoll, J. L.**, The coordination of Cu(II) in a nearly saturated solution of $CuCl_2$ in hydrochloric acid, *J. Inorg. Nucl. Chem.*, 36, 3713, 1974.
128. **Ichihashi, M., Wakita, H., Mibuchi, T., and Masuda, I.**, Coordination structure of Cu^{2+} ion in highly concentrated aqueous $CuBr_2$ solution determined by X-ray diffraction analysis, *Bull. Chem. Soc. Jpn.*, 55, 3160, 1982.
129. **Licheri, G., Musinu, A., Paschina, G., Piccaluga, G., Pinna, G., and Sedda, A. F.**, Coordination of Cu(II) in $Cu(NO_3)_2$ aqueous solutions, *J. Chem. Phys.*, 80, 5308, 1984.
130. **Maeda, M., Maegawa, Y., Yamaguchi, T., and Ohtaki, H.**, X-ray diffraction studies on the structures of diammine- and aqua-silver(I) complexes in aqueous solution, *Bull. Chem. Soc. Jpn.*, 52, 2545, 1979.
131. **Yamaguchi, T., Johansson, G., Holmberg, B., Maeda, M., and Ohtaki, H.**, The coordination and complex formation of silver (I) in aqueous perchlorate, nitrate and iodide solutions, *Acta Chem. Scand.*, A38, 437, 1984.

132. **Maeda, M., Ohtaki, H., and Johansson, G.**, An x-ray investigation of the structure of the tetrachloro- and tetrabromoaurate(III) ions in aqueous solution, *Bull. Chem. Soc. Jpn.*, 47, 2229, 1974.
133. **Dagnall, S. P., Hague, D. N., and Towl, A. D. C.**, X-ray diffraction study of aqueous Zn(II) nitrate, *J. Chem. Soc. Faraday Trans. 2*, 78, 2161, 1982.
134. **Licheri, G., Paschina, G., Piccaluga, G., and Pinna, G.**, X-ray diffraction study of aqueous solutions of $ZnSO_4$, *Z. Naturforsch.*, 37a, 1205, 1982.
135. **Radnai, T., Pálinkás, G., and Caminiti, R.**, X-ray diffraction study on hydration and ion-pairing in aqueous $ZnSO_4$ solution, *Z. Naturforsch.*, 37a, 1247, 1982.
136. **Kruh, R. F. and Standley, C. L.**, An x-ray diffraction study of aqueous zinc chloride solutions, *Inorg. Chem.*, 1, 941, 1962.
137. **Wertz, D. L., Lawrence, R. M., and Kruh, R. F.**, X-ray diffraction study of zinc bromide solutions, *J. Chem. Phys.*, 43, 2163, 1965.
138. **Wertz, D. L. and Bell, J. R.**, On the existence of pseudotetrahedral tetrachlorozincate(II) in hydrochloric acid solutions, *J. Inorg. Nucl. Chem.*, 35, 137, 1973.
139. **Wertz, D. L. and Bell, J. R.**, Solute species and equilibria in concentrated zinc chloride/HCl solutions, *J. Inorg. Nucl. Chem.*, 35, 861, 1973.
140. **Kálmán, E., Serke, I., Pálinkás, G., Johansson, G., Kabish, G., Maeda, M., and Ohtaki, H.**, Complex formation in an aqueous $ZnBr_2$ solution based on electron diffraction, X-ray scattering and Raman Spectra, *Z. Naturforsch.*, 38a, 225, 1983.
141. **Goggin, P. L., Johansson, G., Maeda, M., and Wakita, H.**, The structure of zinc bromide complexes in aqueous solution, *Acta Chem. Scand.*, A38, 625, 1984.
142. **Ahrland, S., Kullberg, L., and Portanova, R.**, On the heats of solvation of the zinc(II), cadmium(II), and mercury(II) ions, and of their neutral halide complexes, in water and dimethyl sulfoxide at 25°C, *Acta Chem. Scand.*, A32, 251, 1978.
143. **Caminiti, R. and Johansson, G.**, On the structure of cadmium sulfate complexes in aqueous solutions, *Acta Chem. Scand.*, A35, 373, 1981.
144. **Paschina, G., Piccaluga, G., Pinna, G., and Magini, M.**, Chloro-complexes formation in a $ZnCl_2$-$CdCl_2$ aqueous solution : an x-ray diffraction study, *J. Chem. Phys.*, 78, 5745, 1983.
145. **Caminiti, R.**, An x-ray diffraction study on an aqueous solution of $CdSO_4$ at 9° and 62°, *Z. Naturforsch.*, 36a, 1062, 1981.
146. **Caminiti, R., Cucca, P., and Radnai, T.**, Investigation on the structure of cadmium nitrate aqueous solutions by x-ray diffraction, *J. Phys. Chem.*, 88, 2382, 1984.
147. **Johansson, G.**, An x-ray investigation of the structure of the Hg_2^{2+} ion in solution, *Acta Chem. Scand.*, 20, 553, 1966.
148. **Johansson, G.**, An x-ray investigation of the hydrolysis products of mercury(II) in solution, *Acta Chem. Scand.*, 25, 2787, 1971.
149. **Sandström, M., Persson, I., and Ahrland, S.**, On the coordination around mercury(II), cadmium(II), and zinc(II) in dimethyl sulfoxide and aqueous solutions. An x-ray diffraction, Raman and infrared investigation, *Acta Chem. Scand.*, A32, 607, 1978.
150. **Kurnetsov, V. V., Trostin, V. N., and Krestov, L. A.**, An x-ray structure study of aqueous solutions of mercury(II) nitrate, *Zh. Strukt. Khim.*, 24, 132, 1983.
151. **Sandström, M. and Johansson, G.**, An x-ray diffraction study of iodide and bromide complexes of mercury(II) in aqueous solutions, *Acta Chem. Scand.*, A31, 132, 1977.
152. **Sandström, M.**, An x-ray diffraction and Raman study of mercury(II) chloride complexes in aqueous solution. Evidence for the formation of polynuclear complexes, *Acta Chem. Scand.*, A31, 141, 1977.
153. **Narten, A. H.**, Diffraction pattern and structure of aqueous ammonium halide solutions, *J. Phys. Chem.*, 74, 765, 1970.
154. **Pálinkás, G., Radnai, T., Szász, Gy. I., and Heinzinger, K.**, The structure of an aqueous ammonium chloride solution, *J. Chem. Phys.*, 74, 3522, 1981.
155. **Musinu, A., Paschina, G., Piccaluga, G., and Magini, M.**, The sulphate ion in aqueous solution: an x-ray diffraction study of a $ZnSO_4$ solution, *J. Appl. Cryst.*, 15, 621, 1982.
156. **Caminiti, R., Licheri, G., Piccaluga, G., and Pinna, G.**, On NO_3^--H_2O interactions in aqueous solution, *J. Chem. Phys.*, 68, 1967, 1978.
157. **Caminiti, R., Paschina, G., Pinna, G., and Magini, M.**, Experimental evidence of interactions SO_4^{2-}-H_2O in an aqueous solution, *Chem. Phys. Lett.*, 64, 391, 1979.
158. **Musinu, A., Paschina, G., and Piccaluga, G.**, On the structure of the NH_4^+ ion in aqueous solution, *Chem. Phys. Lett.*, 80, 163, 1981.
159. **Caminiti, R., Cucca, P., and Atzei, D.**, Phosphate-H_2O interactions in concentrated aqueous H_3PO_4 solutions, *J. Phys. Chem.*, 89, 1457, 1985.
160. **Brown, I. D. and Shannon, R. D.**, Empirical bond-strength-bond-length curves for oxides, *Acta Cryst.*, A 29, 266, 1973.
161. **Brown, I. D. and Kang Kun Wu**, Empirical parameters for calculating cation-oxygen bond valences, *Acta Cryst.*, B 32, 1957, 1976.

Chapter 4

STRUCTURAL INFORMATION ON AQUEOUS SOLUTIONS FROM NEUTRON DIFFRACTION, EXAFS, AND SIMULATION METHODS

I. INTRODUCTION

The great deal of information on the structure of electrolyte solutions, available from the analysis of the X-ray radial distribution functions, has been widely illustrated in the previous chapter. However, it must be always kept in mind that in X-ray diffraction (XRD) studies deconvolution of radial distribution data into the pair interaction peaks is always difficult, often ambiguous, and never unique. In Figure 1 an example taken from a $FeCl_3$ solution shows how complicated the deconvolution can be. Here, the single contributions to the total radial curve have been, of course, derived from the model and could not correspond to the real ones. Nevertheless, whatever the real situation is, the figure gives a precise idea of how many partial contributions participate in the total radial curve calculated from the experimental data.

Then it follows that, whenever possible, the distribution curves obtained from XRD must be compared with those obtained from other techniques. It follows also that techniques suitable to directly provide the single pair correlation functions must be developed. In this chapter, attention will therefore be focused on those structural techniques able to provide total and, above all, partial pair correlation functions (that is, neutron diffraction, ND; extended X-ray absorption fine structure, EXAFS; and computer simulation methods), disregarding other techniques, such as NMR, which can also provide more or less direct structural information. Details on experimental techniques and data treatments will not be given, and only the basic principles necessary to frame the experimental results will be recalled; attention will be rather focused on the structural results and on their comparison, whenever possible, with those obtained by XRD.

II. NEUTRON DIFFRACTION

A. Single Experiments

1. General Considerations

In a neutron diffraction (ND) experiment,[1] similar to what happens with X-ray (see Chapter 1), the experimental diffraction intensities $I_T(s)$ must be converted into mean **absolute intensities** (or **scattering cross-sections**), usually referred to as $(d\sigma/d\Omega)$, by a general formula of the type:

$$I(s) = \alpha(s)\, I_T(s) - \beta(s) \tag{1}$$

where $\alpha(s)$ and $\beta(s)$ are calibration parameters to be determined either experimentally or theoretically for the particular experiment undertaken; major contributions to $\alpha(s)$ are absorption and polarization, and the term $\beta(s)$ is due largely to incoherent and multiple scattering.

In turn, the absolute intensities are related to the **total structure function** i(s), usually referred to as $F_T(s)$, by the equation:

$$I(s) = i(s) + \sum_i x_i f_i^2 \tag{2}$$

where

$$i(s) = F_T(s) = \sum_i \sum_j x_i x_j f_i f_j [S_{ij}(s) - 1] \tag{3}$$

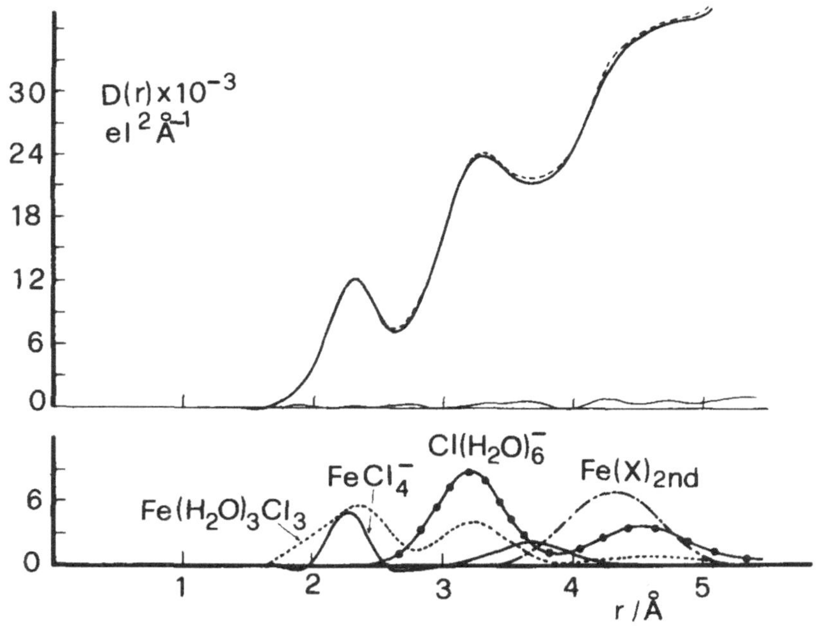

FIGURE 1. Experimental (solid strong line) and calculated (dashed line) distribution curves for a $FeCl_3$ solution. The individual contributions to the calculated curve are shown for the main interactions reported in Table 16, Chapter 3. (From Magini, M. and Radnai, T., *J. Chem. Phys.*, 71, 4255, 1979. With permission.)

The meaning of x_i and f_i is the same of the analogous quantities appearing in XRD equations in Chapter 1.

Therefore, by a single ND experiment (like by XRD), a **total distribution function**, $G^N(r)$, may be obtained by Fourier transforming $F_T(s)$. However, inherent physical differences distinguish X-ray and neutron diffraction phenomena; as a consequence, important differences in data treatment exist, and, moreover, the $G^N(r)$ functions can meaningfully differ from the $G^X(r)$s. It is therefore necessary to point out several aspects of the ND, mainly in order to understand advantages and disadvantages of ND with respect to XRD.

1. The neutron scattering amplitudes of the various chemical elements are of the same order in their magnitude. Neutrons are therefore an excellent probe for the location of light atoms scarcely seen by X-rays, like hydrogen (usually substituted by deuterium in neutron experiments for reasons that will soon be explained). This opportunity allows, theoretically, the investigation of the orientational correlations of the hydrogenated solvent molecules, like H_2O, through the determination of the hydrogen (deuterium) positions. On the other hand, the $G^N(r)$ functions are even more complicated than the corresponding $G^X(r)$s. In fact, in an XRD experiment, a metal halide solution may be thought as a three-component system (cation, anion, and water), while four components are seen in the analogous neutron experiment (cation, anion, oxygen, and deuterium). Ten partial correlation functions are therefore convoluted in the total $G^N(r)$, while only six are convoluted in $G^X(r)$.
2. The neutron scattering factors do not depend, in the static approximation (see below), on the variable s. High-angle ND data should be, therefore, more accurate than corresponding XRD data. In practice, however, due to the weakness of the neutron sources when compared to conventional X-ray tubes, the angular range covered by X-ray has

been, for a long time, wider than that explorable by ND. The ND situation, however, is in rapid evolution. High-flux nuclear reactors (e.g., at Brookhaven, Oak Ridge, Grenoble, etc.) are today available and, even more important, the time-of-flight (TOF) neutron diffraction technique, using a linear accelerator (LINAC), has appeared in the late 1970s. As a consequence, structure functions for liquids, glasses, and solutions, extended to much higher s limits than those obtainable by X-rays, have come to appear in literature. Nevertheless, the complexity of such facilities and their consequent scarce accessibility limit their use strictly to those experiments which necessarily require such a high accuracy.

3. In ND, some corrections are more critical than in XRD. Multiple scattering is usually large and its contribution to the total scattering must be correctly evaluated. The absorption from some isotopes of certain elements (e.g., ^{10}B) is so strong that good diffraction data are difficult to obtain in their presence. Finally, the incoherent contribution to the total scattering is sometimes so important that the extraction of the coherent contribution from the measured intensities would be quite difficult; namely, the ND experiments with aqueous solutions are carried out with D_2O, instead of H_2O, because of high incoherent scattering of H atoms.
4. In ND, a special effect must be accounted for, which comes from an **inelastic** contribution to total scattering. In fact, when a radiation collides with an atomic system, the exchange of energy between radiation and sample is negligible if the mass M of the sample particles is very large compared to the mass M′ of the scattered photons or neutrons, and if the energy E_0 of the incident radiation is much larger than the energy transfer in scattering processes. Under these conditions the scattering particles may be considered rigidly bound (**the static approximation**). These conditions are fulfilled in X-ray scattering, while in neutron scattering, a transfer of both momentum and energy takes place, and a dynamic correction is significant. This can be calculated with good accuracy for heavy nuclides. For aqueous systems, the presence of H or D atoms gives rise to a characteristic drop of $F_T(s)$ at high s, that must be eliminated (**the Placzek correction**). Unfortunately, there is no generally accepted method to take the Placzek effect into account.

As a matter of fact, it is difficult to say whether or not the ND advantages with respect to XRD overcome the disadvantages. It must be emphasized, however, that point 1 represents a basic difficulty for a single ND experiment that cannot be eliminated. The $G^N(r)$, in fact, is a complex function whose resolution into the component pair functions is more difficult than for the analogous $G^X(r)$ function. According to Triolo and Narten, "it is extremely difficult to extract information on ion-water interactions from a single neutron scattering experiment; the interpretation of X-ray results is not quite as hopeless." To the same conclusion we come observing that, for very concentrated solutions, the ion-water contributions to the total scattering pattern are typically 30% in XRD and only 10% in ND.

We can therefore conclude that **ND single experiments**, even if excellent, do not represent a progress with respect to XRD studies. Few experiments of this kind have indeed been carried out and will be shortly examined in the following.

2. Results from Single Experiments

A series of 1-*M* alkali halide solutions has been investigated by Ohtomo and co-workers.[2,3] The diffraction data have been analyzed by comparing the experimental data with those calculated by a great number of possible ionic coordination models. The free water contribution was accounted for by subtracting the weighted intensities obtained from the pure water, in a similar way to that followed in the method FNM1, reported in Chapters 2 and 3. Obviously, the interactions involving the D atoms were considered, so that different possible orientations of the D_2O molecules around the ions (according to the sketches below) were tested.

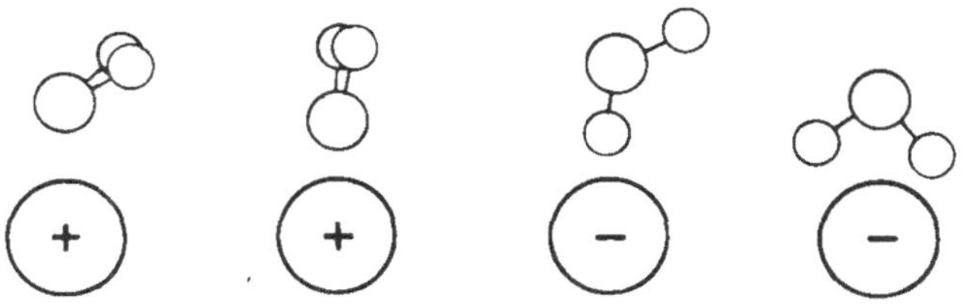

Coordination numbers, ion-oxygen distances, and configurations around the ions determined by Ohtomo and co-workers, are given in Table 1. The results generally agree with those obtained from XRD, except for the Na^+ ion, for which both Na – O distance and coordination number are greater than the corresponding X-ray results. As far as the configuration is concerned, the comparison with XRD is not possible. However, the bifurcated solvent configuration, in some cases suggested around the Cl^- ions, seems somehow surprising, while, we can anticipate, a linear configuration has been always found in the investigations by Enderby and co-workers.

3. Single Experiments Used in Connection with XRD

Although single ND experiments do not guarantee great progress in the knowledge of the structure of ionic solutions, the difference in the neutrons and X-rays scattering factors for the various elements suggests the coupling of both techniques in the study of a given system, in order to **see** it with two quite different **eyes**.[4,5] Figure 2 shows the $G^N(r)$ and $G^X(r)$ obtained from DCl (or HCl) aqueous solutions.[5] It is evident that, while O – O and ion – O correlations dominate the XRD experiment, the correlations involving the D atoms, and within those, the intramolecular ones falling between 1 and 2 Å, dominate the neutron experiment.

The advantages of coupling ND and XRD have been proved by Narten and co-workers in their work on LiCl and HCl (DCl) solutions (see Chapter 3).[4,5] X-ray data were used to fix O – O and ion – O correlations; the transfer of the information to ND data allowed to obtain the ion – D correlations and to propose complete hydration models.

B. Neutron Diffraction with Isotopic Substitution (NDIS): General Principles

As described in Chapter 1, it is possible to accede to single pair structure and correlation functions if different diffraction experiments can be carried out on the same system by varying the scattering factors of the atomic species involved. Since the different isotopes of a given element have different neutron scattering factors, in an ND experiment it is possible, by proper **isotopic substitution**, to perform $M(M + 1)/2$ separate data collections from which derive single pair structure functions. When fewer experiments than those really needed can be performed, linear combinations of a reduced number of pair functions are at least obtained. As a matter of fact, a great progress in structural studies by ND has been actually achieved by the introduction of isotopic substitution method.[1]

Short outlines of the basic equations are given in the following. If we collect data from two MX_n solutions (M = metal, X = halide), which are identical in all respects, except that the isotopic M (or X) state has been changed, a $\Delta_M F(s)$ (or Δ_X F[s]) function can be evaluated. This function is the sum of the four **partial structure functions** $S_{ij}(s)$ describing the interactions between M (or X) and the four atomic species of the solution (O, D, M, X), that is,

$$\Delta F_i(s) = A[S_{iO}(s) - 1] + B[S_{iD}(s) - 1] + C[S_{iM}(s) - 1] + D[S_{iX}(s) - 1] \quad (4)$$

Table 1
IONIC COORDINATION PARAMETERS FROM NEUTRON DIFFRACTION STUDIES

Ion	Coordination number	Ion-oxygen distance	Configuration around ions
Li^+	4	1.90 ± 0.05	Linear
Na^+	8	2.50 ± 0.10	Undetermined
K^+	8	2.70 ± 0.10	Undetermined
Cs^+	8	2.95 ± 0.10	Linear
Cl^-	6	3.10 ± 0.05	Undetermined (for LiCl and CsCl solutions)
Cl^-	6	3.10 ± 0.05	Bifurcated (for NaCl and KCl solutions)

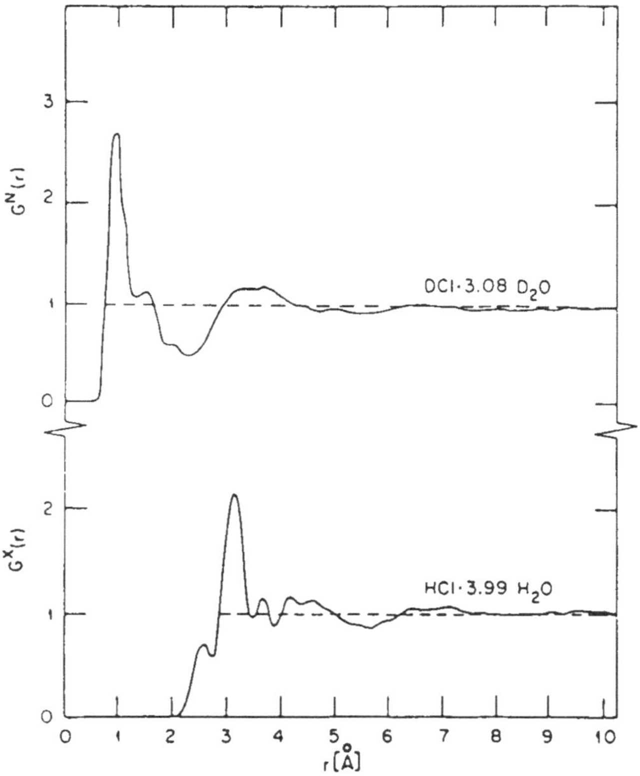

FIGURE 2. Correlation functions for two concentrated hydrochloric acid solutions. $G^X(r)$ and $G^N(r)$ are derived from X-ray and neutron diffraction experiments, respectively. (From Triolo, R. and Narten, A. H., *J. Chem. Phys.*, 63, 3624, 1975. With permission.)

where i stands for M or X and, correspondingly, the coefficients A, B, C, and D will assume different expressions. These coefficients can be derived by algebrical manipulation on the $F_T(s)$ functions reported in Equation 3 and, in short, they are defined by species concentrations and by the scattering factors f_M and $f_{M'}$ (or f_X and $f_{X'}$) of the two isotopic states of M (or X) used in the experiment. The $\Delta F_i(s)$ functions defined in Equation 4 represent the so-called **first order difference scattering functions**.

By Fourier transforming $\Delta F_i(s)$, a radial function of the form,

$$\overline{G}_i(r) = Ag_{i0}(r) + Bg_{iD}(r) + Cg_{iM}(r) + Dg_{iX}(r) + E \quad (5)$$

is obtained, where $E = -(A + B + C + D)$. Since usually $A,B \gg C,D$, $\overline{G}_i(r)$ can be used to investigate the coordination of water around a given ion, without the overlapping of the other interactions and, namely, of the dominant solvent-solvent interactions.

Similarly, it is possible to show that a three-pattern experiment would allow to obtain S_{MM}, S_{XX} and g_{MM}, g_{XX}, and with a four-pattern experiment, also S_{MX} and g_{MX} could be extracted. The ion-ion functions are called **second order difference scattering functions**; they are important for the theories of solutions and of interionic interactions.

The basic principles, here briefly recalled, were known for a long time. However, it was only in the seventies that the powerfulness of isotopic substitution in the field of electrolyte solutions had begun to be exploited, due to the pioneering, precious work of Enderby and co-workers, to which the reader can refer for detailed description of the method.[1] This delay was probably determined by the serious experimental difficulties that must be faced in order to obtain reliable difference functions. In fact, the differences to be evaluated are very small, as shown in Figure 3 which reports four normalized intensity functions taken from four $NdCl_3$ samples, differing in the isotopic state of the Nd ions.[6] The four curves are almost indistinguishable by eye. Only superimposition of pairs of curves on the same scale evidentiates the differences we are looking for (Figure 4).[7]

This is the main reason why it is difficult to obtain reliable first order differences and why it is almost hopeless to obtain second order differences that can be Fourier transformed. In the following, therefore, we will limit our discussion to first order differences by which ionic hydration — the prevailing subject of this book — can be determined.

The experimental accuracy required can be stressed according to Soper et al.:[8] "Great care must be taken with the preparation of the samples as the technique relies on the exact cancellation of the water terms. Accordingly, the difference in light water content of each pair of samples was held to within ± 0.1% by preparative techniques involving the use of an infrared spectrometer. Activity measurements ensured that the concentration scale was reproducible to within ± 1%. Finally, mass spectrographic techniques were employed to determine to within 0.1% the abundances of the various isotopes characteristic of the M and X used in the preparation of the samples."

As far as the procedures of data treatment are concerned, a great advantage is inherent to the difference method, since the Placzek effects are negligible on the ΔF curves. This point is thoroughly discussed by Enderby and co-workers;[1,8] here we limit to observe that the previous statement is experimentally demonstrated by the absence of the characteristic Placzek drop in the ΔF functions. An example referring to a $NiCl_2$ solution is given in Figure 5. The problems posed by inelastic effects are therefore overcome by the isotopic substitution method, and the $\overline{G}_i(r)$ functions can be correctly obtained.

In a practical experiment, conditions may be more or less favorable, depending on the A, B, C, D coefficients values, i.e., depending on the concentrations and on the differences of the neutron scattering factors of the isotopes used. Figure 6 shows an example less ideal than the one just reported.

In the following, results obtained by NDIS on ionic hydration are given and compared with XRD information on the same systems. For the sake of convenience, the results will be reported in the same order of Chapter 3.

C. NDIS: Results and Comparison with XRD Studies

1. Li^+

Li^+ coordination has been determined using ^6Li and ^7Li isotopes.[9] Experiments are complicated by the high absorption of ^6Li; however, at a qualitative level, a well-defined Li^+

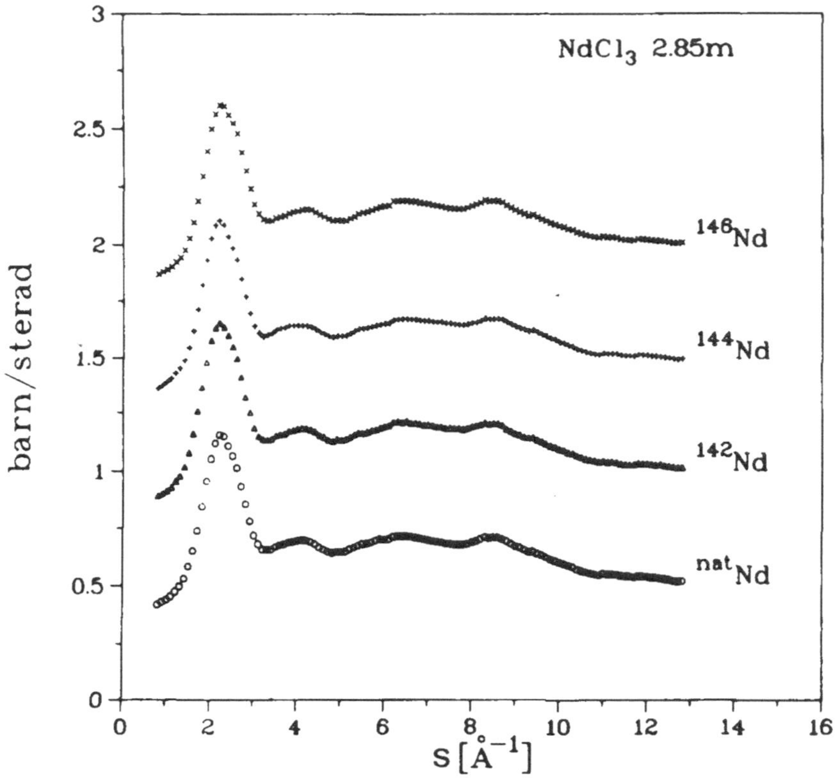

FIGURE 3. Neutron scattering cross sections of 2.85-*m* $NdCl_3$ solutions, identical in every respect, except for the isotopic state of the Nd ions. (From Narten, A. H. and Hahn, R. L., *J. Chem. Phys.*, 87, 3193, 1983. With permission.)

coordination shell has been revealed, as shown in Figure 7. The $\overline{G}_{Li}(r)$ function shows two sharp and resolved peaks at 1.95 and 2.50 Å, where Li – O and Li – D contacts are expected. At about 3.2 Å, the $\overline{G}_{Li}(r)$ tends towards the E value (see Equation 5); this proves that all the pair correlation functions are zero at this r value and, hence, that the lithium ion does have a sharply defined first coordination shell. The first neighbor water molecules are obtained by integrating $4\pi r^2\overline{G}_{Li}(r)$ between 1.7 and 3.2 Å; further, assuming a given D_2O geometry (e.g., the one of the gas phase D_2O molecule, r_{O-D} = 1.0 Å, D – Ô – D = 105.5°), the tilt angle between the Li – O axis and the D_2O plane (see Figure 7) can be evaluated. Table 2 gives the results obtained on two LiCl solutions.[9]

According to these results, the hydration number of Li^+ is strongly concentration dependent, the tilt angle varies a little, while Li – O distance does not vary at all. The change of coordination at different concentrations has been confirmed in two recent investigations.[10-11] Both studies suggest that when the H_2O/LiCl ratio is less than four, a fraction of Li^+ first shell positions is occupied by Cl^- ions. The neutron results are somewhat different from the XRD ones. Narten et al.[4] suggested that the Li^+ ion is four-coordinated in all the concentration ranges explored, while some difference on the Li^+ – O distance was proposed.

2. K^+

The $\overline{G}_K(r)$ derived from two KCl solutions is markedly different from the one of Li^+.[12] Figure 8 shows that K – O and K – D peaks are smeared out and are replaced by a broad feature with a maximum near 3.1 Å; furthermore, this broad peak is not well-separated from

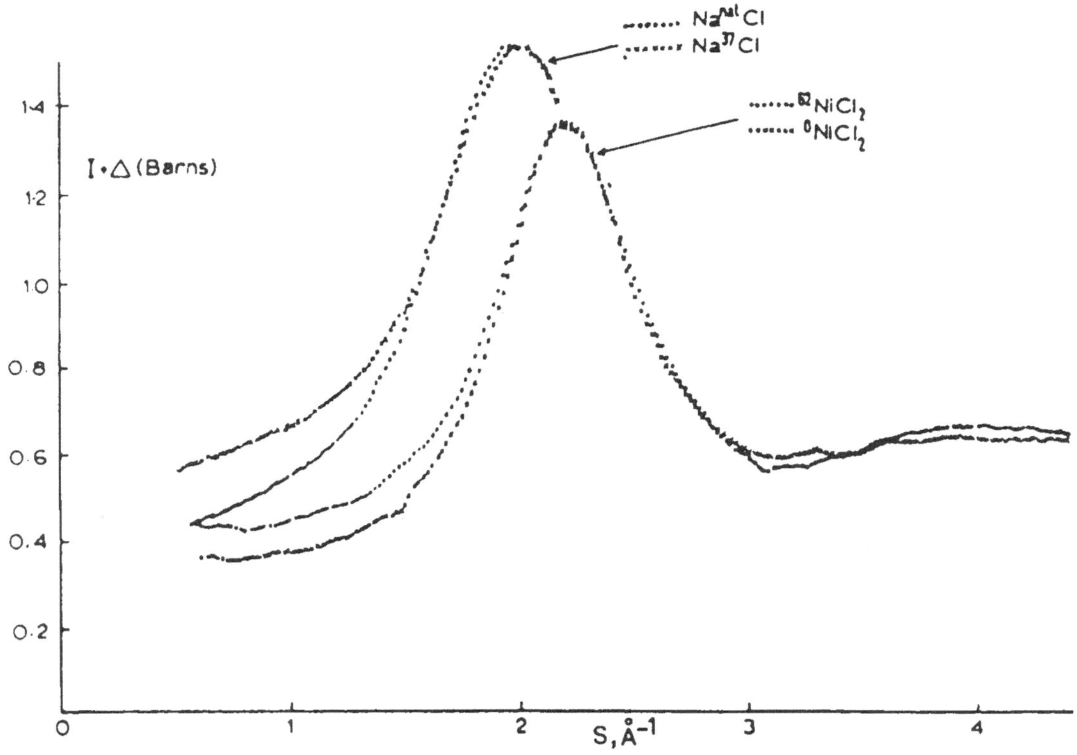

FIGURE 4. Neutron scattering cross sections for pairs of solutions of NaCl and $NiCl_2$, differing in the isotopic state of Cl and Ni, respectively. (From Enderby, J. E., Howells, W. S., and Howe, R. A., *Chem. Phys. Lett.*, 21, 109, 1973. With permission.)

the rest of the curve. All this indicates, in good agreement with the XRD results, that the K^+ ion is weakly coordinated in aqueous solution, and that it is difficult, and probably physically meaningless, to define a coordination number for this cation.

3. Ca^{2+}

Three pairs of $CaCl_2$ solutions have been examied by Enderby and co-workers.[13,14] The $\overline{G}_{Ca}(r)$ functions, reported in Figure 9, show rather well-defined Ca − O and Ca − D peaks, mainly in the most concentrated solutions. Coordination parameters were then determined and are given in Table 3.

Like for Li^+, the Ca − O distance does not vary significantly with concentration, while the coordination number is strongly concentration dependent. The widths of the first peaks decrease with increase in concentration, and this behavior is ascribed to the weak hydration tendency of Ca^{2+} ions in solution. This observation agrees with the difficulties of the XRD to define exactly the Ca^{2+} coordination, mainly in the most dilute solutions. The Ca − O distances from XRD are in very good agreement with the NDIS results, but no evidence of the coordination variation with concentration has been reported in XRD.

4. Rare Earth Ions

Nd(III) and Dy(III) coordination in chloride solutions has been determined by NDIS.[6,15,16] Four different rare earth isotopic states were used in four different solutions (Figure 3). This allowed to calculate difference radial curves without any M-M contribution. An example is given in Figure 10 for the Nd(III) ion.

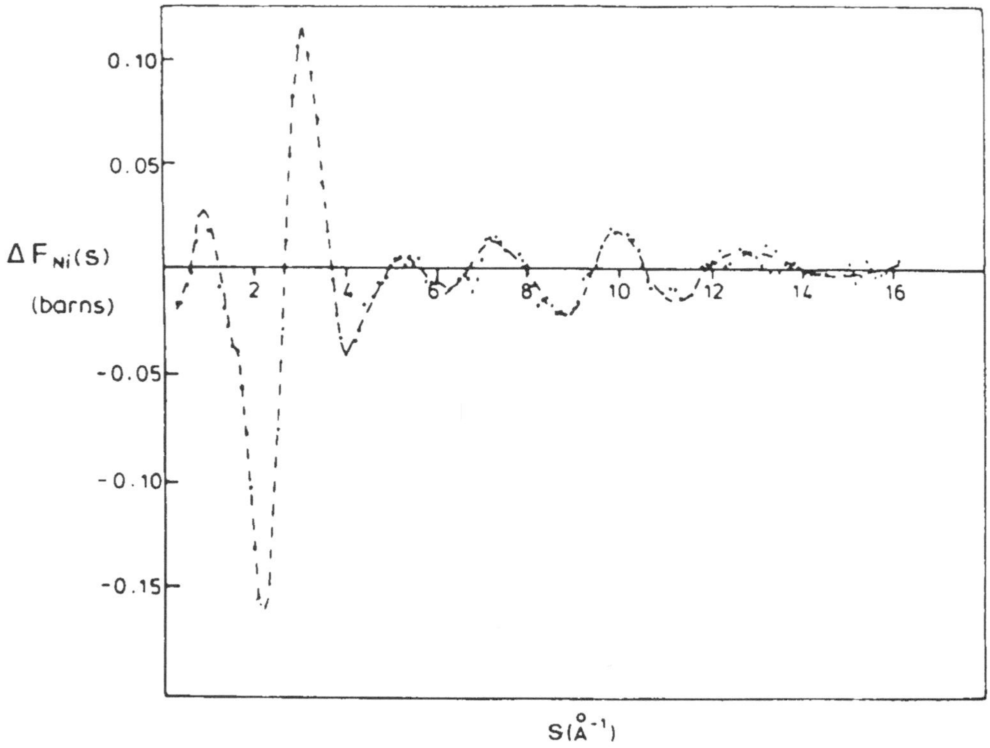

FIGURE 5. $\Delta F_{Ni}(s)$ for a 4.35-*m* solution of $NiCl_2$ in D_2O. (From Neilson, G. W. and Enderby, J. E., *Proc. R. Soc. London,* A390, 353, 1983. With permission.)

It is clear that the coordination shells are sharply defined and the coordination numbers can be easily obtained by area evaluation beneath the M – O and M – D peaks. The values obtained have been confirmed by fitting the $\Delta F_M(s)$ functions in the high s region. The structural parameters obtained in this way are given in Table 4.

For the Dy(III) ion, the possibility of inner-sphere chloride ions was tested in the fitting procedure and the results ruled out that Cl^- ions contribute significantly to the first hydration sphere of this ion. An agreement within 0.03 Å on distances and 0.5 units on coordination numbers is found with XRD results. The agreement is very good and, further, it must be emphasized that the decrease of one neighbor in going from Nd(III) to Dy(III), evidenced by Habenschuss and Spedding[17] by XRD has been fully confirmed by NDIS.

5. *Ni^{2+}*

Ni^{2+} ion is the species more widely investigated by NDIS. First order differences have been obtained for six $NiCl_2$ solutions and for one $Ni(ClO_4)_2$ solution.[8,18,19] Two examples of $\overline{G}_{Ni}(r)$ functions are given in Figures 11 and 12. The two peaks located at 2.07 and 2.67 Å were identified as Ni – O and Ni – D correlations on the grounds that the ratio of the areas beneath them is almost exactly 1:2.

Recently, for a 4.35-*m* $NiCl_2$ solution, the collection of seven distinct scattering patterns allowed the extraction not only of the pair structure functions between $Ni^{2+} - D_2O$ and $Cl^- - D_2O$, but also of the interionic ($Cl^- - Cl^-$, $Ni^{2+} - Ni^{2+}$, $Ni^{2+} - Cl^-$) pair functions.[20] For the first time the quality of experimental data was suitable for the calculation of the radial interionic functions so that really complete structural data could be obtained for this system.

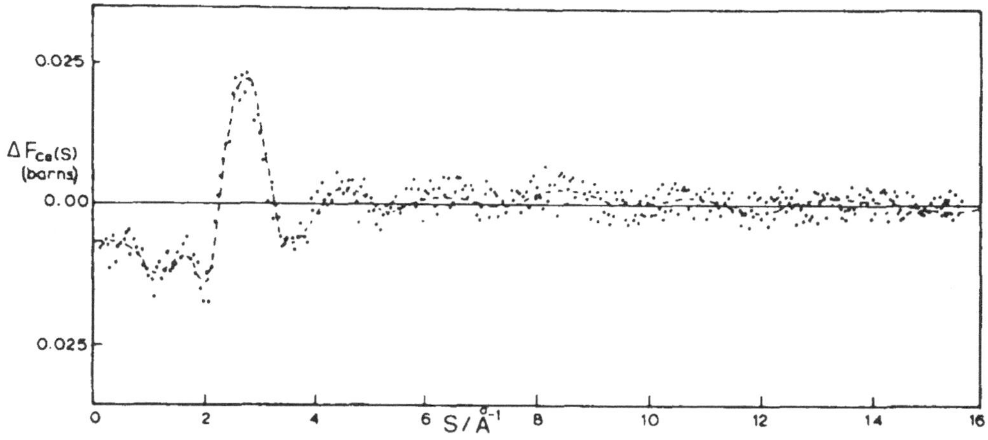

FIGURE 6. $\Delta F_{Ca}(s)$ for a 4.49-*m* solution of $CaCl_2$ in D_2O. (From Cummings, S., Enderby, J. E., and Howe, R. A., *J. Phys. C*, 13, 1, 1980. With permission.)

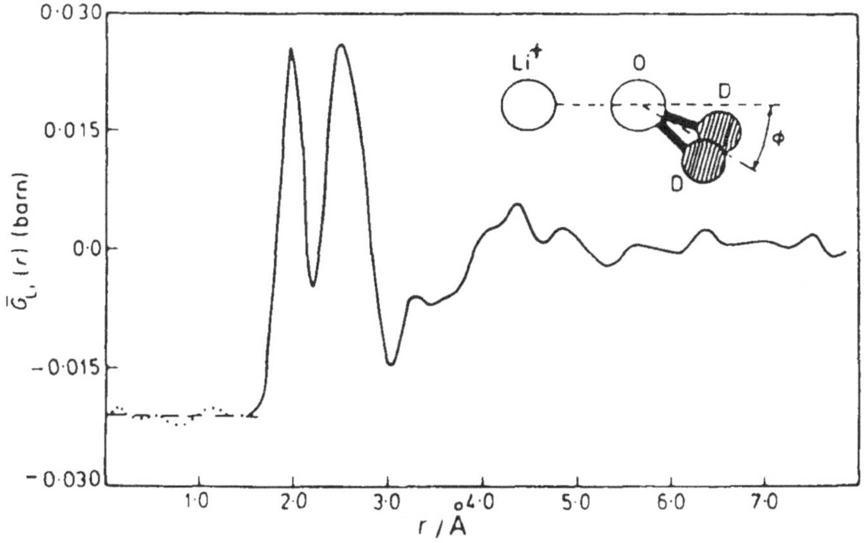

FIGURE 7. $\overline{G}_{Li}(r)$ for a 3.57-*m* solution of LiCl in D_2O. Top right: the Li-D_2O conformation consistent with the $\overline{G}_{Li}(r)$. (From Newsome, J. R., Neilson, G. W., and Enderby, J. E., *J. Phys. C*, 13, L923, 1980. With permission.)

Table 2
HYDRATION OF Li^+

Molality	Lithium-oxygen distance (Å)	Lithium-deuterium distance (Å)	φ	Hydration no.
9.95	1.95 ± 0.02	2.50 ± 0.02	52 ± 5	3.3 ± 0.5
3.57	1.95 ± 0.02	2.55 ± 0.02	40 ± 5	5.5 ± 0.3

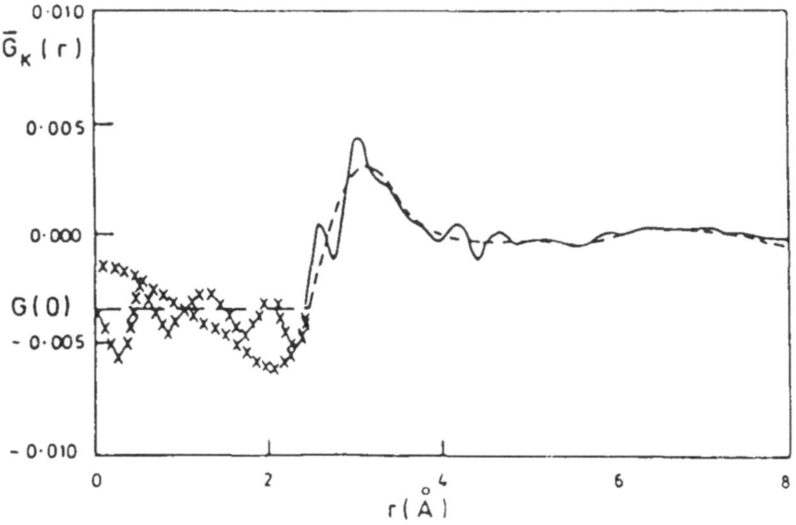

FIGURE 8. $\overline{G}_K(r)$ for a 4-*m* KCl solution in D_2O. The full curve is the result of a point by point transformation of $\Delta F_K(s)$, with the points at low r edited after direct transformation (see crosses between $0 < r < 2.4$ Å) in order that $\overline{G}_K(r) = 0$ for r less than some cut off distance. The dashed curve is the result of smoothing $\Delta F_K(s)$ and truncating at $s = 8$ Å^{-1}. (From Neilson, G. W. and Skipper, N., *Chem. Phys. Lett.*, 114, 35, 1985. With permission.)

The values proposed for Ni^{2+} hydration parameters are given in Table 5. Distances and coordination numbers almost do not change with concentration, neither are they influenced by the different counterion. As far as the Θ angles are concerned, whatever is the geometry assumed for the D_2O molecules, there is a significant angle of tilt which gradually disappears at low concentrations. The stability of the coordination shell, as well the values of the parameters $r_{Ni^{2+}-O}$ and $N_{Ni^{2+}-O}$, are in excellent agreement with XRD determinations. However, the suggestion by the most recent XRD studies of a small amount of Cl^- ions in the first coordination shell of Ni^{2+} is not present in NDIS results.

6. Cu^{2+}

The $\overline{G}_{Cu}(r)$ function from a 4.32-*m* $CuCl_2$ solution,[21] shown in Figure 13, is more complicated than the ones examined so far. In fact, on one hand, the occurrence of Cl^- ions in the first coordination shell of Cu^{2+} is very probable, on the other hand, the coordination polyhedra of Cu^{2+} usually undergo the so-called Jahn-Teller distortions.

The first peak at 2.05 Å has been ascribed to $Cu^{2+}-O$ distances and its magnitude corresponds to about 2.3 contacts. After the subtraction from $\overline{G}_{Cu}(r)$ of the contribution of 2.3 D_2O molecules, a residual curve was obtained, which could be interpreted in different ways. The only model, which was considered consistent with both NDIS and the best XRD data,[22] is shown in Figure 13.

The parameters describing the various interactions occurring in this model are given in Table 6. They differ in some important points from the analogous parameters from XRD; for instance, the $Cu^{2+}-Cl^-$ distance is significantly longer and even the frequency factor $N_{Cu^{2+}-Cl^-}$ is greater than in XRD study.

A variant of the isotopic substitutions can be considered the isomorphous substitution method. Here, the ND data are collected for pairs of solutions of MX_n (or $M[XO_m]_n$) salts, in which, at a given concentration, the cation is changed. If the cations are isostructural (that is, they have equal or very similar $g_{Mi}[r]$), a difference procedure leads again to $\overline{G}_M(r)$ functions, whose meaning is similar to the $\overline{G}(r)$ functions defined by Equation 5. The same method was used by Bol et al.[23] in XRD studies.

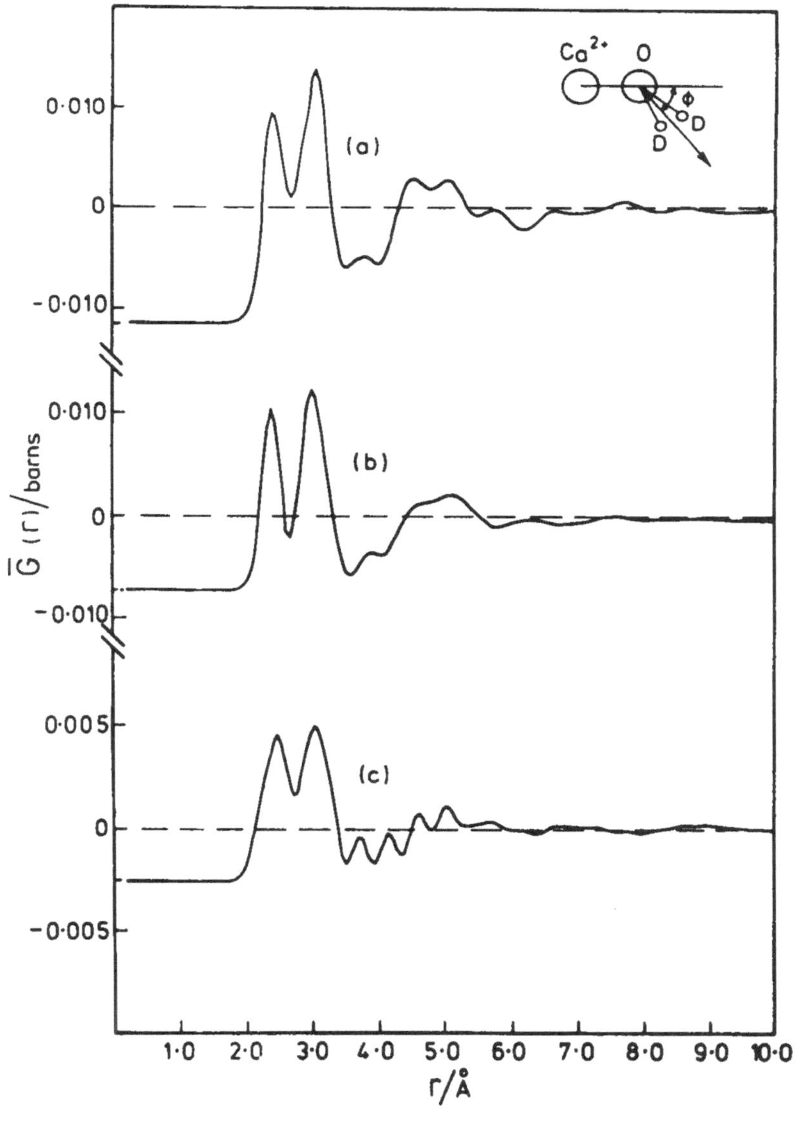

FIGURE 9. The distribution $\overline{G}_{Ca}(r)$ for three $CaCl_2$ solutions in D_2O (a) 4.5 *m* (b) 2.8 *m* (c) 1 *m*. The ion-water geometry is shown at the top right. (From Hewish, N. A., Neilson, G. W. and Enderby, J. E., *Nature, (London)*, 297, 138, 1982. With permission.)

Table 3
Ca^{2+}-WATER COORDINATION

m	r_{CaO}(Å)	r_{CaD}(Å)	ϕ	N_{CaO}
1	2.46 ± 0.03	3.07 ± 0.03	38° ± 9	10.0 ± 0.6
2.8	2.39 ± 0.02	3.02 ± 0.03	34° ± 9	7.2 ± 0.2
4.5	2.41 ± 0.03	3.04 ± 0.03	34° ± 9	6.4 ± 0.3

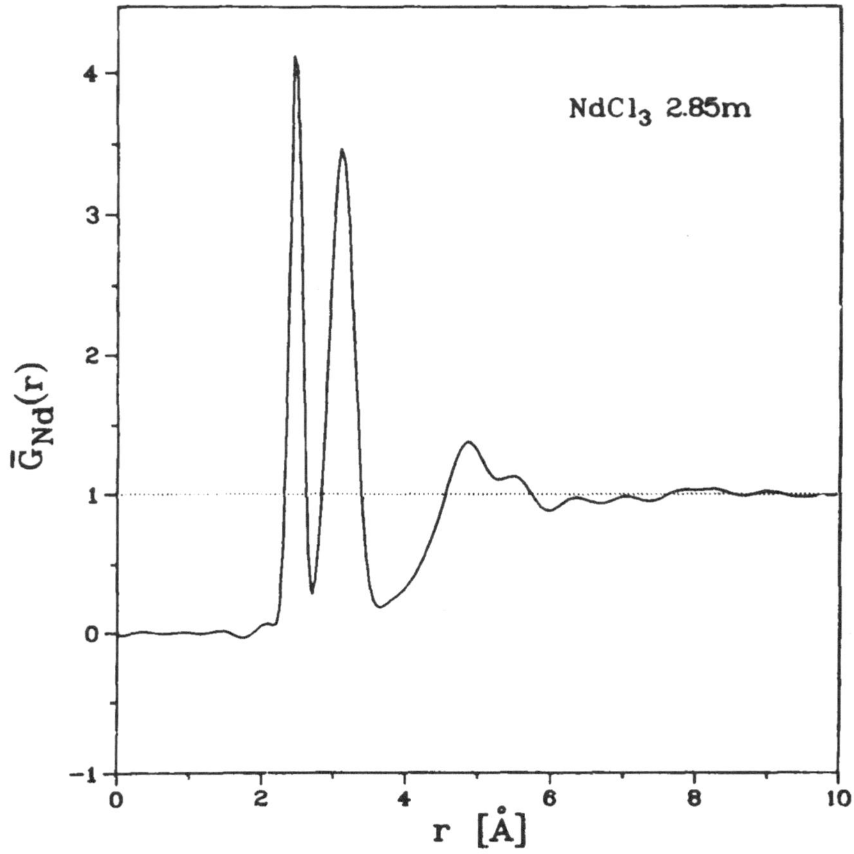

FIGURE 10. Radial distribution of water molecules around the Nd^{3+} ion in a 2.85-*m* solution of $NdCl_3$ in D_2O. (From Narten, A. H. and Hahn, R. L., *Science*, 217, 1249, 1982. With permission.)

Table 4
RARE EARTH IONS COORDINATION

Salt	*m*	r_{Me-O}	r_{Me-D}	ϕ	N_{Me-O}
$NdCl_3$	2.85	2.48	3.13	24	8.4
$DyCl_3$	2.38	2.37	3.04	17	7.4

By this way, the environment of Cu^{2+} in a $Cu(ClO_4)_2$ solution has been investigated, and Ni^{2+} was adopted as an isomorphous cation and its isotopic state was fixed so as to have $f_{Ni^{2+}} = 0$.[24] From the likeness of the $\bar{G}_{Cu}(r)$ thus obtained and the $\bar{G}_{Ni}(r)$ from NDIS, the authors infer that Cu^{2+} and Ni^{2+} should have the same coordination number of 6, and add that "there is no reason to suggest that the water is in two sites." This statement is in total disagreement with XRD results.

7. Ag^+

The first order difference method has been applied to a pair of $AgClO_4$ solutions.[25] In the $\bar{G}_{Ag}(r)$, the peaks Ag − O (at = 2.40 Å) and Ag − D (at = 3.0 Å) give distinct maxima, but are not completely resolved. The peaks were therefore analyzed by the curve fitting

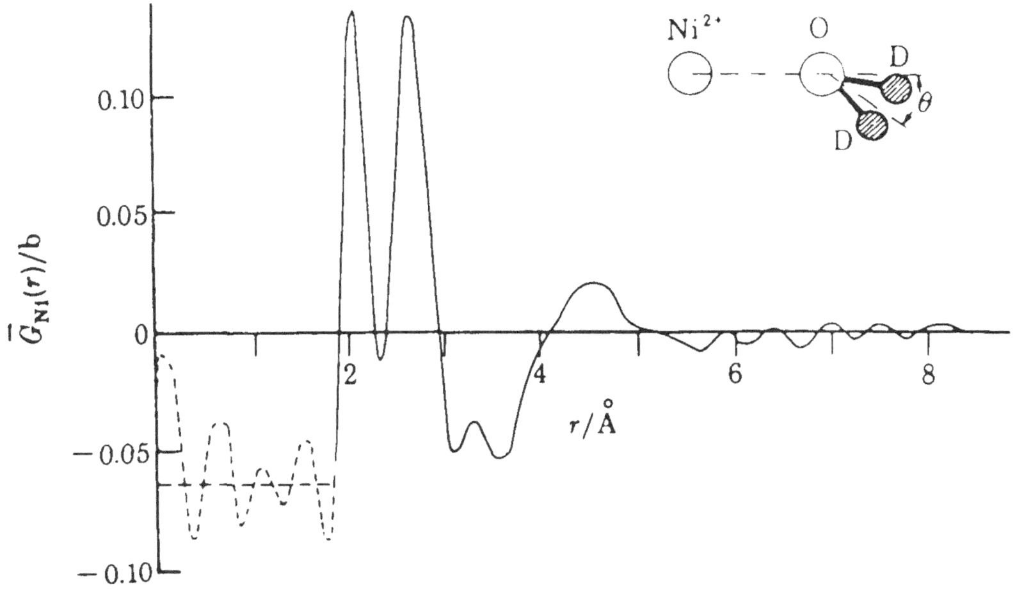

FIGURE 11. $\bar{G}_{Ni}(r)$ for a 4.35-*m* solution of $NiCl_2$ in D_2O. (From Neilson, G. W. and Enderby, J. E., *Proc. R. Soc. London*, A390, 353, 1983. With permission.)

method, since a direct integration of the overlapping peaks would be ambiguous. The average hydration number was estimated to be 4.1 and the distances Ag − O and Ag − D were found to be 2.41 and 2.97 Å. The consistency of these results with those from XRD is remarkable.

8. Cl^-

By preparing pairs of $M^{35}Cl_n$ and $M^{37}Cl_n$ solutions, first order differences have been obtained by Enderby and co-workers for a great number of chloride solutions, for which the $\bar{G}_{Cl}(r)$ functions have been deduced.[8,13,20,26] Two typical examples are given in Figures 14 and 15. In these curves, two well-defined peaks and a third one, less or not at all resolved, are constantly present, centered at 2.25, 3.25, and 3.50 Å, respectively. This is explained by the $Cl^- - D_2O$ conformation sketched in Figure 16.

The parameters describing the Cl^- coordination shell on the various cases examined by Enderby and co-workers are given in Table 7. The general form of $\bar{G}_{Cl}(r)$ is remarkably similar in the various cases; the hydration parameters confirm that there is an apparent lack of sensitivity of anionic hydration to counter-ion and concentration. Only a last result from a very concentrated (14.9 *m*) LiCl solution indicated $N_{Cl^- - H_2O} = 4.4$,[11] but for this solution also, $Li^+ - Cl^-$ contacts were suggested.

The existence of a defined hydration shell for Cl^- ions, having constant characteristics in different systems, has been suggested also by XRD studies. Quantitatively, the agreement among NDIS and XRD results is very good as regards the coordination number, but less good as far as $Cl^- - O$ distances are concerned; in fact, neutron studies give distances systematically longer than those coming from XRD.

Recently, Biggin et al.[27] have proposed a coordination number near 4 for Cl^- in 2.85 *m* $NdCl_3$ solution. The coordination of Cl^- ions in $NdCl_3$ solutions has not been determined by XRD; however, in chloride solutions of other trivalent ions, coordination numbers of Cl^- different from 6 were never suggested. Therefore, the result by Biggin et al. looks a little surprising and deserves confirmation.

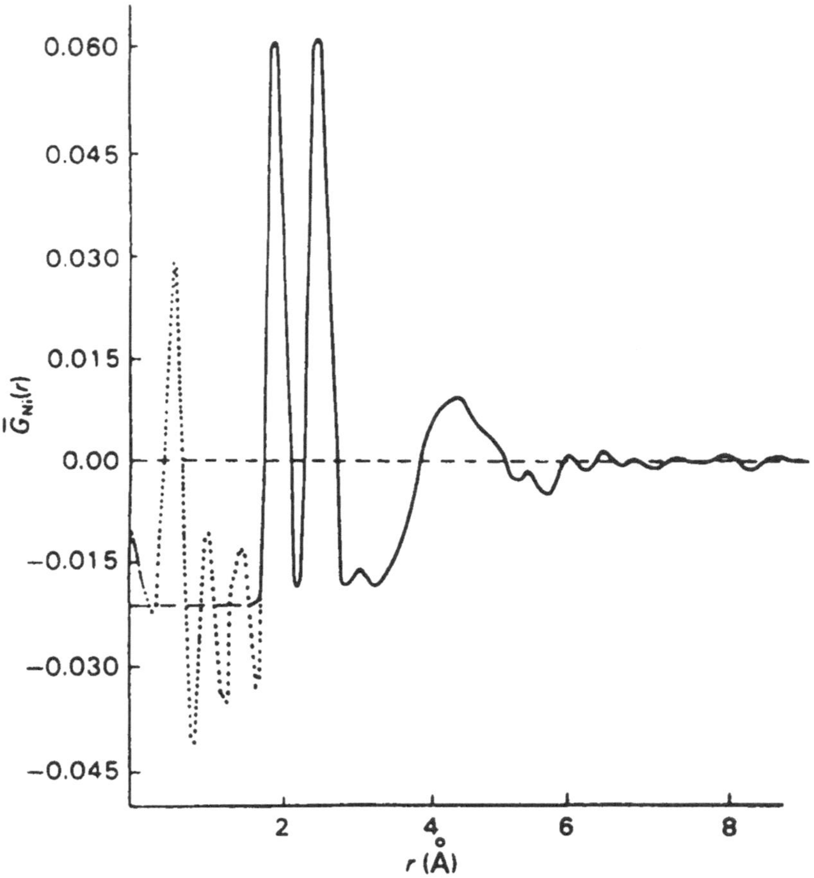

FIGURE 12. $\bar{G}_{Ni}$ (r) for a 1.46-*m* solution of $NiCl_2$ in D_2O. (From Neilson, G. W. and Enderby, J. E., *J. Phys. C*, 11, L625, 1978. With permission.)

9. NO_3^-

The study of a very concentrated $NaNO_3$ solution is the first example of investigation about oxyanions coordination carried out by NDIS.[28] The radial curve coming from first order differences using N isotopes (Figure 17) points out the discrete distances between N and the surrounding atoms (in addition to the O in the NO_3 groups) in a more direct way than in XRD results. However, also in this case, the various peaks are consistent with more than one coordination model, and the final model of Figure 18 has been adopted, since it is the only one consistent with XRD data as well.

D. Final Comments

The results reported in the previous sections cover the literature up to 1985 and clearly demonstrate the high quality level reached by NDIS experiments in the ionic hydration field. We have seen that generally the first order differences and their transforms are very **clean** and allow a reliable description of the structural arrangement of the ions. The structural parameters are often directly derived from the difference curves without any model assumption; only for the determination of tilt angle assumptions about the D_2O molecule geometry are required. The great scientific credit acquired by the NDIS technique is therefore largely justified.

In this respect, two main questions still remain open. To date, is the information obtainable

Table 5
HYDRATION OF Ni^{2+}

Ion	Solute	Molality	Ion-oxygen distance (Å)	Ion-deuterium distance (Å)	θ[a] ($\vec{OD}$ = 0.94 Å)	θ[a] ($\vec{OD}$ = 1.00 Å)	Hydration no.
Ni^{2+}	$NiCl_2$	4.41	2.07 ± 0.02	2.67 ± 0.02	34° ± 8°	42° ± 8°	5.8 ± 0.2
		3.05	2.07 ± 0.02	2.67 ± 0.02	34° ± 8°	42° ± 8°	5.8 ± 0.2
		1.46	2.07 ± 0.02	2.67 ± 0.02	34° ± 8°	42° ± 8°	5.8 ± 0.3
		0.85	2.09 ± 0.02	2.76 ± 0.02	10° ± 10°	27° ± 10°	6.6 ± 0.5
		0.46	2.10 ± 0.02	2.80 ± 0.02	0° ± 10°	17° ± 10°	6.8 ± 0.8
		0.086	2.07 ± 0.03	2.80 ± 0.03	0° ± 10°	0° ± 20°	6.8 ± 0.8
	$Ni(ClO_4)_2$	3.80	2.07 ± 0.02	2.67 ± 0.02	—	42° ± 8°	5.8 ± 0.2

[a] θ is the angle between the plane of the water molecule and the MO axis calculated for two values of $\vec{OD}$. $\widehat{DOD}$ is assumed in both cases to be 105.5°.

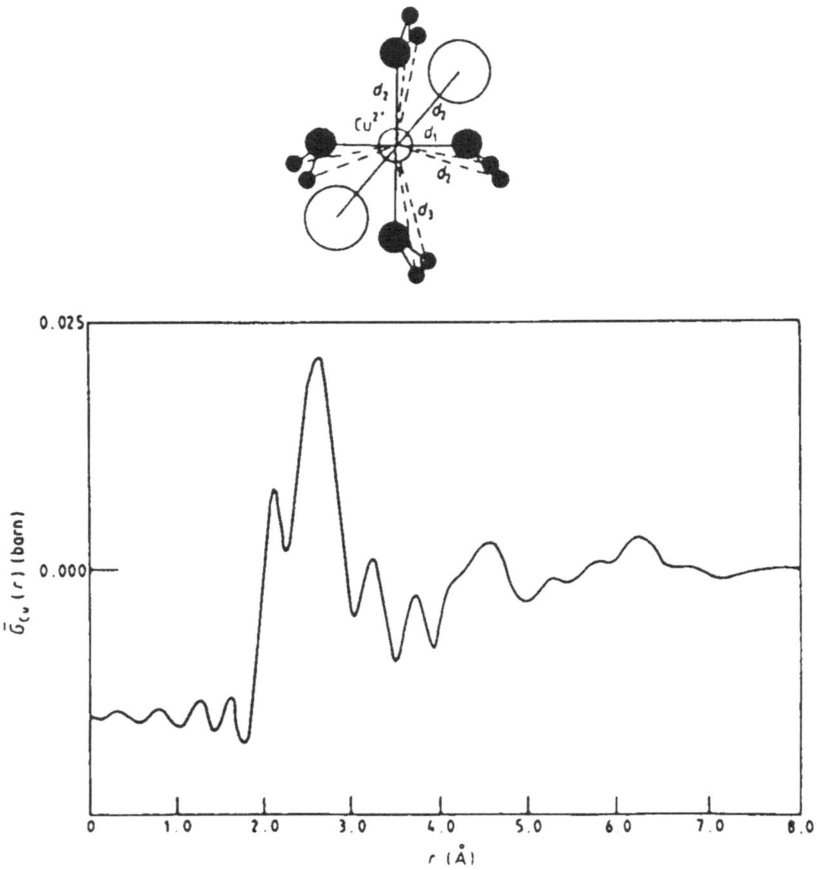

FIGURE 13. $\bar{G}_{Cu}$ (r) for a 4.32-*m*- solution of $CuCl_2$. In the top, the model proposed for Cu^{2+} coordination is shown: d_1 = 2.05 Å, d_2 = 2.56 Å, and d_3 = 3.2 Å. (From Neilson, G. W., *J. Phys. C*, 15, L233, 1982. With permission.)

Table 6
Cu^{2+} COORDINATION IN A $CuCl_2$ SOLUTION

Ion	Solution	*m*	$r_{Cu\text{-}O}$	$N_{Cu\text{-}O}$	$r_{Cu\text{-}D}$	$r_{Cu\text{-}Cl}$	$N_{Cu\text{-}Cl}$	$r_{Cu\text{-}O}$	$N_{Cu\text{-}O}$
Cu^{2+}	$CuCl_2$	4.32	2.05	2.3	2.56	2.56	2.8	2.5	2

by NDIS always completely reliable? Should NDIS replace any other structural technique? It is not difficult answering the second question. NDIS is evidently limited by the availability of the needed number of isotopes required in a given experiment and, when available, by the required characteristics of the isotopes (e.g., enough scattering factors difference). Improvement of the instrumentation (like high flux reactors, LINAC, etc.) and increase of the available isotopes (e.g., by using radio nuclides) will surely increase in the future the number of practicable experiments. Nevertheless, it is easy to predict that a large number of atomic species will be excluded anywhere from this kind of structural analysis. Other structural techniques will therefore always be necessary; it is then very important to verify the consistency of their results with those obtained by NDIS. This test is also the key to answer the first question; the comparison between NDIS and XRD results, already included in the previous sections, will therefore be discussed in greater detail in the following.

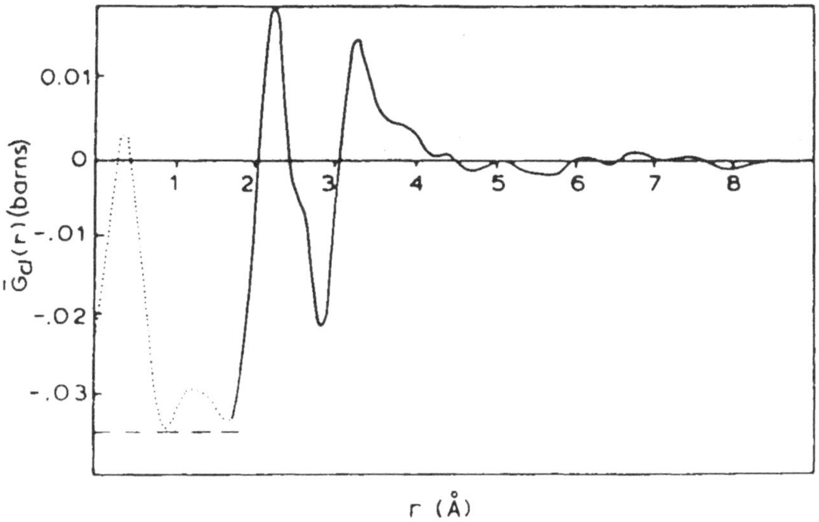

FIGURE 14. $\bar{G}_{Cl}$ (r) for a 5.32-*m* solution of NaCl in D_2O. (From Soper, A. K., Neilson, G. W., Enderby, J. E., and Howe, R. A., *J. Phys. C*, 10, 1793, 1977. With permission.)

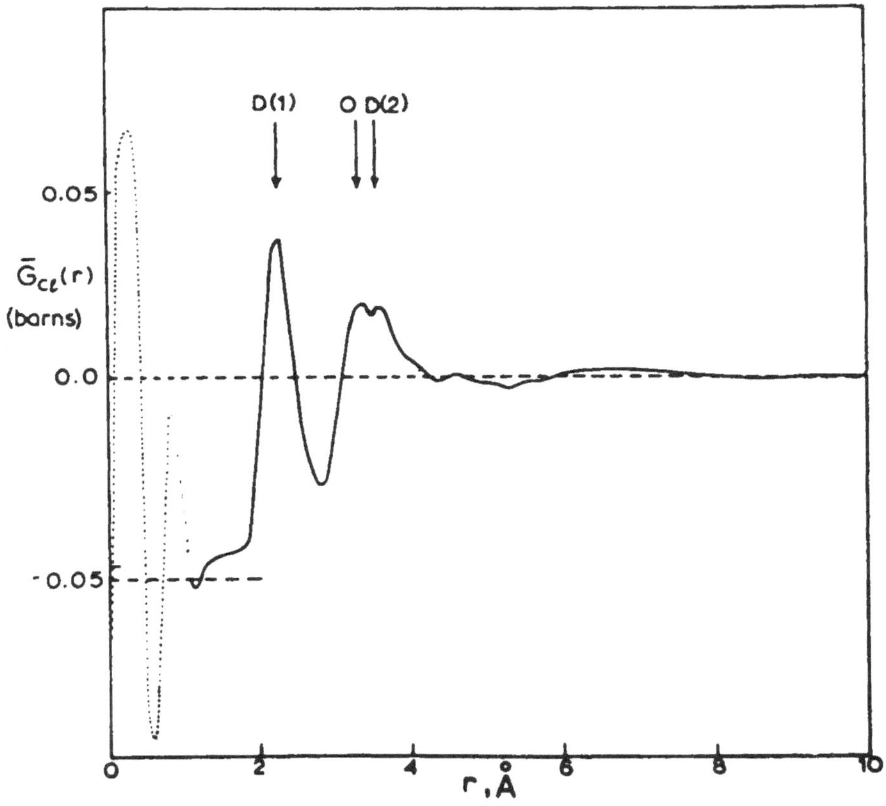

FIGURE 15. $\bar{G}_{Cl}$ (r) for a 4.49-*m* solution of $CaCl_2$ in D_2O. (From Cummings, S., Enderby, J. E., and Howe, R. A., *J. Phys. C*, 13, 1, 1980. With permission.)

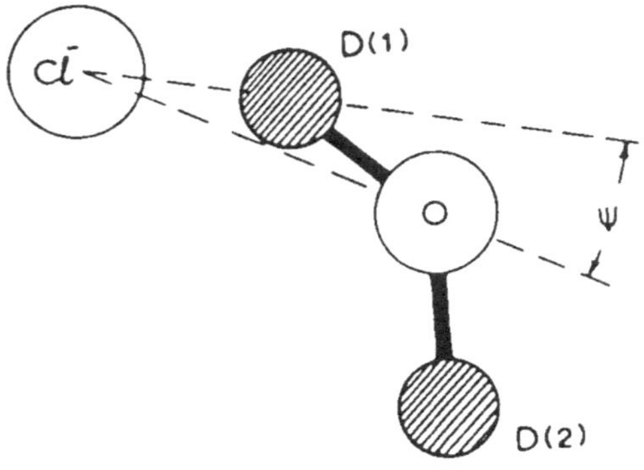

FIGURE 16. The anion-water conformation consistent with the $\overline{G}_{Cl}$ (r)s shown in Figures 14 and 15.

Table 7
Cl-WATER COORDINATION

Solute	*m*	$r_{Cl\text{-}D_1}$	$r_{Cl\text{-}O}$	r_{Cl-D_2}	ψ	$N_{Cl\text{-}O}$
LiCl	3.57	2.25 ± 0.02	3.34 ± 0.04	—	0	5.9 ± 0.2
LiCl	9.95	2.22 ± 0.02	3.29 ± 0.04	3.50—3.68	0	5.3 ± 0.2
NaCl	5.32	2.26 ± 0.03	3.20 ± 0.05	—	0—10	5.5 ± 0.4
RbCl	4.36	2.26 ± 0.03	3.20 ± 0.05	—	0—10	5.8 ± 0.3
$CaCl_2$	4.49	2.25 ± 0.04	3.25 ± 0.04	3.55—3.65	0—6	5.8 ± 0.2
$NiCl_2$	4.35	2.29 ± 0.02	3.20 ± 0.04	3.40—3.50	6—11	5.7 ± 0.2
$NiCl_2$	3.00	2.23 ± 0.03	3.25 ± 0.05	—	0—6	5.5 ± 0.4
$BaCl_2$	1.10	2.24 ± 0.04	3.26 ± 0.05	—	0—6	6.2 ± 0.4

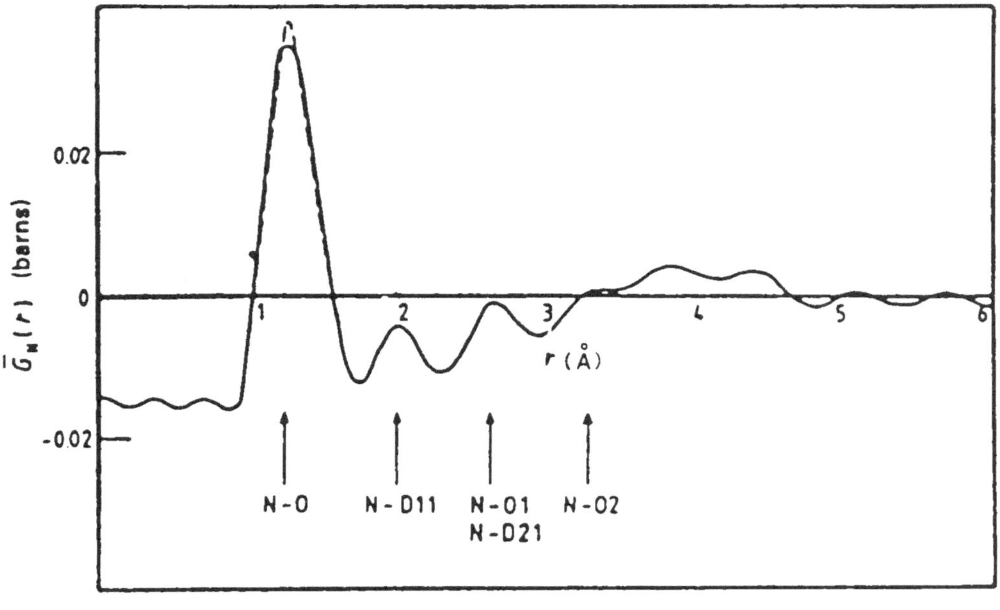

FIGURE 17. $\overline{G}_N$ (r) for a 7.8-*m* solution of $NaNO_3$ in D_2O. The arrows indicate the most probable positions of neighboring oxygen and deuteron atoms with respect to nitrogen. The broken curve shows the effect of extending the data to 50 $Å^{-1}$. (From Neilson, G. W. and Enderby, J. E., *J. Phys. C*, 15, 2347, 1982. With permission.)

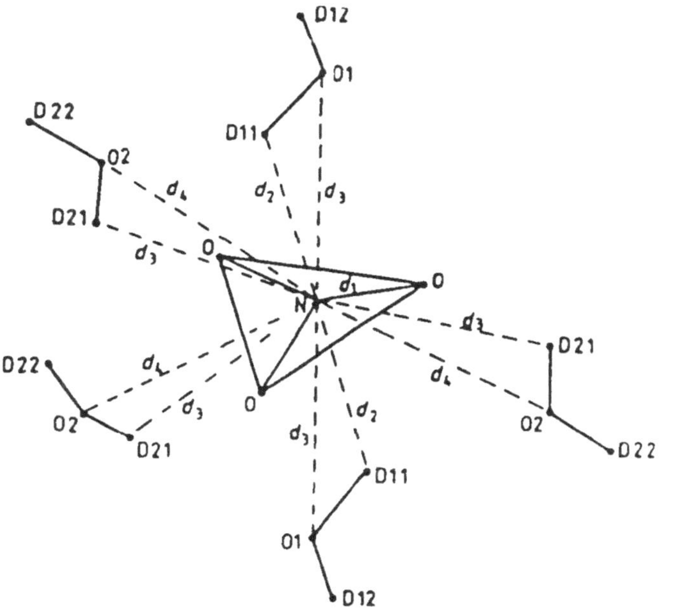

FIGURE 18. Planar projection of the local coordination of the nitrate ion based on the $\overline{G}_N$ (r) shown in Figure 17. In Å, d_1 = 1.23, d_2 = 2.05, d_3 = 2.65 and d_4 = 3.40. (From Neilson, G. W. and Enderby, J. E., *J. Phys. C*, 15, 2347, 1982. With permission.)

The agreement between XRD and NDIS is very often more than satisfactory and two (quite different) examples for all will be recalled. In the case of Ni(II) salt solutions, the distances and coordination numbers for the $Ni^{2+}-O$ interaction are in excellent agreement, as expected since the $Ni^{2+}-O$ peaks in both NDIS and XRD distribution curves are well-resolved. In the case of chloride solutions, the $Cl^{-}-H_2O$ peaks in the $G^X(r)$ were usually poorly resolved and the indication of defined (sixfold) coordination mainly came from fitting procedures. However, also in this case, the indications from XRD have been substantially confirmed by NDIS investigations.

Nevertheless, minor and not minor disagreements exist, and the discrepancies have been systematically attributed by NDIS workers to the "inherently poor discrimination of total X-ray patterns from solutions". This statement is, in our opinion, somewhat acritic and superficial and can lead to underestimation of the XRD results. We believe that all the matter regarding the discrepancies in discussion must be carefully examined without any prejudice against or in favor of any technique with respect to the other. The observations, given in the following, intend to be a contribution in this direction.

Difficulties of resolution are present also in NDIS work, although to a minor extent as compared to XRD studies. In the case of chloride solutions, the first order $\overline{G}_i(r)$ are a linear combination of four terms, while six terms are present in the total $G^X(r)$. When ion-ion pair correlations can be neglected, the terms in $\overline{G}_i(r)$ become two and, similarly, three terms are contained in $G^X(r)$. It is therefore not surprising that sometimes different interatomic distances are strictly mixed up even in the partial $\overline{G}^i(r)$. This was clear in Figures 13 and 15, representing real space distribution functions around Cu^{2+} and Cl^-, respectively.

Another point worthy of attention is that the partial $\overline{G}^i(r)$ functions of Equation 5 are evidently a part of the total, poorly discriminating picture offered by the total $G^X(r)$ or $G^N(r)$. Compatibility between partial neutron $\overline{G}_i(r)$ and total $G^X(r)$ functions **must** exist, provided they are all reliable functions. This means that the structural arrangements proposed by

NDIS should never be inconsistent with the total $G^X(r)$ function. If incompatibility occurs, it cannot be explained with the undoubted difficulty of XRD data analysis; indeed, the only conclusion is that at least one of the two experiences is wrong and, in our opinion, it cannot be assumed an *a priori* major reliability of the NDIS data, due to the by far greater difficulties of this technique compared with the relatively simple and well-established XRD.

The reasonableness of this judgment is demonstrated by the fact that in some controversial cases NDIS results are in contrast with the behavior predictable on the basis of the chemical knowledge rather than with XRD results. For instance, the strong variation of Ca^{2+} coordination number with concentration in $CaCl_2$ solutions does not agree with the almost constant $Ca^{2+}-O$ distance values (2.40 to 2.45 Å); in coordination chemistry, it is usually found that metal-ligand distances increase with the coordination number. In the Ca^{2+} case, values of coordination number equal to 9 have been found in crystalline hydrates, such as $CaCl_2 \cdot 6H_2O$ and $Ca(NO_3)_2 \cdot 4H_2O$, where the $Ca^{2+}-O$ average distances lengthen to 2.55 Å. Another example is given by the coordination polyhedron proposed around the Cu^{2+} ion in perchlorate medium. It is hard to believe that the Jahn-Teller distortions, revealed by X-ray analysis in similar solutions, do not exist, as suggested by the neutron studies.

The reported examples are, in our opinion, sufficient to say that, at present, NDIS results cannot be accepted blindly. The high accuracy required in the difference method could be the origin of some nonconvincing results; an inspection of the $\Delta F_i(s)$ functions shows that the data noise is sometimes still too high (see Figure 6). Another point could be still more important. A careful examination of the discrepant cases has convinced us that the divergent conclusions could be explained by a different way of analyzing the experimental data, rather than by the different quality of the experiments. In this connection we recall that in XRD the structural parameters are usually derived by a least square fitting procedure, while in NDIS investigations they are directly evaluated from the radial functions, with the exception of some very recent works.[6,16,25] Then, in NDIS, ion $-$ O distances are fixed at the maxima positions of the partial radial curves and the coordination numbers are given by the peaks area. Overlapping of adjacent peaks can originate some trouble in this procedure; for instance, the determination of the $Cl-O$ distance certainly suffers by the superimposition of the $Cl-O$ and $Cl-D_2$ peaks (Figures 14 and 15). Even worse is the case of $CuCl_2$, because of the complexity of the Cu^{2+} environment; three interatomic distances, $Cu-D$, $Cu-Cl$, and $Cu-O$, were assumed to be coincident with the position of the second maximum in the $\overline{G}_{Cu}(r)$ (2.50 Å) and, at least for the $Cu^{2+}-Cl^-$ pair, the distance proposed is not consistent neither with the values found in crystallographic investigations, nor with the results obtained by XRD in solution.

The fitting methods, although less direct and apparently artful, allow to understand the complete meaning of the small variations in the radial curves. We recall that, in the studies of halo-complexation performed by XRD, often the $M-X$ interactions did not give rise to distinct peaks in the radial curves, but only to minor variations of the distribution functions, which were quantitatively defined by the fitting work. It is possible that in NDIS studies the meaning of some small variations in a series of $\overline{G}_i(r)$ has been disregarded. For instance, some different details of the $\overline{G}_{Ni}(r)$ for the $NiCl_2$ solutions at different concentrations (Figures 11 and 12) could be indication of a small presence of Cl^- in the first Ni^{2+} coordination shell; likewise, the small differences of $\overline{G}_{Cu}(r)$ and $\overline{G}_{Ni}(r)$ in Reference 24 might be diagnostic of some distortion in the coordination polyhedron of Cu^{2+}.

In conclusion, NDIS studies on ionic hydration represent one of the most advanced contributions, but some improvement in experimental data and interpretative procedures is still possible; anyway, integration with other structural techniques is highly recommendable. In this connection, we recall the NDIS investigations on $CuCl_2$ and $NaNO_3$ solutions, where XRD information has been utilized to choose, among various possible interpretations, the one consistent with both NDIS and XRD.

III. EXAFS SPECTROSCOPY

A. X-Ray Absorption Spectra

X-ray absorption spectroscopy (XAS) is nowadays becoming a powerful tool for obtaining chemical and structural information about the near environment of a particular atomic species in a complex system.[29,30] This is mainly due to the recent availability of tunable and high flux X-ray beams from synchrotron radiation sources, which allow X-ray absorption spectra to be measured with high accuracy.[30-32] Moreover, the rapid theoretical progresses in understanding the physical origin of the spectral features observed near and beyond an inner shell absorption edge, are extending the field in which XAS may find attractive applications.[33]

An absorption spectrum is a measure of the X-ray absorption coefficient as a function of the incident photon energy. Typically, it is obtained through a transmission experiment by measuring the intensities I_0 and I of the incident and transmitted beams and determining the absorption coefficient according to:

$$\mu d = \ln(I_0/I) \tag{6}$$

where d is the sample thickness. The absorption spectrum can also be obtained indirectly by measuring the decay products of the photoabsorption events, such as the fluorescent X-rays or the Auger electrons. Some of these indirect detection procedures are particularly appropriate for certain types of samples or for approaching particular problems; for example, the fluorescence method is suitable for dilute species, while nonradiative schemes (Auger or secondary electrons) probe the structure of surfaces.

Since X-ray photons are absorbed through the excitation of inner shell electrons, a sharp step is produced in the absorption coefficient whenever the X-ray energy crosses an inner shell ionization threshold, and fine structures appear in connection with electronic transitions from an atomic core level to unoccupied final states. In general, the binding energies of inner shell electrons, either in the same atom or in different atoms, are well-separated one from another; therefore, an X-ray absorption spectrum is characteristic of a specific core level, in a specific atom, embedded in a specific environment. This accounts for the special property of XAS of probing selectively the local situation around each atomic species in multicomponent materials.

Figure 19 shows a typical X-ray absorption spectrum with the conventional and somewhat arbitrary partition between XANES (X-ray absorption near edge structure) and EXAFS (extended X-ray absorption fine structure) regimes. Actually, these two spectral regions are dominated by different physical processes and hence contain different and complementary information. The first narrow region within a few eV around the high absorption jump is essentially determined by chemical bonding effects like transitions to empty molecular orbitals; therefore, it provides information about the electronic structure of the absorbing atom.[34-36] The edge position itself is sensitive to the local chemistry; as an example, Figure 20 shows the K-edge shift of molybdenum ions in aqueous solution, which can be considered indicative of an increasing amount of covalency in metal-ligand bonds as the Mo oxidation state increases.[37]

XANES is the low energy side of the absorption spectrum and extends up to some tens of eV above the edge. In XANES region, the electron ejected by the absorber in the photoionization process has low kinetic energy; consequently, it is strongly scattered by the neighboring atoms through multiple scattering processes.[38-42] A more subtle partitioning of the spectrum distinguishes a first region of **full multiple scattering** regime, where all the multiple pathways which begin and end at the absorber contribute to the spectrum, followed, before the transition to EXAFS, by a region where only the contributions from low order multiple scattering pathways are important. The interpretation of this **intermediate multiple**

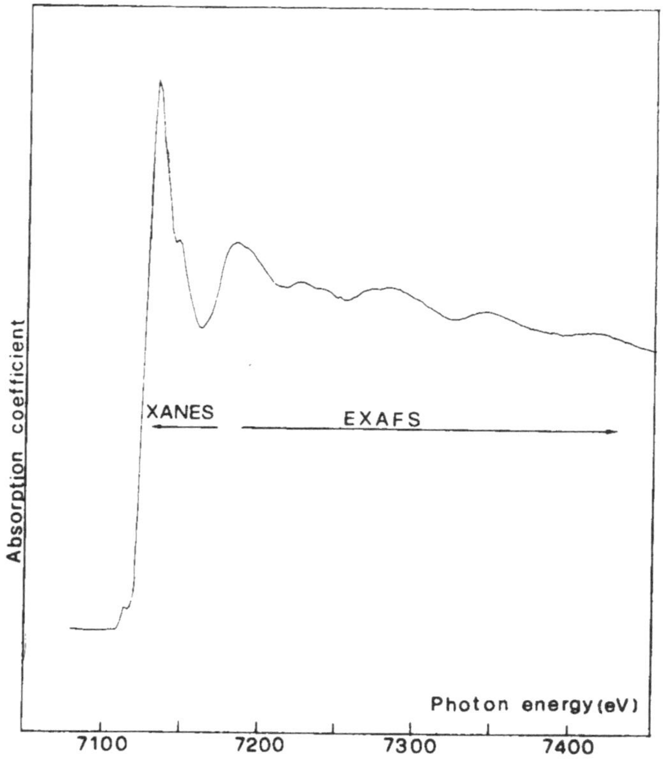

FIGURE 19. Typical X-ray absorption spectrum showing the near edge region and the EXAFS oscillations.

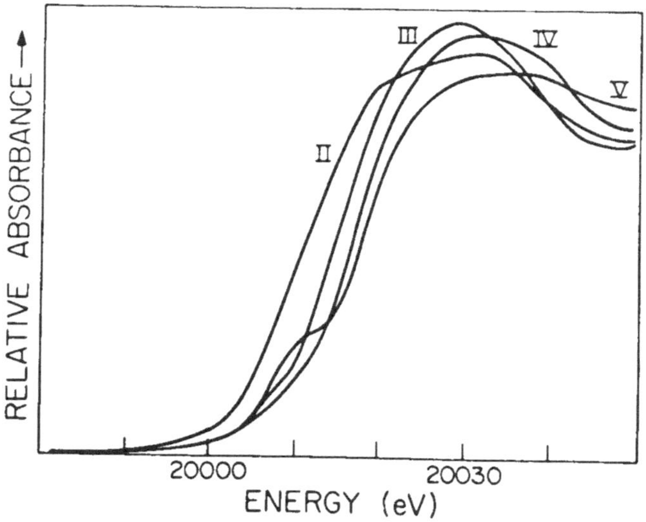

FIGURE 20. Absorption edge for aqueous molybdenum ions in different oxidation states (the spectra were scaled to yield the same amplitude at 20 KeV). (From Cramer, S. P., Eidem, P. K., Paffet, M. J., Winkler, J. R., Dori, Z., and Gray, H. B., *J. Am. Chem. Soc.*, 105, 799, 1983. With permission.)

scattering regime would open the possibility to probe high order correlations.[43,44]

In the EXAFS region, extending up to about 1 KeV above the edge, the ejected electrons are weakly scattered, so that only single scattering processes are important. The term EXAFS refers to the oscillations in this spectral region which originate because of interference effects between the electron wave outgoing from the absorbing atom and the waves backscattered by its neighbors. The EXAFS theory, formulated within a **short-range, single-electron, single-scattering** approach,[45-49] is today generally accepted, and its application to structural determinations involves almost standard procedures.[50-55] In the following, we will deal only with the EXAFS technique, as practically only the EXAFS spectral region has been used in structural studies on aqueous solutions.

B. EXAFS: Basic Concepts and Relations

The physics of the EXAFS phenomenon can be outlined as follows.[46,51-55] As shown schematically in Figure 21(A), the absorption of an X-photon induces the transition of an electron from the initial state $|i\rangle$ to a final unoccupied continuum state $|f\rangle$. By conservation of energy:

$$E = E_f - E_i = E_0 + \hbar^2 k^2/2m \quad (7)$$

where $E = h\nu$ is the X-photon energy, k is the photoelectron wave vector and E_0 is the threshold energy, that is the energy value corresponding to $k = 0$.

The photoabsorption process can be treated semiclassically within the **dipole approximation** for the interaction between an atom and the electromagnetic field. In the single-electron approximation, the absorption coefficient μ can be then calculated through the **Fermi Golden Rule**:

$$\mu \propto |\langle f | \vec{\epsilon} \cdot \vec{\mathbf{r}} | i \rangle|^2 \rho(E_f) \quad (8)$$

which involves the transition matrix element of the dipole operator and the density of states at the final energy E_f.

It has been shown that the oscillatory behavior of μ can only originate from the matrix element, because $\rho(E_f)$ behaves monotonically as the photon energy changes, and can be assumed to be that of a free electron.[46] Besides, changes in the matrix element can only be produced by final-state effects, since the initial state $|i\rangle$ does not change with the photon energy E. The initial state is localized in the deep core of the absorbing atom assumed as origin. As a consequence, EXAFS depends on how the wave function of the final state is modified near the origin by the presence of other atoms.

The behavior of $|f\rangle$ can be qualitatively visualized as in Figure 21. In the absence of neighbors, the final state $|f\rangle = |f_0\rangle$ is a spherical wave outgoing from the origin and describing the propagation of the electron outside the central atom (Figure 21[B]). In this case, the absorption coefficient $\mu = \mu_0$ varies monotonically with the energy of the incident photon. On the contrary, when the absorber is surrounded by other atoms, the outgoing photoelectron travels in presence of potentials centered at the atomic sites which act as scatterers. The final state $|f\rangle = |f_0\rangle + \Sigma|f_{sc}\rangle$ can be considered a combination of an unperturbed outgoing wave and of those backscattered by each surrounding atom towards the central one (Figure 21[C]). The interference between these two parts changes $|f\rangle$ near the origin, where $|f\rangle$ overlaps with the initial state, and consequently changes the matrix element and the absorption coefficient μ. Depending on the photon energy, the interference can be constructive or destructive, producing, respectively, an increase or a decrease of μ with respect to the monotonic behavior of μ_0.

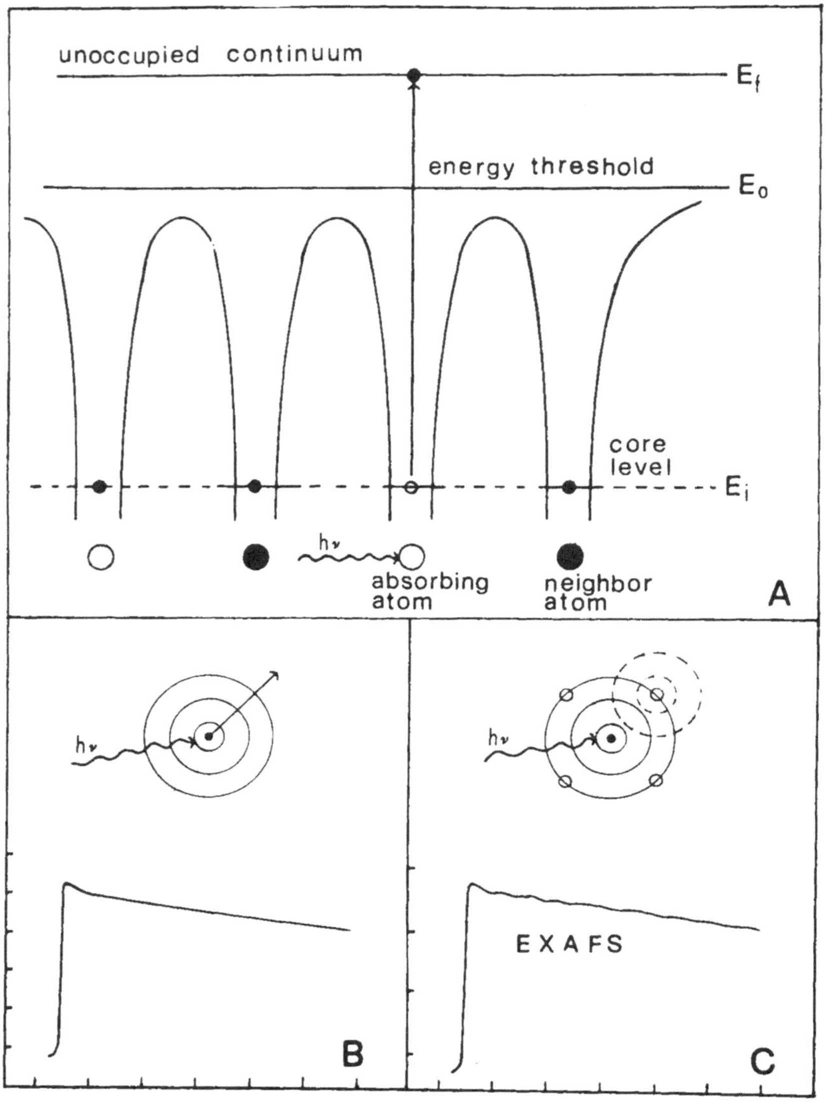

FIGURE 21. Pictorial view of EXAFS origin.

We can write the absorption coefficient according to

$$\mu = \mu_0[1 + \chi + \text{multiple scattering terms}] \tag{9}$$

where χ is the single-scattering term, called EXAFS. Thus, in the spectral region where multiple scattering effects are absent or unimportant, EXAFS can be easily extracted from the measured absorption as the oscillatory part of μ normalized to the atomic background μ_0, i.e.

$$\chi(E) = [\mu(E) - \mu_0(E)]/\mu_0(E) \tag{10}$$

Within a plane wave (small atom) approximation, the EXAFS interference function for a system with randomly oriented interatomic distances can be expressed by a simple mathematical form as function of the photoelectron wave vector K. The EXAFS for the photoexcitation of the atomic species i is given by the superposition of contributions from all the backscattering atoms surrounding i,[45-49] i.e.,

$$\chi_i(k) = \sum_b A_{ib}(k, r_{ib}) \sin[2kr_{ib} + \phi_{ib}(k)] \tag{11}$$

Each atom contributes a sinusoidal oscillation with a frequency depending on the distance r_{ib} from the central atom, and a total phase shift

$$\phi_{ib}(k) = 2\eta_i(k) + \delta_b(k) \tag{12}$$

which accounts for the fact that the photoelectron experiences the atomic potentials of both the absorber and the backscatterers. The term $\delta_b(k) = \arg[t_b(k)]$ is due to the complex scattering amplitude $t_b(k)$ of the backscatterer, while the term $2\eta_i$ comes from the central atom which is crossed twice by the photoelectron.

The sinusoidal behavior is modulated by the amplitude

$$A_{ib}(k) = S_i(k)\, F_b(k) \exp[-2r_{ib}/\lambda(k)]/(kr_{ib}^2) \tag{13}$$

where $F_b(k) = |t_b(k)|$, while $S_i(k)$ and $\exp[-2r_{ib}/\lambda(k)]$ are phenomenological terms introduced to account for inelastic scattering processes which reduce the EXAFS amplutide. The reduction term, $S_i \leq 1$, is due to many-body effects (multiple-electron excitations) associated with the absorbing atom, which are not accounted for in the single-electron approximation. The exponential damping term approximates the effects of inelastic losses due to the environment, by introducing the photoelectron mean free path $\lambda(k)$.

Equation 11 corresponds to an instantaneous snapshot of the atomic configuration around an absorbing atom. It must therefore be averaged over all configurations, that is, over the radial pair correlation functions $g_{ij}(r)$ of the system. Considering that $4\pi r^2 x_j\, \rho_0\, g_{ij}(r)dr$ represents the number of j-type atoms at distance between r and r + dr from an i-type one, Equation 11 can be written as:

$$k\chi(k) = 4\pi\rho_0 S_i \sum_{j=1}^{M} x_j F_j(k) \int_0^\infty g_{ij}(r) \exp(-2r/\lambda) \sin[2kr + \phi_{ij}(k)]\, dr \tag{14}$$

where the sum is over the species present in an M-component system, x_j is the composition coefficient of the j-type atoms, and ρ_0 the bulk number density of the sample.

If we use a picture involving coordination shells around the central atom, Equation 14 can be rewritten as:

$$k\chi_i(k) = S_i \sum_j F_j(k)\, N_{ij} \int_0^\infty [P_{ij}(r)/r^2]\, e^{-2r/\lambda} \sin[2kr + \phi_{ij}(k)]\, dr \tag{15}$$

where j numbers are now the succession of coordination shells, N_{ij} is the **coordination number** of the jth shell, and $P_{ij}(r)dr$ is the probability of finding an atom of the jth shell in the range r to r + dr.

When the thermal and static disorder can be described by gaussian distributions with mean distances r_{ij} and r.m.s. deviations σ_{ij}, Equation 15 can be approximated to:

$$k\chi_i(k) = S_i \sum_j F_j(k)(N_{ij}/r_{ij}^2) \exp(-2r_{ij}/\lambda) \exp(-2k^2\sigma_{ij^2}) \sin[2kr_{ij} + \phi_{ij}(k)] \quad (16)$$

which is the most popular EXAFS expression, as it explicitly relates $\chi(k)$ to structural parameters; these can be extracted straightforwardly from the experimental spectra if the other terms appearing in Equation 16 are known independently. It must be observed, however, that the distance distribution may not always be gaussian, and Equation 16 may not be appropriate.

The approximations underlying the above EXAFS equations limit their validity at photon energy values greater than 30 to 40 eV beyond the threshold (typically, $k \geq 3$ Å^{-1}). The limitatitions on the structural information we can hope to extract from EXAFS data, will be discussed shortly.

C. Data Analysis

Only a brief description of the main points of EXAFS data analysis will be given here; the reader interested in more details may refer to the many books and reviews that extensively cover the subject.[30,50-55] Although most of the following matter has general meaning, we here only refer to transmission measurements at K-edge.

The first step in data reduction is the extraction of the function $\chi(E)$ as defined in Equation 10. This implies the removal from the measured spectrum of the **pre-edge absorption** (coming from lower energy edges of the absorber itself or of other atoms), and the determination of the **atomic absorption** $\mu_0(E)$. Both contributions are smooth functions, so they can be approximated by slowly varying functional forms or splines. The first contribution is determined in the preedge region and is then extrapolated above the edge. The function approximating μ_0 is fitted in the region above the edge up to the limit of the data range; the EXAFS oscillations are in fact of sufficiently high frequency as not to be washed out by the use of a slowly varying function.

The next step consists in obtaining $\chi(k)$ as function of the photoelectron momentum k,

$$k = [2m/\hbar^2(E - E_0)]^{1/2} \quad (17)$$

which implies the knowledge of the threshold energy E_0. Generally, it is almost impossible to obtain the true value of E_0 from theoretical calculations, or to identify it in the experimental spectrum. It is thus common practice to fix E_0 preliminarily at a some feature of the edge (e.g., the maximum of the first derivative) and to adjust it to a more proper value in a subsequent step of the analysis. The E_0 problem is less important if the analysis involves the comparison of the EXAFS spectrum with that of a reference compound of known structure. By fixing E_0 in a consistent way for the two samples, the absolute distortion in the k scale does not affect the comparison between the two data sets.

Once the function $\chi(k)$ has been calculated, the structural information can be obtained by using Equations 14 to 16, if the "nonstructural functions" (amplitude, phase shift, mean free path) are known or can be determined independently.

For simplicity, let us first consider the case of a one-component system, where Equation 14 becomes:

$$k\chi(k) = 4\pi\rho_0 SF(k) \int_0^\infty g(r)\, e^{-2r/\lambda} \sin[2kr + \phi(k)]\, dr \quad (18)$$

This equation is not a Fourier transform, and it cannot be formally inverted using the Fourier integral theorem to obtain the pair correlation function g(r) directly, as it can be done in X-ray diffraction. However, the main problem is not the inversion of the integral equation, which could be approached by numerical regularization techniques,[56-58] but rather the boundaries of the accessible k range, typically from k_{min} ~3 $Å^{-1}$ to k_{max} ~15 $Å^{-1}$. The most severe limitation for noncrystalline materials comes from the lower limit. In fact, the lack of low k information suppresses the slowly varying and broad components in g(r) mainly coming from the high r region, so that only sharp features at low r can be actually observed.

In the case of multicomponent systems, further problems derive from the presence of different atomic species acting as backscatters. To go on in the analysis, it is common practice to visualize the structural information in real space through a complex Fourier transform in the complementary variables 2k and r:[45,47]

$$F(r) = (1/2\pi)^{1/2} \int_{k_{max}}^{k_{min}} k\chi(k)\ W(k) \exp(-2ikr)\ d(2k) \tag{19}$$

where W(k) is a window function which damps the data to zero at k_{min} and k_{max}. The magnitude |F(r)| is a **modified radial distribution function**, which exhibits peaks corresponding to the various interatomic distances between the absorber and the backscattering atoms, shifted from the true values because of the k dependence of the phase shift in the sine argument of the EXAFS equations.[47] When more than one shell of neighbors contribute significantly to the EXAFS spectrum, the Fourier transform can isolate, as distinct peaks, the different contributions which are mixed together in $\chi(k)$, allowing the analysis of each of them independently. However, this analysis is not usually made in r-space; the common practice consists in obtaining a **Fourier filtered spectrum** and analyzing it in k-space. This procedure involves singling out a part of F(r), e.g., the peak of interest, and Fourier transforming it back into k-space. The result represents the contribution to $\chi(k)$ from the shell (or shells) determining the r region backtransformed.

Figure 22 shows an example of the Fourier transform of the Cr K-edge EXAFS spectrum of a 0.5-*M* $Cr(NO_3)_3$ aqueous solution, where the main peak is due to Cr-O nearest neighbor distances.[59] The appearing of only one peak corresponds to the disappearance of longer distance and more disordered contributions, as discussed before. The peak position is shifted about 0.40 Å below the known value; a few tenths of angstroms are in fact typical values for this shifting effect in F(r).

The bottom part of Figure 22 shows the filtered spectrum corresponding to the peak in F(r). It can be seen that the filtered spectrum reproduces the original one without the noise, thus confirming that the total signal is determined by only the nearest neighbor shell.

If the peak backtransformed corresponds to only one coordination shell, the Fourier filtering has the capability of decomposing the modulated sine wave

$$\chi_{shell} = B(k) \sin\psi(k) \tag{20}$$

into the total amplitude B(k) and the total phase shift $\psi(k)$ functions, which can be analyzed separately. In the case of gaussian disorder, the structural interpretation can be performed in terms of the structural parameters, R, N, and σ through the equations:

$$\psi(k) = 2kR + \phi(k) \tag{21}$$

$$B(k) = (N/kR^2)\ S(k)\ F(k) \exp(-2R/\lambda) \exp(-2k^2\sigma^2) \tag{22}$$

The distance R could be obtained from ψ if φ is known independently from theoretical

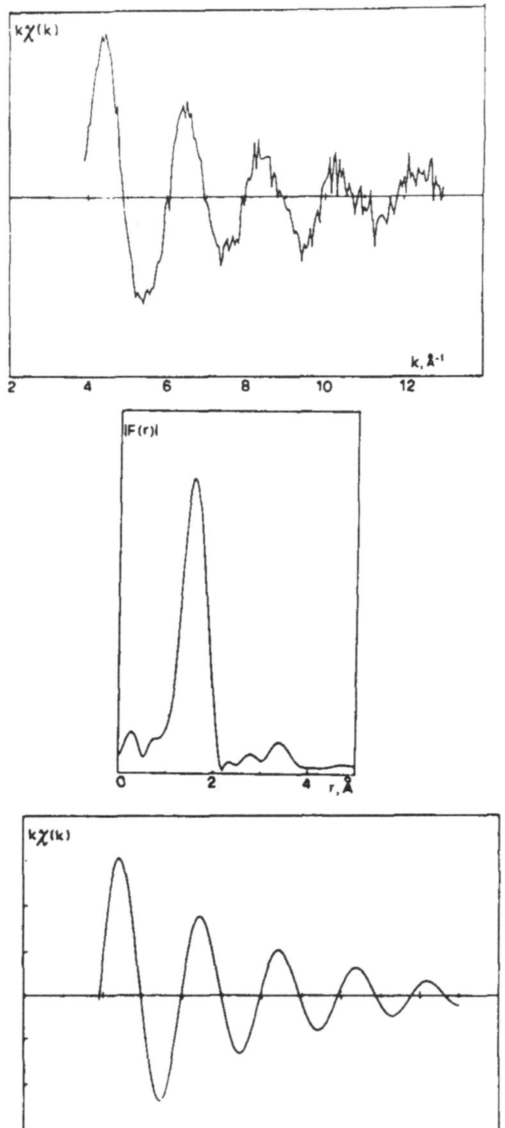

FIGURE 22. kχ(k) for a 1-*M* aqueous solution of $Cr(NO_3)_3$ (top), Fourier transform of kχ(k) (middle), Fourier filtered spectrum of the main peak (bottom).

calculations.[60] Unfortunately, the values of phase shift tabulated in literature,[61,62] cannot be used straightforwardly. In fact, a phase shift value is defined for a given value of the threshold energy E_0 which generally does not correspond to that used in fixing the k scale of the measured spectrum. For this reason, a common practice consists in treating E_0 as an adjustable parameter in this stage of the data analysis.[62] Alternatively, a comparison can be made between the unknown (u) system and a model (m) system with known distance R_m between the same atom pair as in u.[63-66] Assuming chemical transferability of the phase shift, we have

$$\Delta\psi(k) = \psi_u - \psi_m = 2k(R_u - R_m) \quad (23)$$

so that R_u is obtained by fitting a straight line to $\Delta\psi$. If the k scale has been calculated with the correct value of E_0 the fitted straight line must cross the origin; this property can be used to adjust the E_0 value properly, before obtaining ΔR from the slope of Equation 23.

The total amplitude B(k) in Equation 22 can be used to obtain the coordination number and σ, once R_u and E_0 have been determined with the above procedure. The tabulated values reported for backscattering amplitudes are not sufficiently reliable because of the approximations necessary for their calculation.[62,63] So, the use of a suitable model compound of known structure becomes a necessary procedure. If it can be assumed that $S_u = S_m$, $F_u = F_m$, $\lambda_u = \lambda_m$, and $|R_u - R_m| \ll \lambda$, the behavior of $\ln(B_u/B_m)$ vs. k^2

$$\ln(B_u/B_m) = \ln(N_uR_m^2/N_mR_u^2) - 2(\sigma_u^2 - \sigma_m^2)\,k^2 \quad (24)$$

can be fitted to a straight line to obtain σ_u^2 and N_u from the slope and the intercept, respectively.[47] Unfortunately, the assumption of chemical transferability of amplitudes is much more critical than for phase shifts.[68-70] Consequently, it should be necessary to have a reference compound with very similar chemical environment at about the same distance as in the unknown.[70] Note that when studying metal ion aqueous solutions, the corresponding hydrated crystals can be generally considered very good standards.

When two peaks in Fourier transforms are not well separated, or a single peak can be ascribed to two not resolved contributions at different distances, the above procedures break down. In these cases, the filtered spectrum involving the two contributions must be fitted with a proper expression like Equation 16 based on some structural model. Phase shifts and amplitudes can still be extracted from one or more model compounds. A common procedure consists in parameterizing these functions using appropriate analytical forms;[61,67] these parameters are obtained from models and then transferred to the unknown system. Combinations and/or variations of these procedures can obviously be employed in connection with the specific problem one has to deal with.

The accuracy obtainable for the values of structural parameters depends on the particular situation. In general, the distance value of a near neighbor shell can be obtained with high accuracy: up to 0.01 Å in favorable cases of low noise to signal ratio, and depending on good knowledge of phase shifts, or on availability of appropriate model compounds. Much more critical problems come out when determining coordination numbers and r.m.s. deviations. This is mainly due to the poor knowledge of the EXAFS amplitudes and their bad transferability, and also to a high sensitivity of EXAFS amplitude to incorrect experimental setups and sample preparation. For example, effects due to sample inhomogeneities, sample thickness,[71-73] and size of powder particles,[74] produce significant reduction and distortion in the measured amplitudes. So, when studying aqueous solutions, where sample effects can be easily minimized, care should be taken in preparing the standards, which usually are powdered hydrated crystals.

D. EXAFS vs. XRD

The characteristics of EXAFS can be usefully discussed through a comparison with diffraction techniques.[52-54,75-80] We refer here only to X-ray diffraction, but most of the discussion applies to neutron diffraction as well.

When studying the structure of crystalline materials, the comparison is clearly in favor of XRD which generally determines the complete tridimensional atomic arrangement of a single-crystal sample through sophisticated, but well-founded techniques. However, EXAFS can play an important role in the cases of complicated structures with large number of atoms in the unit cell when the emphasis is on the local structure around a particular species, or when the species of interest is diluted in a crystalline matrix.

The comparison is more appropriate in the case of noncrystalline materials where both techniques only provide a one-dimensional representation of the structuring. Let us compare analogous expressions relating experimental data to structural information. The EXAFS interference function given in Equations 14 and 15 shows evident similarities with the XRD structure function written according to

$$
\begin{aligned}
si(s) &= 4\pi\rho_0 \sum_{i=1}^{M} \sum_{j=1}^{M} x_i x_j f_i f_j \int_0^{\infty} r[g_{ij}(r) - 1] \sin(sr)\, dr \\
&= \sum_{i=1}^{M} \sum_{j} x_i f_i f_j N_{ij} \int_0^{\infty} (1/r)[P_{ij}(r) - 4\pi r^2 x_j \rho_0] \sin(sr)\, dr
\end{aligned}
\tag{25}
$$

where, in the second expression, j runs over distinct coordination shells around each atomic species taken as origin. Extending the analogy to cases where P_{ij} is a gaussian distribution of distances, the EXAFS Equation 16 can be compared with the XRD equation,

$$
si(s) = \sum_i x_i \sum_j f_i f_j (N_{ij}/r_{ij}) \exp(-s^2\sigma_{ij}^2/2) \sin(sr_{ij}) \tag{26}
$$

in which the structural parameters characteristic of the coordination shells appear explicitly. Let us observe that the modulation induced by gaussian disorder involves r.m.s. deviations around mean values of interatomic distances. It means that, in the case of noncrystalline materials, σ has the same meaning in EXAFS and XRD, in contrast with crystallography which considers r.m.s. deviations of atoms around their mean position.

The comparison can be further extended to r-space. If Equation 14 is rewritten according to [75]

$$
k\chi_i(k) = 4\pi\rho_0 \sum_j x_j \int_0^{\infty} g_{ij}(r)\, \mathrm{Re}\{\exp(ikr)\, \Lambda_{ij}(k, r)\}\, dr \tag{27}
$$

where

$$
\Lambda_{ij} = -iS_i(k)\, F_j(k) \exp[-2r/\lambda + i\phi_{ij}(k)] \tag{28}
$$

it can be shown that the Fourier transform (Equation 19) of $k\chi(k)$ can be written as[75]

$$
F(r) = 4\pi\rho_0 \sum_j x_j \int_0^{\infty} g_{ij}(u)(1/2)\{\Gamma_{ij}(r - u) + \Gamma_{ij}^*(-r - u)\}\, du \tag{29}
$$

where

$$
\Gamma_{ij}(r, r') = (1/2\pi)^{1/2} \int_{k_{min}}^{k_{max}} W(k)\, \Lambda_{ij}(k, r') \exp(-2ikr)\, d(2k) \tag{30}
$$

Equation 29 shows that the radial function F(r) is a perturbed image of a linear combination of the subset of g_{ij} describing the structuring around the absorbing species. The perturbation is given by the convolution with the peak-shape function Γ_{ij} which involves terms inherent to the physics of the electron scattering process and effects of data treatment. This representation is the analogue of the XRD equation

$$D(r)/r = 4\pi\rho_0 \sum_i x_i \sum_j x_j \int_0^\infty u g_{ij}(u)\{T_{ij}(r - u) - T_{ij}(r + u)\}\, du \qquad (31)$$

where the total radial function D(r) appears to be a convolution perturbed image of the true partial correlation functions (see Chapter 1).

This formal comparison reveals the existence of analogies, but also of important differences between EXAFS and XRD. The most evident analogy regards the kind of structural information which consists of pair correlation functions in both techniques. The most obvious difference regards the number of partial correlation functions g_{ij} which contribute to the experimental data for an M-component system. As discussed in Chapter 2, the presence of all the $M(M + 1)/2$ independent partial correlation functions complicates dramatically the interpretation of XRD spectra as the chemical complexity of the system increases. Moreover, each g_{ij} is weighted by the additional composition coefficients x_i, so that the signal from diluted species disappears. In contrast, EXAFS is a few steps forward towards the objective of separating correlation functions, and this is a significant element which advantages EXAFS over XRD. In fact, each EXAFS measurement singles out a subset of M g_{ij} functions (i = absorber, $j = 1,M$) and it is possible to change the absorber in different measurements, even choosing dilute species as the observation point of the surrounding structure. Moreover, EXAFS can take advantage from a better distinguishability of atomic species, as scattering amplitude functions of different atoms have different shapes.

There are, however, other significant differences in the way in which the structural information appears in analogous equations. It can be seen that the information about a shell centered at R_i enters multiplied by $1/R$ in XRD and by $\exp(-2R/\lambda)/R^2$ in EXAFS; consequently the signal from shells beyond the first are more quickly lessened in EXAFS than in XRD. Let us consider, for example, two shells at distances of about 2 and 4 Å, which are typical chromium-water distances in aqueous solutions. The EXAFS signal from the second shell is lessened, relatively to the first, by a factor of $(R_1/R_2)^2\exp[-2(R_2-R_1)/\lambda] \sim 0.15$ for a typical value of $\lambda \sim 8$ Å. The corresponding XRD signal is lessened by only $(R_1/R_2) \sim 0.5$.

Going beyond the formal comparison between equations, a significant difference regards the range of momentum space actually accessible by EXAFS and XRD. In fact, the quantity and the quality of the structural information can depend critically from the experimental range available for its extraction. The momentum regions to be compared are those of s in XRD and 2k in EXAFS,[75] because the corresponding of the XRD scattering vector s (momentum transfer between the incident and scattered photon), is the momentum transfer of the photoelectron 2k. The XRD region starts practically at $s = 0$ Å^{-1} and ends at about $s = 16$ Å^{-1} because of experimental constraints. On the contrary, EXAFS covers approximately the region $6 \lesssim 2k \lesssim 30$ Å^{-1}; so EXAFS displays a better resolution in detecting the nearest neighbor environment, but suffers the lack of low k data, as discussed before. Let us consider this point in some detail by using the mentioned example involving two Cr-O distances. Suppose that the guassian broadening of the two shells is ~0.08 and ~0.26 Å respectively, as found in a 1-*M* $Cr(NO_3)_3$ aqueous solution (see Chapter 3). The signal from any shell is damped by $\exp[-\sigma^2 s^2/2]$ as s increases both in XRD and in EXAFS ($s = 2k$ in EXAFS). So, the second shell is lessened, relatively to the the first, by the factor $\exp[-(\sigma_2^2-\sigma_1^2)s^2/2]$ which is ~0.75 at $s = 3$ Å^{-1}, ~0.33 at $s = 6$ Å^{-1} where EXAFS data are just starting, and becomes ~0.05 at $s = 10$ Å^{-1}. The accessible EXAFS k range is essentially insensitive to such a broadened shell. More in general, this effect adds to the limit due to the short mean free path $\lambda(k)$ in dropping the information about long range order from EXAFS spectra. In disordered systems, as aqueous solutions are, the consequence is that only the first shell can be detected. The comparison above, between the respective momentum regions, stresses why EXAFS and XRD must be considered complementary, as they are respectively more sensitive to different aspects of the pair correlations. Therefore, the combined use of both techniques may lead to a more detailed description of a system. Perhaps the most attractive perspective in this direction consists in involving both EXAFS and XRD measurements to separate single g_{ij} functions as suggested by Babanov et al.[56]

E. Review of the Results

The first EXAFS spectra of solutions date back to 1975, to a paper by Eisenberger and Kincaid,[81] who reported EXAFS data of a dilute $CuBr_2$ aqueous solution taken at Cu and Br edges. This study provided the first direct EXAFS evidence of the hydration of ions in solution. Many investigations have followed, in the attempt to exploit the special characteristics of this technique for a better understanding of the near environments of the ions and related order phenomena.

1. Ion – Oxygen Interactions

In Table 8 we report first shell ion-oxygen distances and coordination numbers (when available) from studies carried out in about a decade. Only distance values are reported in many cases, particularly in the early studies. As pointed out before, the distance between the absorber and its first neighbors is the structural parameter that can be obtained by EXAFS data with greater accuracy. The evaluation of coordination numbers proved to be a more delicate problem, and only in some cases reliable results have been obtained. Many investigations were carried out on systems already examined by XRD, with the explicit aim of testing the new technique; nowadays, the EXAFS technique is considered an almost standard tool for structural investigations, so it has been used confidently to approach some interesting problems concerning the state of ions in solution.

The local environment of the Ni^{2+} ion in concentrated $NiCl_2$ solutions has been a point of discussion. In fact, an aqueous environment was supported by XRD and ND data even at high concentrations, while Raman results were interpreted in terms of $NiCl_4^{2-}$ complexes,[95] in amount dependent on solution concentration. EXAFS investigations on this system contradicted the latter hypothesis. Figure 23 shows the filtered spectra of the first shell of Ni^{2+} obtained by Licheri et al.[87,89] for 2- and 4-*M* $NiCl_2$ and $Ni(NO_3)_2$ solutions. The direct comparison between chloride and nitrate solutions demonstrates that essentially the same near environment of oxygens is present around the cation in the two systems. The same picture about Ni^{2+} coordination was given by Lagarde et al.[94] for $NiCl_2$ and $NiBr_2$ solutions.

The probably distorted environment of the Cu^{2+} ion in solution was probed by the EXAFS technique. In the mentioned study on a dilute solution of $CuBr_2$, the shortest Cu – O distance was detected.[81] In a following investigation on a 0.32-*M* solution of $Cu(ClO_4)_2$, Sano et al.[90] proposed two values for Cu – O distances, based on the peaks appearing in the radial function. The difficulties in finding clear indications of longer Cu – O distances in EXAFS data were pointed out by Joyner[96] who examined the crystalline $CuSO_4 \cdot 5H_2O$. In fact, he was able to determine the distance value (1.95 Å, in good agreement with ND and XRD results) of the four Cu – O interactions due to equatorial water molecules in the octahedral polyhedron tetragonally distorted, but unable to detect the contribution from the axial oxygens, pertaining to the SO_4^{2-} (he estimated this contribution to be about 7% with respect to the one due to the short distances). A detailed analysis of the experimental data of a 10-*M* solution of copper perchlorate was carried out by Sham et al.[84] A small peak appearing at the right side of the main one, due to the equatorial Cu – O interactions, was thought as coming from axial contributions, and the entire peak region was interpreted by a two-shell fit. Actually, the authors realized that a number of factors influence the quantitative interpretation of the data. The combined effects of different σs and distances on the radial

Table 8
STRUCTURAL PARAMETERS FOR ION-OXYGEN INTERACTIONS AS CALCULATED BY EXAFS MEASUREMENTS[a]

Ion	Sample	Conc. (*M*)	$r_{Me\text{-}O}$ (Å)	$N_{Me\text{-}O}$	$r'_{Me\text{-}O}$[b] (Å)	Ref.
Cr^{3+}	$Cr(NO_3)_3$	0.1	2.01	—		82
	$Cr(ClO_4)_3$	1	1.966	—		83
Cr^{2+}	$Cr(ClO_4)_2$	1	1.984	—		83
	$Cr(ClO_4)_2$	1	1.985	—	2.30	84
Mn^{2+}	$Mn(ClO_4)_2$	1	2.177	—		83
Fe^{2+}	$Fe(ClO_4)_2$	1	2.095	—		83
Fe^{3+}	$Fe(ClO_4)_3$	1	1.990	—		83
Co^{2+}	$Co(NO_3)_2$	0.1	2.08	—		82
Ni^{2+}	$Ni(NO_3)_2$	0.1	2.05	—		85
	$Ni(NO_3)_2$	0.1	2.06	—		82
	$Ni(NO_3)_2$	0.5	2.05	6.4		86
	$Ni(NO_3)_2$	1.1	2.06	7.1		86
	$Ni(NO_3)_2$	2	2.06	6.5		86
	$Ni(NO_3)_2$	4	2.06	6.6		86
	$Ni(NO_3)_2$	4	2.06	6.6		87
	$NiCl_2$	2.78	2.06	7		88
	$NiCl_2$	3.74	2.07	6.2		88
	$NiCl_2$	2	2.06	6.5		89
	$NiCl_2$	4	2.06	6.8		87
Cu^{2+}	$CuBr_2$	<0.1	1.97	—		81
	$Cu(ClO_4)_2$	0.32	(1.94)	(4)	2.46	90
	$Cu(ClO_4)_2$	1	1.94	—		83
	$Cu(ClO_4)_2$	1	1.955	(4)	2.60	84
Ag^+	$AgClO_4$	3	2.31	2.9		91, 93
	$AgNO_3$	3	2.36	3.9		91, 93
	$AgNO_3$	9	2.34	2.6		91, 93
Sn^{2+}	$Sn(ClO_4)_2$	3	2.21	3.8		92, 93
Br^-	NaBr, $CuBr_2$	<0.1	3.14	—		81
	$SrBr_2$	0.35	3.40	—		94
	$SrBr_2$	3.5	3.40	—		94

[a] Values in parentheses were kept fixed.
[b] Ion-oxygen second distance values.

functions were explored in detail, and it was shown how the small peak is affected. The conclusion reached was that the structural parameters defining the shorter interactions are quite reliable (4 oxygens at 1.955 ± .010 Å, with σ = 0.036 ± .005 Å), while a large uncertainty is associated with the second distance (r = 2.60 ± .05 Å and σ = 0.12 ± .02).

XRD studies on concentrated solutions of Ag^+ perchlorate and nitrate gave results of about

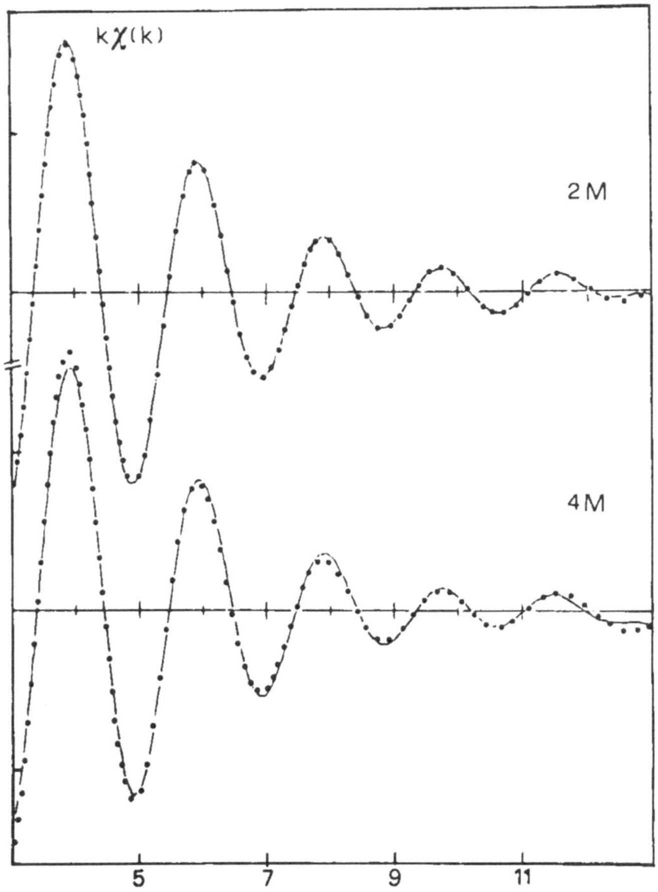

FIGURE 23. First shell filtered spectra of 2- and 4-M solutions of $Ni(NO_3)_2$ (full line) and $NiCl_2$ (dots).

two oxygens (from water molecules) close to the ion at a distance of 2.41 to 2.45 Å,[97] value which seems rather long if coupled with the low coordination number. Crystalline structures of silver salts display a variety of geometrical arrangements around this ion (linear, distorted tetrahedron, square planar, distorted octahedron, antiprism), each one characterized by different distance values. Though the EXAFS technique cannot be expected to evidence the spatial arrangement of the atoms, the accurate determination of the $Ag^+ - O$ distance would support one specific geometry. Yamaguchi et al.[91,93] examined by EXAFS 3-*M* solutions of silver nitrate and perchlorate and a 9-*M* solution of $AgNO_3$, besides several solid samples used as reference compounds. As a result of their analysis, they found coordination numbers in the range 3 to 4 and distance values of 2.31 to 2.36 Å. On this basis, the authors suggested the tetrahedral coordination geometry. Also recent XRD investigations led to similar numerical results.[98]

Also in the case of Sn^{2+} ions, XRD studies met with the usual difficulties due to overlapping interactions. On the other hand, crystallographic investigations on Sn^{2+} salts had revealed the key role of the Sn^{2+} active lone pair, which is reflected in an asymmetric environment of this ion. The EXAFS investigation[92,93] of a 3-*M* solution of $Sn(ClO_4)_2$ agreed with the general view of the stereochemically active Sn^{2+} ion; the data were interpreted in terms of one hydration shell of about 4 neighbors, at a distance of about 2.24 Å. Longer distances were not revealed.

2. Ion – Halide Interactions

EXAFS has been applied to the study of the halo-complexation, on account of its special characteristics. The structural investigation of this phenomenon has proved to be a not easy task, and contrasting pictures have often been provided by different techniques. As extensively discussed in Chapter 3, the interpretation of XRD data is not straightforward, and only systematic investigations and special situations have given satisfactory answers. When the complexation degree is low, the metal-halide contribution to the experimental function is hidden by the dominant ones and the very existence of the phenomenon may turn out questionable. In these conditions, not even EXAFS has always provided conclusive answers.

Let us consider, for example, the case of chloro-complexation. EXAFS data on the absorption edge of the metal show to be not very sensitive to the presence of small amounts of Me-Cl contacts. Cation-chloride distances are in fact longer than cation-oxygen ones, and thus they are more strongly dampened in the EXAFS signal. Besides, although oxygen and chlorine are distinguishable as backscatterers on the basis of their different phase shift functions, they are both light scatterers, with amplitudes quickly decaying as k increases. Their effect on the medium and high k region is small in general, and minor terms are poorly detectable. However, another route can be followed here; the chlorine atom can be used as absorber, becoming the observation point of the structural information; the species acting as backscatterers are now quite different and they can be identified more easily; moreover, also the distances involved, Cl-metal and Cl-water, have fairly different values and can give rise to distinct peaks in the radial functions.

This type of measurements has been made recently by Sandstrom,[99] who examined several solutions of $FeCl_3$, in the concentration range from 0.40 to 5.75 *M*, and the $FeCl_3$ hydrated salt. In the radial functions reported in Figure 24, iron-chloride and water-chloride distances are clearly detectable as relatively well-resolved peaks, which heights vary in accord with the increasing expected degree of complexation depending on concentration. The same indication is found in the pre-edge region of the spectrum, in which a peak of amplitude approximately proportional to the amount of direct Fe-Cl bonding (Figure 25) tends to disappear with increasing dilution. As this pre-edge feature is ascribed to transitions to antibonding molecular orbitals derived from the iron 3d and chlorine 3p atomic orbitals, it concurs in supporting extensive and concentration dependent formation of iron-chloride complexes. Other EXAFS measurements on solutions of $FeCl_3$ have been carried out by Asakura et al.[100] at the iron K-edge. The absorbing atom is here the center of the complexes $FeCl_x(H_2O)_y$, and the objective of the study was their description. The Fourier transforms of dilute (less than 1 *M*) and concentrated (3 to 4 *M*) solutions (some of them acidified with nitrate or chloride acid), were compared with those of crystalline compounds where the iron ions are in the form of tetrahedral complexes, $[FeCl_4]^-$, and octahedral complexes, $[FeCl_2(H_2O)_4]^+$ and $[Fe(H_2O)_6]^{3+}$. The comparison indicated that the complex, $[Fe(H_2O)_6]^{3+}$, is the predominant species in the dilute solutions, and the complex, $[FeCl_2(H_2O)_4]^+$, in the 3 to 4-*M* solutions. Structural parameters describing the chloro-complexation were obtained by fitting procedures (Table 9), where the sum of the coordinated Cl and O atoms were assumed to be 6 on the basis of XANES and UV-visible spectra. This might be the reason why the extent of chloro-complexation in the concentrated solutions is in systematic contrast with XRD findings (see Table 16 in Chapter 3).

Measurements at the Cl K-edge were also used by Sandstrom to study nickel-chloro complexes in solutions of $NiCl_2$.[99] In the radial functions reported in Figure 26, the crystalline $NiCl_2 \cdot 6H_2O$ shows clearly two peaks due to Cl – Ni and Cl – O distances. The functions of the solutions show a progressive decrease of the first peak, indication of diminishing direct cation-anion interactions with increasing dilution. The same evidence comes from the inspection of the pre-edge region, where a small but clear hump, diagnostic of Ni-Cl interactions appears, though less evident than in the case of $FeCl_3$ solutions. Although the evaluation of

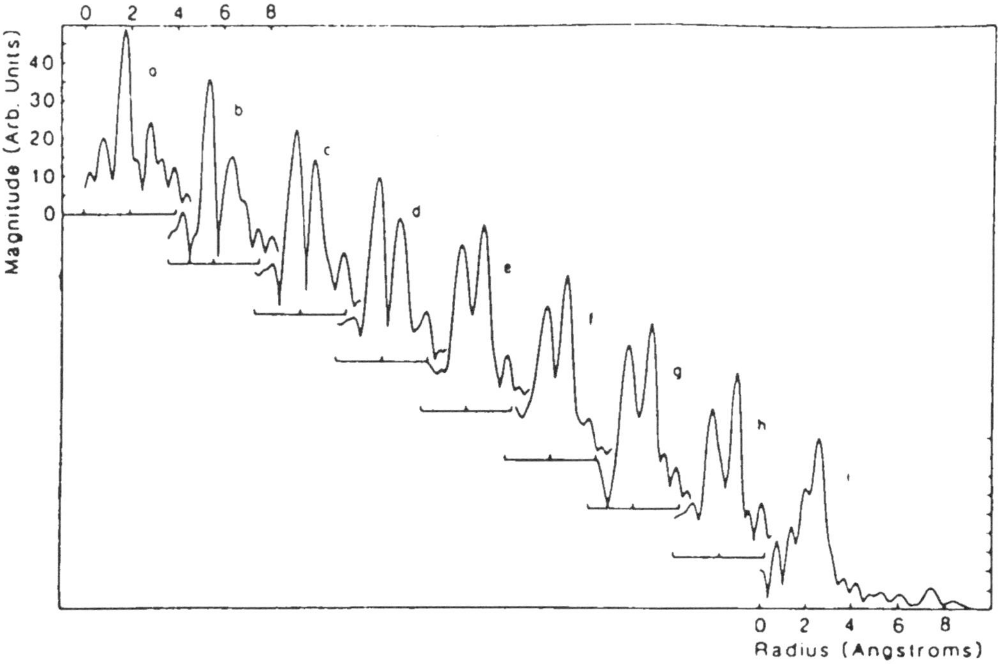

FIGURE 24. Fourier transforms of Cl K-edge EXAFS spectra for $FeCl_3$ hydrated salt (a) and $FeCl_3$ solutions in order of decreasing Fe^{3+} concentration in range 5.75 to 0.4 *M* (b-i). (From Sandström, D. R., in *EXAFS and Near Edge Structure III*, Hodgson, K. O., Hedman, B., and Penner-Hahn, J. E., Eds., Springer-Verlag, Berlin, 1985. With permission.)

chloro-complexation extent at the various concentrations was not attempted, these results definitely show that Ni – Cl contacts are present at high concentrations, in accordance with the most recent XRD results.[101] On the contrary, and as discussed before, Ni K-edge spectra were not sensitive enough to reveal the same phenomenon. A more favorable situation was found in Ni K-edge measurements of 2-*M* $NiCl_2$ solutions, with Cl/Ni ratios greater than the stoichiometric value.[89] Figure 27 shows the radial functions of three solutions and of the crystalline $NiCl_2{\cdot}6H_2O$. It is apparent that the solutions display some evolution towards the crystal (4 oxygens at 2.05 Å and 2 chloride ions at 2.36 Å), suggesting an increasing degree of complexation as the Cl/Ni increases. The values of the structural paremeters obtained by best fitting procedures are reported in Table 10, together with the results obtained in XRD studies.[101,102] The comparison reveals complete agreement as regards the trend of the complexation with the chloride concentration, while the Ni – Cl distance values obtained by the two techniques are somewhat different. An investigation where the EXAFS unique selectivity is exploited is reported in a paper by Corrias et al.[103] aimed at the understanding of the competitive behavior of nickel and cobalt towards the chloro-complexation. A mixed solution of $NiCl_2$ and $CoCl_2$, 0.5 *M* in each of the cations, and with a chloride/metal ion ratio of 7 was examined by XRD and EXAFS. In this case, XRD could only provide information about an "average" metal species, because of the similar response of the two metal ions to the diffraction. The absorption data collected in the proper energy range show clearly the absorption edges of the two metals (Figure 28), and the two interference functions can be analyzed separately. The best quantitative results were obtained assuming that only oxygen atoms surround the nickel, while two types of distances, namely Co – O and Co-Cl, had to be accounted for to obtain a good fit to the experimental Co function. The conclusion was therefore reached, that in these conditions the Co^{2+} ion tends to form chloro-complexes in greater extent than the Ni^{2+} does. These results are reported in Table 11.

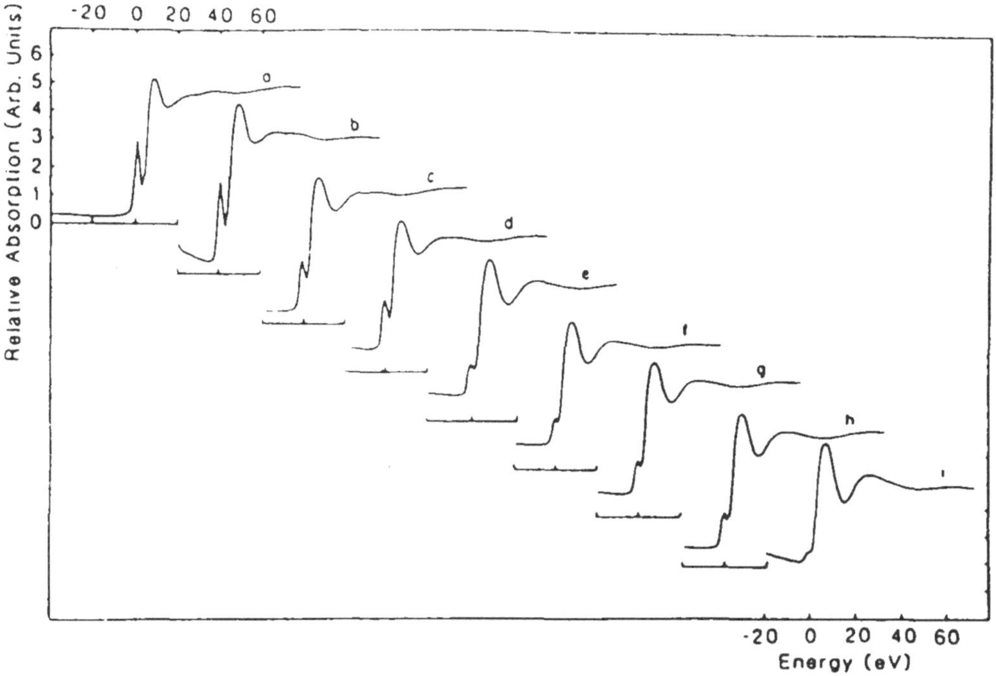

FIGURE 25. Cl K-edge XANES spectra for the same samples as in Figure 24. (From Sandström, D. R., in *EXAFS and Near Edge Structure III*, Hodgson, K. O., Hedman, B., and Penner-Hahn, J. E., Eds., Springer-Verlag, Berlin, 1985. With permission.)

Table 9
RESULTS OF CURVE-FITTING ANALYSIS OF EXAFS DATA OF $FeCl_3$ SOLUTIONS

Sample	Fe^{3+} conc. (M)	$(Cl^-)/(Fe^{3+})$	$r_{Fe\text{-}O}$ (Å)	$r_{Fe\text{-}Cl}$ (Å)	$N_{Fe\text{-}O}$	$N_{Fe\text{-}Cl}$
1	<1	3	1.98—2.00	2.26—2.35	5.2	0.8
2	3.0	3	2.02	2.30	4	2
3	4.3	3	2.02	2.30	4.2	1.8
4	3.0[a]	3	2.01	2.30	4	2
5	3.1	4	2.04	2.30	4	2

[a] Solution acidified with HNO_3.

Me – Br bonding formation in $MeBr_2$ solutions (Me = Cu, Zn, Cd) in a wide range of concentration was demonstrated through measurements on cation and anion edges. In these studies, the emphasis was on the involvement of bromo-complexes in the formation of extended structures. Solutions of $CuBr_2$ in the concentration range from 0.05 to 4.5 M were investigated by Fontaine et al.[104] and Lagarde et al.[94] On the basis of numbers and types of neighbors independently found around central Cu and Br, the following picture for local ordering around the ions was proposed: (1) at low concentrations (<0.5 M), half of the Cu and Br ions are linked together, but with a large degree of disorder; the hydration number of the Cu ions is about 7, revealing prevalence of hydration; (2) at high concentrations (>1 M), the local ordering between Cu and Br ions becomes definite and similar to that found

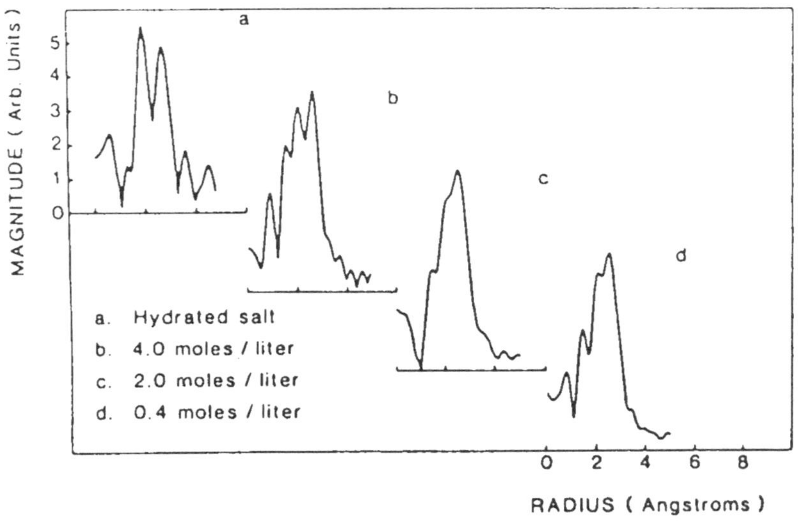

FIGURE 26. Fourier transforms of Cl K-edge EXAFS spectra for $NiCl_2$ hydrated salt and solutions. (From Sandström, D. R., in *EXAFS and Near Edge Structure III*, Hodgson, K. O., Hedman, B., and Penner-Hahn, J. E., Eds., Springer-Verlag, Berlin, 1985. With permission.)

in the anhydrous $CuBr_2$ (parallel chains of rectangles, with Cu set at the center and Br at the corners). These findings might suggest the existence of long chains even in concentrated solutions; however, the authors themselves observe that EXAFS, being a local probe, cannot tell how far the order actually extends. The same interpretative scheme was used by Lagarde et al.[94] in a study on solutions of $ZnBr_2$ (from 0.1 *M* to the saturation). It was proposed that an increasing number of ions depending on the concentration (about 85% at saturation) is engaged in $ZnBr_4^{=}$ tetrahedral units linked together in a solid-like structure. At concentrations as low as 0.1 *M*, there is still evidence of metal-halogen bondings, but the overall structure is increasingly dominated by hydrated ions. Measurements taken on solutions of $CuBr_2$ and $ZnBr_2$ at different temperatures did not reveal changes in the metal-halogen coordination.[94] On the contrary, a general decrease in the hydration number was observed at increasing temperature, with a sudden change between 45 and 60 °C. This effect was also observed by Galli et al.[105] on a 2-*M* $CuBr_2$ solution. In contrast to the behavior of Cu^{2+} and Zn^{2+} ions, measurements at Br K-edge of $SrBr_2$ solutions indicated complete absence of Sr – Br bonding.[94] Combinations of Cu, Zn, and Sr bromide in solution were investigated, with the aim of comparing the ability of Cu^{2+} and Zn^{2+} ions to form bromo-complexes. Alliota et al.[106] examined several mixed solutions of composition $MeBr_2$ + $Me'Br_2$ (with Me, Me' = Zn, Cu, Sr) by EXAFS measurements taken at Cu and Zn edges. They found that Sr^{2+} ions behave as "bromine donors" towards Cu^{2+} and Zn^{2+}, while the Cu^{2+} ion plays the same role towards Zn^{2+}. Another system which seems to display extended order phenomena was examined by Sadoc et al.[107] Seven solutions of $CdBr_2$ of concentration ranging from the saturation to one diluted 20 times were examined at Cd and Br edges. The average coordination numbers obtained were interpreted in terms of units like pyramidal $CdBr_3^-$ or tetrahedral $CdBr_4^{=}$ forming polymeric networks, with some hydrated Cd^{2+} ions along. Cd – Br bondings were still evident at very low concentration in amount even greater than in the case of $CuBr_2$ and $ZnBr_2$ solutions. A comparison with XRD results about the extent of ordering in Cu, Zn, Cd bromide solutions can be made only for Cu and Zn cations. Ichihashi et al.[108] studied $CuBr_2$ solutions in a wide range of concentrations and Br/Cu ratios; neither the coordination numbers they obtained, nor the general picture of the bromo-

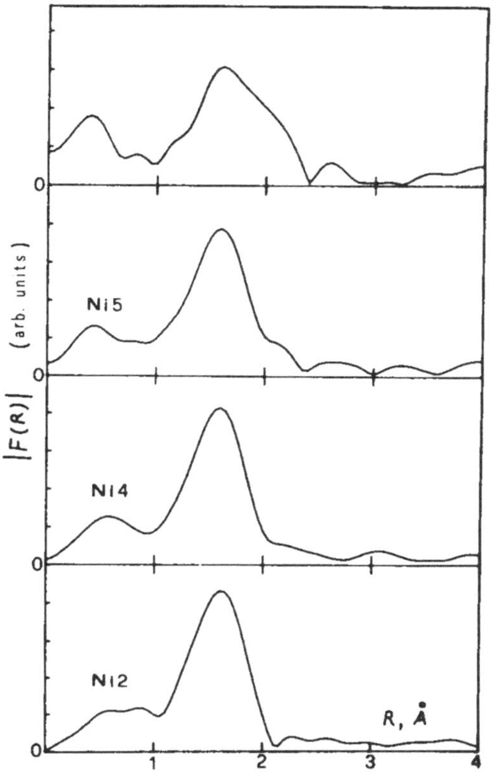

FIGURE 27. Fourier transforms of Ni K-edge EXAFS spectra for $NiCl_2$ hydrated salt (top), and for 2-*M* solutions with $(Cl^-)/(Ni^{2+})$ ratios equal to 2, 4, and 5 (from bottom). (From Licheri, G., Paschina, G., Piccaluga, G., and Pinna, G., *J. Chem. Phys.*, 79, 2168, 1983. With permission.)

Table 10
RESULTS OF CURVE-FITTING ANALYSIS OF EXAFS AND XRD DATA OF $NiCl_2$ SOLUTIONS

			r_{Ni-O} (Å)		r_{Ni-Cl} (Å)		N_{Ni-O}		N_{Ni-Cl}	
Sample	(Ni^{2+})	$(Cl^-)/(Ni^{2+})$	EXAFS	XRD	EXAFS	XRD	EXAFS	XRD	EXAFS	XRD
NI2	2 *M*	2	2.06	2.06	—	—	6.5	6	—	—
NI4	2 *M*	4	2.05	2.06	2.38	2.47	6.0	5.1	0.7	0.9
NI5	2 *M*	5	2.05	2.07	2.38	2.44	5.3	4.6	1.0	1.4

complexes they describe seem to be consistent with polymeric structures. The extensive study of Goggin et al.[109] on solutions of $ZnBr_2$ shows evidence of Zn – Br contacts involving $ZnBr_4^{2-}$, $ZnBr_3^-$, and $ZnBr_2$ complexes, but reports no indication of Zn – Br – Zn bridgings in support of extended structures.

F. Final Comments

The amount of EXAFS results so far obtained on solutions is somewhat limited for an

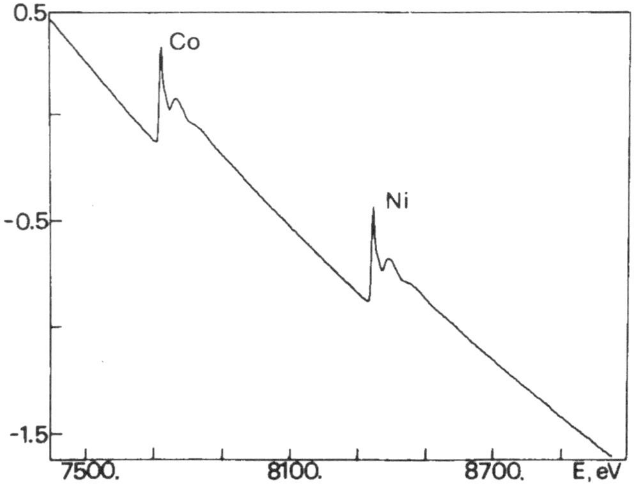

FIGURE 28. XAS spectrum of a mixed solution of $NiCl_2$ (0.5 *M*), and $CoCl_2$ (0.5 *M*), enriched in Cl^- to a ratio $(Cl^-)/(Me^{2+}) = 7$. (From Corrias, A., Musinu, A., and Pinna, G., *Chem. Phys. Lett.*, 120, 295, 1985. With permission.)

Table 11
RESULTS OF CURVE-FITTING ANALYSIS OF EXAFS DATA OF A MIXED SOLUTION OF $NiCl_2$ AND $CoCl_2$ ENRICHED IN Cl^-

	Cobalt	Nickel
$r_{Me\text{-}O}$(Å)	2.11	2.06
$N_{Me\text{-}O}$	5.68	7.00
$r_{Me\text{-}Cl}$ (Å)	2.46	—
$N_{Me\text{-}Cl}$	1.23	—

exhaustive evaluation of the state of the art. One reason for this situation is that the "in principle" potentialities of EXAFS often clash with objective difficulties that prevent their complete exploitation. For example, unfavorable experimental conditions in some energy regions prevent the obtaining of good spectra at the edges of many ions. Moreover, the limited access to synchrotron facilities does not allow many people to acquire the confidence in the experimental data they generally reach when working with their own equipment. The progress in the instrumentation, the development of reliable laboratory spectrometers, and the availability of new **dedicated** synchrotron radiation sources will permit to overcome these difficulties. However, our feeling is that many people have surrendered too quickly, after the initial enthusiasm, when they realized that the EXAFS technique could not be employed as straightforwardly as they had expected.

Going to the papers reviewed, let us consider briefly the reliability of the structural information obtained, in comparison with what is known from the XRD technique, which, as illustrated in Chapter 3, has produced the largest quantity of structural results about solutions. In the case of full hydration of the cation, the first shell distance values reported in Table 8 appear, as expected, to be of good accuracy. These distance values generally agree, within the experimental errors, with XRD ones, which are also evaluated in a rather direct way and with as much accuracy. When the shell is doubly composed, EXAFS has shown to be strongly sensitive to the disorder of the longer distance contribution. This has been the case with the oxygen environment around Cu, where EXAFS was unable to detect $Cu-O_{ax}$ distances, while XRD succeeded in revealing them among the other contributions which determine the experimental spectrum. On the contrary, when the disorder is low, the distance values determined by EXAFS showed to be good estimations, generally better than XRD determinations, as in the case of O and Cl around Ni ions.

The values reported for coordination numbers came out much less reliable than distances, as expected after the discussion in the sections above. Therefore, great caution should be used when sophisticated interpretations are strongly based on the values of coordination numbers. For example, the proposal that Cu – Br ordering in aqueous solutions is similar to that present in anhydrous $CuBr_2$ is critically dependent on the determined values of coordination numbers.[104] The values of N_{Cu-Br} and N_{Cu-O} are not too dissimilar from those obtained by XRD in a study by Ichihashi et al.,[108] but perhaps sufficiently dissimilar to open different interpretations. In particular, the decrease of N_{Cu-Br} with decreasing concentration reported by Ichihashi et al. does not agree with the hypothesis of substantial extent of complexation even at very low concentration. Again, in the case of $ZnBr_2$,[94] the coordination numbers N_{Zn-Br} differ systematically by a factor of about 2 from those reported subsequently in the XRD study of Goggin et al.,[109] raising questions about the model compound used which suffered preparation difficulties, as stated by the authors themselves. Here EXAFS pays for its inherent difficulties in treating the amplitudes. Yet, the general lines of these EXAFS studies are the correct ones, as two independent pieces of information about cations and anions are put together for understanding the overall structure.

Indeed, the field of aqueous solutions seems to be particularly suitable for a large involvement of the specific and somewhat unique potentialities of the EXAFS technique. For example, the study of dilute solutions, inaccessible to XRD, may become the source of new important structural information which cannot be obtained by extrapolation from high concentrations. Moreover, as the structural behavior of ionic solutions consists in a local restructuring around the dissolved ions, the independent access to cations and anions would be a fundamental element for the complete understanding of hydration and complexation phenomena. In particular, when the chemical complexity of the solution increases, as in mixed solutions of different salts, the atomic selectivity of EXAFS becomes irreplaceable.

Finally, we want to observe that EXAFS has the capability to give direct, qualitative evidences which can be much more important than some numerical values. For example, the direct evidence of a small, but certain, amount of Cl ions nearest neighboring to Ni in concentrated aqueous solutions of $NiCl_2$ represents the definite settlement of a long controversy.[99]

IV. COMPUTER SIMULATION STUDIES

A. Introduction

Among the techniques which can give information about the short range order of an ionic solution, computer simulations have a quite interesting position. In fact, detailed descriptions of spatial correlations among ions and water molecules can be obtained through averages of large numbers of different spatial configurations, which are derived one from the other following certain rules. A necessary, though not sufficient, condition why a simulated model can be judged as a good approximation of a real system is that the calculated value of a certain property is in good agreement with the experimental value of the same property, within the limits of the unavoidable errors. As we are interested in the static description of a system, the property of choice may be the correlation function (or a function related to it) and the experimental reference can be obtained by scattering experiments. Neutron diffraction, in fact, through the isotopic substitution technique provides the partial correlation functions describing the interactions in the solution, and comparison of these with the ones deduced from the simulation might be a good test for the latter. However, the available NDIS information is, today, rather limited, and, in some cases, of still debated reliability, so that this kind of comparison is not of general feasibility. Additional testing of the simulated model can be achieved comparing the total correlation function of the simulated solution with the ones obtained by single diffraction experiments (ND or XRD). This comparison, though not as detailed as those mentioned above, is as much meaningful, since the total correlation function of a realistic model must be consistent with its experimental counterpart. Assuming that a satisfactory agreement is reached, computer simulations offer the advantage of a very detailed description of the system. This, in turn, may support one particular solution when more than one interpretation of experimental results is possible, and/or may be of help in planning new experiments.

In the following, a brief description of the simulation techniques and related problems is presented in Section B. Section C deals with the most common expressions of potential functions describing the interactions in the solution, and comparison of these with the ones deduced from the simulation might be a good test for the latter. However, the available

B. Molecular Dynamics and Monte Carlo Techniques

Computer simulation studies have come out with increasing frequency in the last decades, following closely the development of computer hardware and computational methods. The early work was devoted to model systems (hard disks, hard spheres, Lennard-Jones particles), which was followed by applications to molecular systems. Today, many problems in liquid and solid systems have been approached by simulation methods,[110,111] using procedures that still follow the general outlines given in the fifties by two classical papers.[112,113]

Alder and Wainwright in 1959 introduced the simulation method called **molecular dynamics** (MD).[112] The simulated system consists of an assembly of particles confined in a closed space. A starting arbitrary configuration is chosen, where the particles are at rest or move with random distribution of velocities, but the system evolves rapidly to situations with Maxwellian distribution of velocities. The time evolution of the system is followed solving numerically the motion equations of the particles. After the effects of the initial configuration have died out, that is, when the system has "relaxed", the physical properties of the system can be obtained as averages over the time evolution.

Another possible approach is represented by the **Monte Carlo** method (MC). In this case, the system is assumed to be represented by a large number of configurations generated one from the other by random displacements of the particles (a roulette wheel might well be used for such operations), and weighted by the appropriate Boltzmann factor. However, a more efficient scheme has been devised by Metropolis et al.[113] in 1953. Following their "importance sampling" procedure, each new configuration is generated with a probability equal to its Boltzmann factor. The physical properties of the system can be calculated as averages over the configurations.

Both methods suffer from the approximations and limitations that have to be used in dealing with a multibody system. A severe approximation is given by the relatively small number of particles that can be considered simultaneously. For complicated molecules such as water, systems of no more than a few hundred molecules have been considered, with a few ions in case of solutions. A problem connected with the number of particles is given by the finite (and small) volume which contains the system; the ratio surface/volume is then not negligible and the description of the system would be greatly affected. To minimize this boundary effect, periodic conditions are generally introduced by surrounding the fundamental cell with image cells that repeat the system. Interactions among particles within the central cell and among particles pertaining to different cells are then accounted, and the system becomes large, albeit periodic.

C. Interaction Potentials

The fundamental problem in the simulation of a realistic system is still represented by the realistic description of the forces acting between particles. In the absence of external forces and accepting some approximations and simplifications, the **total intermolecular potential energy** U_N of an assembly of N particles can be written as a sum of terms, the first containing the sum of pair interactions, the second of triplet interactions and so on:

$$U_N = \sum_{i<j=1}^{N} u_2(\vec{\mathbf{r}}_i, \vec{\mathbf{r}}_j) + \sum_{i<j<k=1}^{N} u_3(\vec{\mathbf{r}}_i, \vec{\mathbf{r}}_j, \vec{\mathbf{r}}_k) + \ldots \tag{32}$$

With very few exceptions, only the first term of the expansion has been used in practical applications, with the approximation known as **pairwise additive forces**.

Much effort has been devoted to the calculation of water-water pair potentials, following, essentially, two different approaches. In the first theoretical approach, numerical values of the potential energy surface as function of the reciprocal position and orientation of two water molecules are obtained by *ab initio* quantum mechanical calculations. The second class of potentials has been based on empirical models of the water molecule, "waterlike" objects rather than molecules, which present characteristics peculiar to the water itself. In both cases, the pair potentials have to be expressed by analytical formulas as function of the coordinates of the molecules. A number of adjustable parameters appears in these formulas, and their values are fixed through fitting procedures against proper references: dimerization energies obtained by theoretical calculations in the case of theoretical potentials, values of physical properties of the water itself in the case of empirical potentials. Corrections for effects not yet accounted for are often introduced at this stage, particularly in theoretically derived potential expressions, so as to make the pair functions as **effective** as possible. The total potential energy of the system U_N is then assumed to be acceptably represented by sums of effective pair potentials V_{ij}. The same two approaches have been followed for the description of the other interactions which occur in an ionic solution, namely, ion-water and ion-ion interactions. General aspects of the matter can be found in References 114 to 117.

1. Theoretical Approach

The theoretical approach was followed by Clementi and co-workers in a truly massive attempt to describe the interactions between two water molecules and between a water molecule and several simple ions. Popkie et al.[118] calculated the potential energies for 200 configurations of the water dimer. With reference to a simple point charge model of two interacting molecules (Figure 29), they presented an analytical formula which gives the pair potential curve as function of distance and orientation of one water molecule with respect to the other. The pair potential V_{H_2O,H_2O} has the following expression:

$$\begin{aligned} V_{H_2O,H_2O} = {} & q^2(1/r_{13} + 1/r_{14} + 1/r_{23} + 1/r_{24}) + 4q^2/r_{78} \\ & - 2q^2(1/r_{18} + 1/r_{28} + 1/r_{37} + 1/r_{47}) + a_1\exp(-b_1r_{56}) \\ & + a_2[\exp(-b_2r_{13}) + \exp(-b_2r_{14}) + \exp(-b_2r_{23}) + \exp(-b_2r_{24})] \\ & + a_3[\exp(-b_3r_{16}) + \exp(-b_3r_{26}) + \exp(-b_3r_{35}) + \exp(-b_3r_{45})] \end{aligned} \tag{33}$$

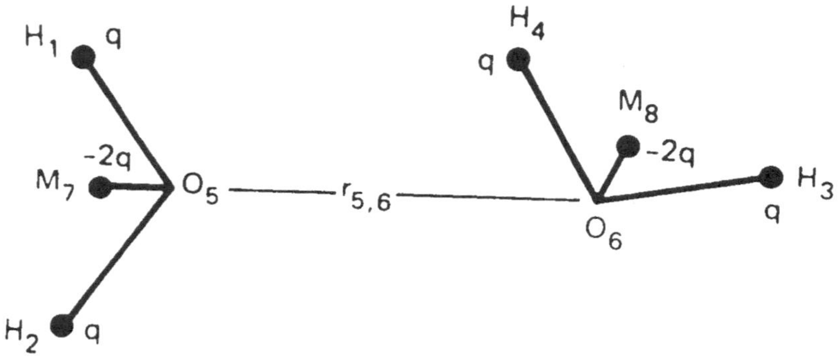

FIGURE 29. Point charge model used in the derivation of analytical water-water interaction potential. (From Kistenmacher, H., Lie, G. C., Popkie, H., and Clementi, E., *J. Chem. Phys.*, 61, 546, 1974. With permission.)

The best values of the constants in Equation 33 were obtained by fitting methods against the numerical values of the potential obtained theoretically. Modifications of this basic formula have been presented in the years, some of which take into consideration effects that were neglected in the original calculations. The influence of triplet and quadruplet terms on the total potential energy of the water (Equation 32) was examined,[119] but more important were other effects deemed: in particular, several empirical forms of correlation energy corrections were critically evaluated,[120] and definite improvements in the simulation results were obtained introducing into the theoretical potential a radial-dependent correlation correction calculated by perturbation methods.[121]

New and more realistic theoretical calculations on the water dimer were obtained in 1976 using the method of **configuration-interaction** (CI).[122,123] Also the angular part of the correlation term was taken into account by this method. An analytical formula formally differing from Equation 33 for a few extra terms was proposed; all the parameters appearing in it (which, of course, do not have any physical meaning) were obtained by fitting procedures against the numerical values of the potential energy relative to 66 different geometries of the two water molecules.

Theoretical calculations on ion-water potentials have been performed by the same group[124,125] who also proposed analytical formulas to represent the numerical values obtained. The simplest ion-water model is given in Figure 30, while Equation 34 represents the ion-water potential, V_{i,H_2O}

$$V_{i,H_2O} = Q(R^{-1}_{i-H_1} + R^{-1}_{i-H_2} - 2R^{-1}_{i-M}) - Q_m(R^{-1}_{i-M_1} - R^{-1}_{i-M_2} - 2R^{-1}_{i-O}) + A_1[\exp(-a_3R_{i-H_1}) + \exp(-a_3R_{i-H_2})] + A_2\exp(-a_4R_{i-O}) \quad (34)$$

The ability of these formulas to represent realistic situations was tested through simulations (see following sections).

For a more complete review of the work in this field see Reference 117.

2. Empirical Approach

The approach to empirical potential energy functions started in 1972 with the work of Ben-Naim and Stillinger.[126] The BNS model of the water molecule (Figure 31) is a tetrahedral object apt to simulate the type of coordination exhibited by the water molecules in bulk water and in aqueous solutions. The interaction potential between two water molecules is given by two terms, a spherically symmetric Lennard-Jones interaction term and a noncentral

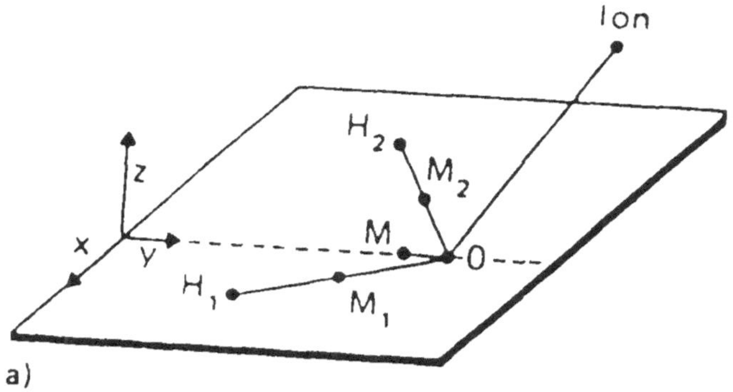

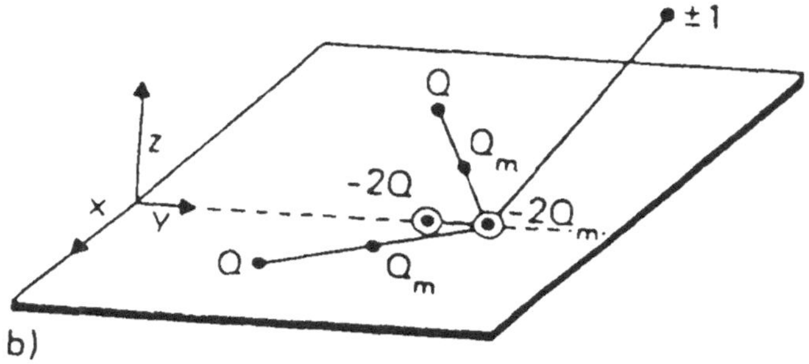

FIGURE 30. Geometrical relative position (a) and value and sign (b) of the point charges used to represent a simple model for analytical ion-water potential. (From Kistenmacher, H., Popkie, H., and Clementi, E., *J. Chem. Phys.*, 59, 5842, 1973. With permission.)

electrostatic contribution. The first term, v_{LJ}, refers to oxygen nuclei as the force centers, and the constants ϵ and σ are assumed to have the same values as in neon which is isoelectronic with water:

$$v_{LJ}(r_{12}) = 4\epsilon[(\sigma/r_{12})^{12} - (\sigma/r_{12})^{6}] \quad (35)$$

The electrostatic contribution, v_{el}, accounts for the interactions between two tetrahedral charge distributions and consists of 16 different charge-pair terms:

$$v_{el}(\vec{\mathbf{r}}_1, \vec{\mathbf{r}}_2) = (\eta e)^2 \sum_{\alpha_1,\alpha_2=1}^{4} [(-1)^{\alpha_1\alpha_2}]/[d_{\alpha_1\alpha_2}(\vec{\mathbf{r}}_1, \vec{\mathbf{r}}_2)] \quad (36)$$

Here, α_1 and α_2 run over the four charges of molecule 1 and 2, so that even and odd values correspond to negative and positive charges, respectively; the quantity $d\alpha_1\alpha_2$ is the distance between the charges α_1 and α_2 and depends on the full set of molecular coordinates $\vec{\mathbf{r}}_1$ and $\vec{\mathbf{r}}_2$; the value of the point charges ηe is considered an adjustable parameter. The pair potential V_{H_2O,H_2O} assumes therefore the expression:

$$V_{H_2O,H_2O} = v_{LJ}(r_{12}) + S(r_{12})\, v_{el}(\vec{\mathbf{r}}_1, \vec{\mathbf{r}}_2) \quad (37)$$

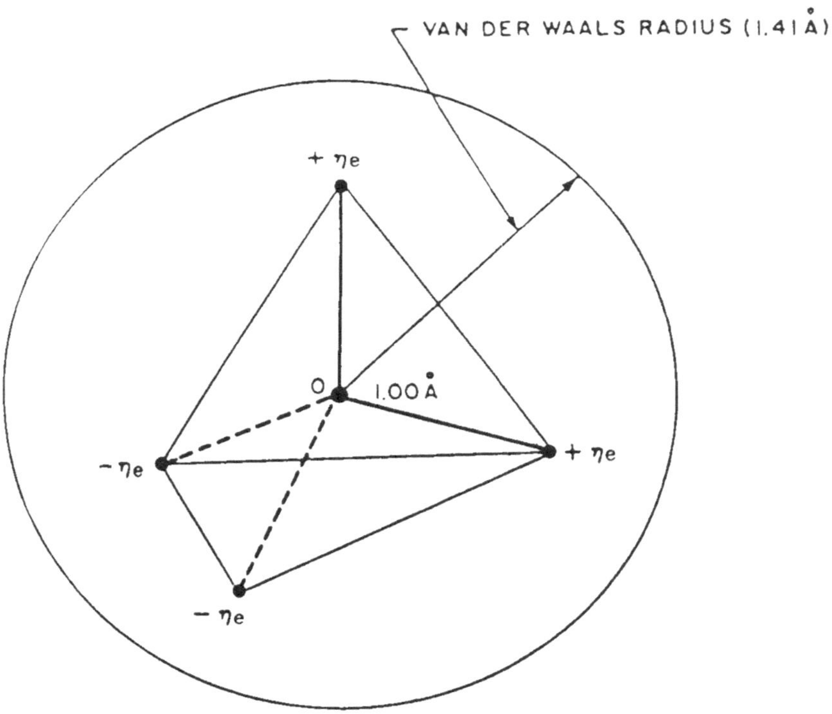

FIGURE 31. Tetrahedral charge distribution for the water molecule. The oxygen nucleus O is at the center of the regular tetrahedron with circumradius 1.00 Å. Positive charges are shielded protons, negative charges simulate unshared electron pairs. (From Stillinger, F. H. and Rahman, A., *J. Chem. Phys.*, 60, 1545, 1974. With permission.)

The term v_{el} is multiplied by a **switching** function S to prevent the collapse of two charges of opposite sign ($d_{\alpha_1 l_{\alpha_2}} = 0$). The BNS model underwent some variations to make it more realistic. The new version, called ST2,[127] is basically the same, except for the oxygen-hydrogen distance within the water molecule; the pair potential function is expressed as in Equation 37, but the constants appearing in Equations 35 and 36 have different values. The capability of the ST2 model to describe the structure of pure water was tested through an MD simulation: the main features displayed by the experimental structure functions obtained by X-ray and neutron diffractions were well reproduced in the same function obtained by simulation.[127]

To describe ion-water and ion-ion interactions, the same basic formulas have been generally used, with the necessary variations to replace four-charge molecules with single charge points; new values were given to the constants appearing in the Lennard-Jones part, and the electrostatic part consisted of fewer terms. Slightly different switching functions were considered for ion-water terms. Besides, other types of switching functions were often introduced in the simulations, to cut-off long range interactions between ions and save computer time.

A new class of models of liquid water was presented in 1975. Lemberg and Stillinger[128] proposed a picture of the water molecule as a three-point effective charge. Interacting through central forces only, from which the appellation of central-force model, three charges automatically assume the molecular structure of the isolated molecule; besides, they are also able to give rise to water dimers of remarkable stability. The description of liquid water can be made using a linear combination of the three water potentials V_{HH}, V_{OH}, and V_{OO}, which are expressed by simple functional forms. Also in this case, a MD simulation of pure water provided the necessary test of the model.[129] As a consequence, the expressions of the three potentials were slightly changed. Further attempts to improve the formulas brought variations only in the constants.[130] The latest expressions of the three potentials are given by:

$$
\begin{aligned}
V_{HH}(r) &= \frac{36.1345}{r} + \frac{18}{1 + \exp[40(r - 2.05)]} \\
&\quad - 17\exp[-7.62177(r - 1.45251)^2], \\
V_{OH}(r) &= -\frac{72.269}{r} + \frac{6.23403}{r^{9.19912}} - \frac{10}{1 + \exp[40(r - 1.05)]} \\
&\quad - \frac{4}{1 + \exp[5.49305(r - 2.2)]} \\
V_{OO}(r) &= \frac{144.538}{r} + \frac{26758.2}{r^{8.8591}} \\
&\quad - 0.25\exp[-4(r - 3.4)^2] - 0.25\exp[-1.5(r - 4.5)^2]. \qquad (38)
\end{aligned}
$$

Other attempts to describe the water molecule as a rigid object made up by several (three, four, and five) interaction sites were made by Jorgensen.[131,132] Good compromise between simplicity and effectiveness was found in 4-site and 3-charge models (TIPS2 and TIP4P),[133] in which the water molecule was described by two point charges set on the hydrogen atoms, and one set on a point displaced away from the oxygen on the C_2 axis. The pair potentials between these monomers were described by coulombic interaction terms between the intermolecular pairs of charges and a Lennard-Jones term between oxygens. The best functions obtained were introduced in a MC simulation of pure water,[133] and the pair correlation functions calculated were found in good agreement with ND experimental results.

D. Molecular Dynamics Simulations of Aqueous Solutions

Much work has been devoted to the study of liquid water, often with the aim of testing the models devised for the water molecule and the interaction potentials proposed for water pairs. More recently, the MD method has been applied to the simulation of aqueous solutions.

The mentioned general limitations of computer simulations still remain, but substantial improvements have been effected over the years. The number of particles is still of the order of a few hundreds, with a number of ions approaching 20. However, the algorithms used have become more differentiated in computing the various interactions, and long-range forces are accounted for using sophisticated procedures. Also the problem of periodic boundary conditions has been dealt with, taking into account, when needed, interactions between particles in the fundamental cell and particles in the image cells. This means that the few hundred particles have become virtually infinite and this has permitted the obtainment of meaningful values of bulk phase properties.

The picture emerging from MD simulations on solutions is today rather various. All the alkali metals have been considered in a variety of situations that go from infinite dilution (only one ion and different quantities of water molecules) to one case of LiCl about 14 *m*. Also, the halide ions have been studied in similar situations. Besides, there are some interesting investigations on divalent metals: Ca^{2+} and Mg^{2+} have been actually studied and also some results have been given for a transition metal ion (Ni^{2+}).

As for the interactions between particles, the existing studies have exploited the many possibilities mentioned above, so that a parallel evolution in theoretical studies and in simulations of real systems can be followed in the years ahead. The results are apparent in the interesting comparisons between simulated and experimental properties often reported in the papers, in particular, between simulated radial functions and data from ND and XRD.

1. One – Ion Studies

Leaving apart early studies on systems composed by too few particles (a "realistic" system should comprise a number of water molecules at least sufficient to form two hydration shells around the ion), mention must be done of a paper by Briant and Burton.[134] They considered clusters of eight water molecules and one monovalent ion: Na^+, Rb^+, Cs^+, F^-, Br^-, and I^-. Larger clusters containing up to 29 water molecules were also simulated in two cases, with Cs^+ and F^-. Water-water and ion-water interactions were described using the BNS potential.[126] The radial functions of the systems showed some evolution at increasing cluster size, and the authors concluded that quantitative information about the structure of large clusters could not be obtained by extrapolation from small clusters.

More recently, three studies have presented cases of large clusters of some tens of water molecules and one ion, Li^+, Na^+, K^+, Cl^-, F^-, Ca^{2+}, and Ni^{2+}.[135-138] The number of water molecules was almost the same in all the systems (64 or 125 in Reference 135, 79 in Reference 136, again 64 in References 137 and 138), but different types of water-water interaction potentials were used. Impey et al.[135] described the interactions between water molecules by the analytical potential function previously obtained by Matsuoka et al.[122] and based on *ab initio* quantum mechanical calculations. Nguyen et al.[136] modeled the water interactions using the revised central field potential developed by Stillinger and Rahman,[130] that is, through an empirical potential which treats both intermolecular and intramolecular interactions. Bounds[137,138] gave preference to the TIP4P model of the water molecule proposed by Jorgensen.[133] Ion-water interactions were described using different analytical formulas fitted to the same theoretical data of Kistenmacher et al.[124] in the first two cases, whereas Bounds used his own quantum mechanical calculations as a basis, to which he fitted simple forms. All pair correlation functions for ion-oxygen and ion-hydrogen reported in these papers display the same main features, with minor differences beyond the nearest neighbor range. The structural data obtained are reported in Table 12. The numerical values of ion-oxygen distances and coordination numbers of columns a and b agree quite well, with slight differences in the cases of light ions; the distance values of column c are higher than the others. Direct comparisons of these results with XRD data are difficult for the alkali ions, for the low reliability of the structural parameters determined by XRD in these cases. Among the results given for the Cl^- ion, those by Bounds (column c) seem the most realistic. For the divalent cations examined, XRD and ND provide good reference data; here the simulation results display a systematic overevaluation of distances and coordination numbers, which the author himself is inclined to attribute to some insufficiency in ion-water potentials. In any case, Table 12 presents the results of one of the few simulations of the hydration of an alkaline-earth ion, and of the only one so far carried out about a transition metal ion in water.

2. Solution Studies

a. The 1:1 Salt Solutions

The attempt to simulate realistic solutions has been made in a series of papers published from 1974. Heinzinger and Vogel[139-143] simulated 2.2-*m* aqueous solutions of LiI, LiCl, NaCl, CsCl, and CsF and a 0.55-*m* solution of NaCl. The water-water interactions were described using the ST2 model for the water molecule (see Figure 31) with the pair potential of the form given in Equations 35 to 37.[127] The ions were treated as point charges at the centers of Lennard-Jones spheres, and their interaction potentials with the water molecules and with the other ions were given by suitable modifications of Equation 37. The description of the ions in solution gave rise to some problems. First, their effect is far reaching, so, in order to prevent too long calculations, a proper switching function had to be introduced,[139] which rapidly reduced the coulombic part of the potential, Equation 36. It turned out that this approximation had to be not too drastic, so different choices were made in the follow-

Table 12
ION-OXYGEN DISTANCES r AND COORDINATION NUMBERS N OBTAINED BY MD SIMULATIONS OF ONE-ION IN WATER SYSTEMS[a]

	$r_{ion\text{-}O}$ (Å)			$N_{ion\text{-}O}$		
	a	b	c	a	b	c
Li^+	1.98	1.92	2.04	5.3	4	6.0
Na^+	2.29	2.24	2.35	6.0	6	6.2
K^+	2.76	2.76	2.86	7.5	7	7.6
F^-	2.67	2.64	—	5.8	6	—
Cl^-	3.29	3.28	3.19	7.2	7	6.0—6.4
Ca^{2+}	—	—	2.54	—	—	9.3
Ni^{2+}	—	—	2.17	—	—	8.0

[a] Results from: (a) Impey, R. W., Madden, P. A., and McDonald, I. R., *J. Phys. Chem.*, 87, 5071, 1983; (b) Nguyen, H. L. and Adelman, S. A., *J. Chem. Phys.*, 81, 4564, 1984; and (c) Bounds, D. G., *Mol. Phys.*, 54, 1335, 1985.

ing simulations.[140-142] Some information began to emerge, that is, ion-water distances, ion coordination numbers, and preferential orientation of the water molecules around the ions.[139-144] However, the quantitative comparison of these data with values obtained by XRD showed that all the anion-water distances tended to be shorter than the corresponding experimental ones.[141] The reason of this systematic discrepancy was essentially attributed to improper evaluation of the parameters appearing in the Lennard-Jones part of ion-water and ion-ion pair potentials whenever the anion was involved. New parameter values were calculated for the halide ions and the simulation of the 2.2-*m* solution of NaCl was performed again.[145] It is now interesting to compare the results of this simulation with experimental data, so as to understand how realistic the description of the various interactions is. However, the available experimental function obtained by X-ray diffraction obviously contains the contributions from all the pairs.[146] The separation of the main contributions, that is cation-water, anion-water, and water-water, was obtained fitting a first neighbor model (FNM2, see Chapter 2) to the experimental data. The partial functions thus calculated were considered as "experimental" references of the corresponding simulated functions. The results of this procedure are reported in Figure 32. The general picture is rather good, although the cation-water peak encompasses the "experimental" one and some discrepancy remains in the water-water correlation function. Leaving apart punctilious comparisons of single structural parameters which cannot be meaningful given the model interpretation of the experimental data,[147] we may obtain an overall impression of the reliability of this simulation by looking at the total structure functions obtained directly from XRD and from MD simulation. They are reported in Figure 33 (points and dashed line, respectively) and show clearly the major point of discrepancy. In fact, the two functions agree quite well in the whole s range, with the exception of the first double-peaked maximum, characteristic of water-water interactions, which appears as a simple oscillation in the MD function. This suggests that the ST2 model of the water molecule should be responsible for the disagreement and may be inadequate to reproduce the behavior of the water in solution.

A completely different approach was adopted by the same group in a following attempt to simulate the same system. Bopp et al.[148] used the latest version of the central force (CF)

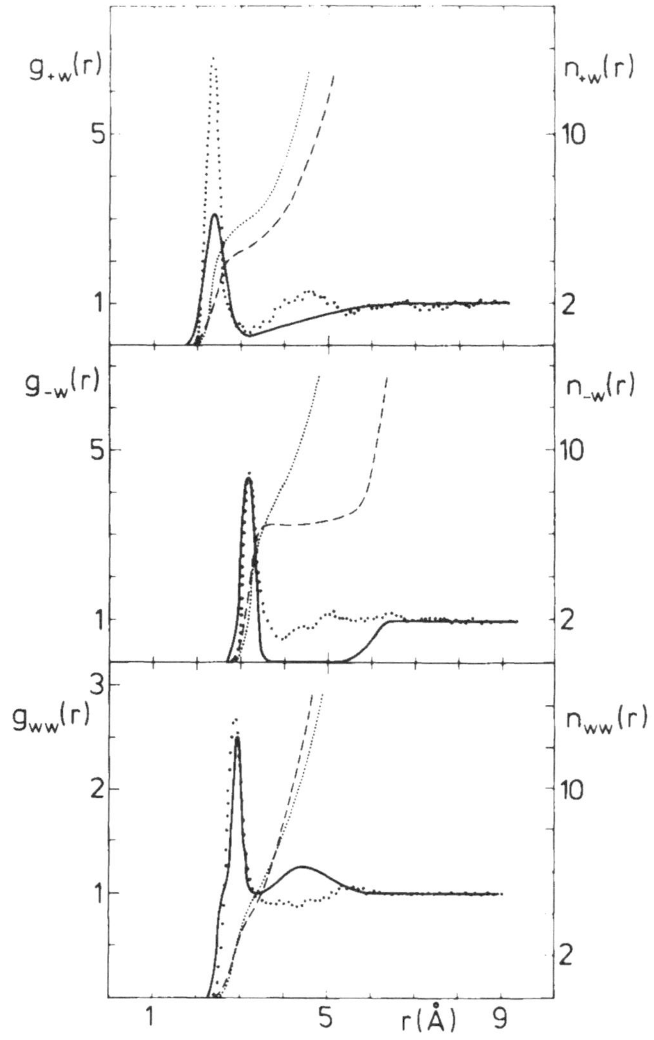

FIGURE 32. Pair correlation functions g(r) and running coordination number n(r) obtained by MD simulation of a 2.2-*m* solution of NaCl (dots) and by FNM2 fits to XRD results: g(r) (full line) and n(r) (dashed line). (From Pálinkás, G., Riede, W. O., and Heinzinger, K., *Z. Naturforsch.*, 32a, 1137, 1977. With permission.)

model of the water to describe the solvent.[130] As it implies the use of three different pair potentials, V_{HH}, V_{OH}, and V_{OO} (see Equations 38), separate contributions ion-O and ion-H had to be introduced. These were obtained fitting simple three-term formulas to *ab initio* numerical results, while ion-ion interactions were described as in the previous simulation.[145] The total structure function obtained in this work is also reported in Figure 33 (full line). It can be seen that the use of the CF model for the water leads to a good reconstruction of the double first oscillation, but, on the whole, the agreement MD vs. experimental function is definitely worse. It may be also inferred from the systematic frequency shift of the MD function that ion-water distances will have values somehow different from the experimental ones (and from those previously obtained as well).

The ST2 model of the water molecule was again used in the simulation of a 2.2-*m* solution

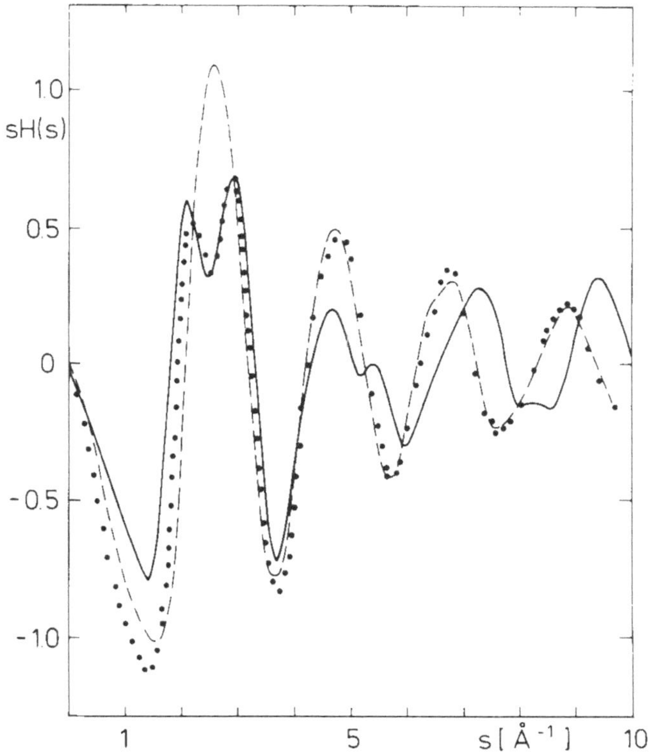

FIGURE 33. Comparison of structure functions of a 2.2-*m* solution of NaCl: XRD results (dots); MD simulation using ST2 (dashed line), and CF (full line) water model. (From Bopp, P., Dietz, W., and Heinzinger, K., *Z. Naturforsch.*, 34a, 1424, 1979. With permission.)

of LiI.[149-151] Using the same procedure applied in the case of NaCl,[145] the parameters to be introduced in the Lennard-Jones part of the potential were recalculated so that the interaction distances of the anion would fall at more realistic values. The total correlation function of the simulation is reported in Figure 34 (full line). It can be seen that the agreement with XRD data (points) is still poor, but the disagreement is essentially to ascribe to inadequate treatment of the water-water contribution (second peak in the G[r]). According to this observation, the treatment of the water was changed again in the simulation of a concentrated solution of LiCl (13.9 *m*).[153] The central force potential was in fact used to describe the interactions between water molecules,[130] while the intramolecular interactions were described adapting a spectroscopical potential.[152] Also ion-oxygen and ion-hydrogen pair potentials were calculated on the basis of theoretical data. The good result of the simulation is apparent in Figure 35, where the total correlation function calculated (full line) is reported together with the experimental one obtained by XRD (dashed line). The Li – O peak centered around 2 Å is still not perfectly reproduced in the simulation, but here also, the reference curve is not completely reliable because of the light weight of the Li^+ ion.

The overall picture emerging from these simulations is quite encouraging. It seems that at a qualitative level, the available pair potentials, and the MD procedure in general, have reached a sufficiently good level so as to reproduce the main features of water-water and ion-water interactions. The quantitative information is obviously more delicate. In fact, it seems that modifications of the adjustable constants appearing in the pair potential functions might lead to correct values of the structural parameters describing the systems. On the

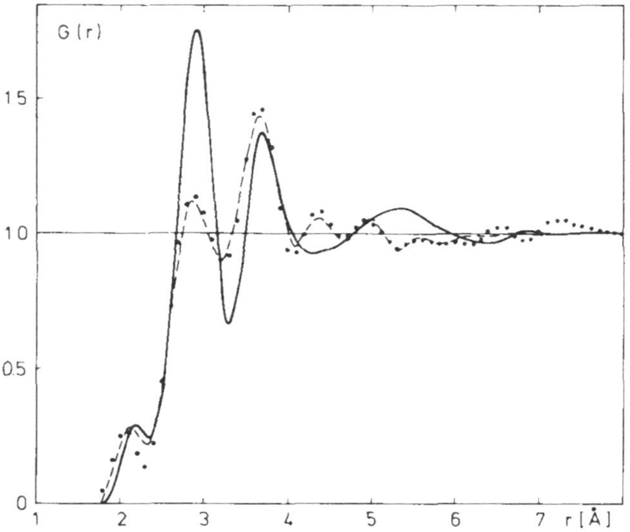

FIGURE 34. Total correlation functions for a 2.2-*m* solution of LiI from XRD data (dots), model fit to XRD data (dashed line) and MD simulation (full line). (From Radnai, T., Pálinkás, G., Szász, G. I., and Heinzinger, K., *Z. Naturforsch.*, 36a, 1076, 1981. With permission.)

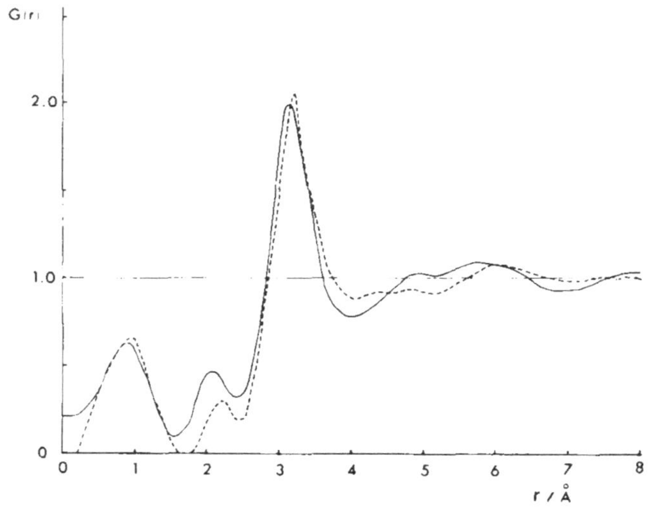

FIGURE 35. Total correlation functions for a 13.9-*m* solution of LiCl from XRD data (dashed line) and from MD simulation (full line). (From Bopp, P., Okada, I., Ohtaki, H., and Heinzinger, K., *Z. Naturforsch.*, 40a, 116, 1985. With permission.)

other hand, reliable experimental references must be available to judge the final simulation results. As pointed out in the previous chapter, alkali ions are detected by XRD not without problems. At the same time, they are the ones for which a great deal of theoretical calculations have been done, so that the solutions of alkali ions have been, and still are, the most convenient systems in the field of solutions to which apply the simulation procedures. Nor can the problem be solved considering the partial functions that are directly obtained by MD calculations. Only for a few cases, in fact, are experimental partial functions from neutron diffraction studies available, while XRD total functions can be separated into contributions only with the help of model interpretation. If the weight of one of the contributions is light, as in the case of Li^+, or the interactions are weak, as in the case of Na^+, the model interpretation of these terms is less reliable.

b. The 2:1 Salt Solutions

The experimental reference problem should not arise in the case of an ion, like magnesium, for which solutions, in fact, XRD radial functions show sharp interaction peaks corresponding to a well-defined first neighbor shells. The simulation of a 1.1-*m* solution of $MgCl_2$ was carried out by Dietz et al.[154] The description of water-water interactions was made using the central force pair potential,[130] on account of the good description of the water obtained in a previous calculation.[148] The ion-hydrogen, ion-oxygen, and oxygen-oxygen interactions were described separately using potentials based on *ab initio* calculations of different ion-water configurations. The formulas for ion-ion potentials had the same form, and were obtained by the former with slight modifications of them. A detailed comparison of the simulation results with experimental data is reported in a paper by Pálinkás et al.[155] Immediate impression of the simulation quality is given by Figure 36 which reproduces MD and XRD correlation functions. On the whole, the comparison is quite satisfactory, and all the experimental features are reproduced in the MD correlation function; the major point of discrepancy is given by a slight shift of the peak around 2 Å, which describes the first coordination shell of the cation: a coordination distance about 0.1 Å shorter than the experimental one can be anticipated from the figure. Considering the high reliability of the experimental results on this particular solution, we may use it as a test case and examine more closely the detailed description of a system as it is offered by the simulation technique.

An interesting aspect is given by the pair correlation functions directly provided by the simulation. Figure 37 reports the pair correlation functions of Mg^{2+} (top) and Cl^- (bottom) with H and O atoms of the water. The structuring of the ions is described in detail. The first shell of the cation is very well defined and separated from the second one; it is made up by 6 water molecules tightly bound to the ion (see dashed lines in Figure 37). The area under the peak centered at 2.75 Å contains the 12 hydrogen atoms of the 6 first shell water molecules; its sharpened shape says that each molecule is symmetrically oriented towards the cation. A second shell of about 15 water molecules is indicated by the second peaks in the g_{Mg-O} and g_{Mg-H}. Traces of weak structuring beyond the second shell are still visible in the range 6 to 7 Å.

The environment of the anion is less structured, as expected. Only one peak centered at 3.18 Å appears in g_{Cl-O}, indicating that about 6 water molecules form the first hydration shell around the anion; the two peaks in g_{Cl-H} show that the water molecules point one of the hydrogen atoms towards the anion.

Information on the symmetry of the coordination shells, on the distribution of the water molecules around equilibrium positions and on their preferential orientations are provided directly by the MD simulation. So, in the case in point, the six water molecules nearest neighbors to the cation form an almost perfect octahedron with very sharp distributions around the geometrical vertices, while a much more spread arrangement is displayed by the water molecules around Cl^-. With the help of a polar coordinate system, in which the ion defines the origin, one oxygen atom determines the z direction, and a second one the xz-plane, the spatial distribution of the water molecules around the ions can be visualized. Figure 38 shows the probability distributions of water molecules around the ions (and around a water molecule) expressed as functions of the azimuth angle φ and of the polar angle ϑ. In the case of Mg^{2+}, the distribution displays sharp maxima around $\vartheta = 90°$ and 180° (xy-plane and negative z-direction), and around $\varphi = 90°$, 180°, and 270° in the xy-plane (the

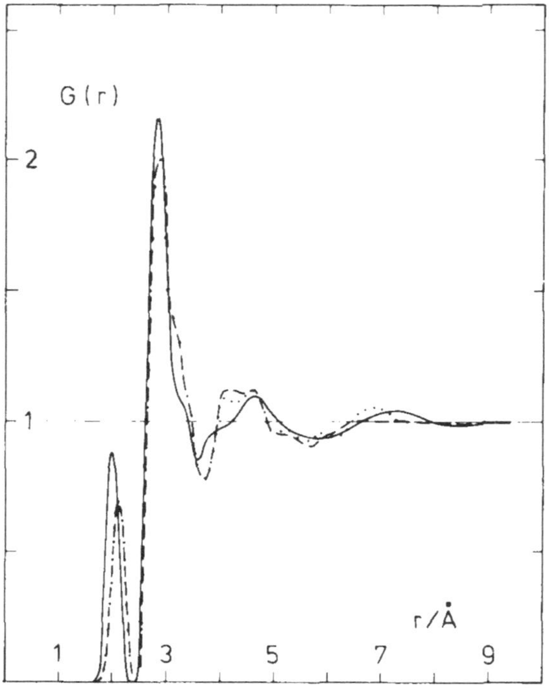

FIGURE 36. Total correlation functions for a 1.1-*m* solution of $MgCl_2$ obtained from XRD data (dots), model fit to XRD data (dashed line), and MD simulation (full line). (From Pálinkás, G., Radnai, T., Dietz, W., Szász, G. I. and Heinzinger, K., *Z. Naturforsch.*, 37a, 1049, 1982. With permission.)

densities relative to the molecules defining the reference system are omitted in the figure). The same functions for Cl^- show that the spatial arrangement of the water molecules around the anion is similar, but the preferential sites are loosely defined.

The orientation of the water molecules around an ion may be visualized in several ways.[154-156] A simple description is given by Szász et al.[156] Defining as ϑ the angle between the dipole moment vector of the water and the vector pointing from the oxygen to the ion, $P(\cos\vartheta)$ distribution shows how the water molecules are oriented towards the ion. Figure 39 shows that the water molecules around Mg^{2+} have almost perfect trigonal orientations, even more evidenced by the comparison with the case of Na^+; on the contrary, a broad distribution around the mean value of 51° is displayed by the water molecules around Cl^-.

Finally, the MD simulation provides us with another interesting aspect regarding the hydration shells of the ions, that is with the probability of finding a given number of coordination site directions occupied in coincidence. The result for the $MgCl_2$ is that in more than 60% of the time the six octahedral directions around magnesium are occupied by water molecules, while the frequency of this event is of about 1% in the case of Cl^-. Figure 40 shows the histograms relative to Mg^{2+}, Cl^-, and water (an occurrence of about 3% of full occupancy of the tetrahedral directions has been found for the latter).

A global image of the structuring around Mg^{2+} and Cl^- is provided by the tridimensional drawings reported in Figure 41.

A 1.1-*m* solution of $CaCl_2$ was simulated by Probst et al.[157] The pair potentials used to describe the ten possible interactions were of the same form as those used in the simulation of the $MgCl_2$ solution.[154] The correlation function obtained by the MD study is reported in

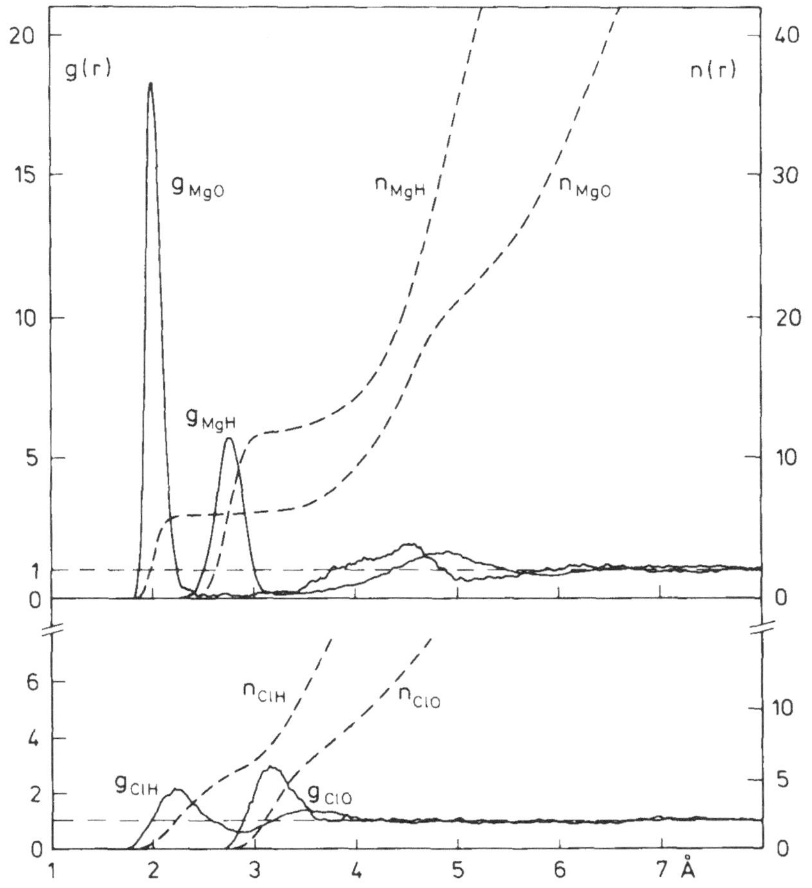

FIGURE 37. Ion-oxygen and ion-hydrogen pair correlation functions g(r) (full line) and running coordination numbers n(r) (dashed lines) for a 1.1-*m* solution of $MgCl_2$. (From Dietz, W., Riede, W. O., and Heinzinger, K., *Z. Naturforsch.*, 37a, 1038, 1982. With permission.)

Figure 42, together with one obtained by XRD data, and a model interpretation of the experimental function. The comparison is encouraging, as all the main experimental features are well reproduced by the simulation. The distance value obtained for Ca^{2+} contacts, 2.39 Å, is in agreement with the experimental results of X-ray and neutron diffraction studies. The experimental verification of the rather high value of 9 obtained by the simulation for the cation coordination is more difficult. In fact, the values obtained by diffraction data range from 5.5 to 10; besides, the interpretation of XRD data of $CaCl_2$ have presented some problems (see Chapter 3). In such a case, the simulation study might suggest possible explanations for a structural situation which appears unexpectedly so different from the case of Mg^{2+}. The projection of density maps of oxygen atoms around the Ca^{2+} are shown in Figure 43A. Figure 43B reports the same functions for Mg^{2+} from the mentioned investigations.[154-156] The water molecules around the Ca^{2+} show a high degree of order, but their arrangement does not seem to fit any regular polyhedron. The most probable candidate, the regular hexahedron, would involve the maxima which are marked with crosses in the figure, and that appear not to coincide with the maxima of the simulation. This particularly complicated arrangement might explain the difficulties found in diffraction studies to assign a definitely preferential coordination number to this ion. Further investigations will perhaps clarify better the overall picture of this system.

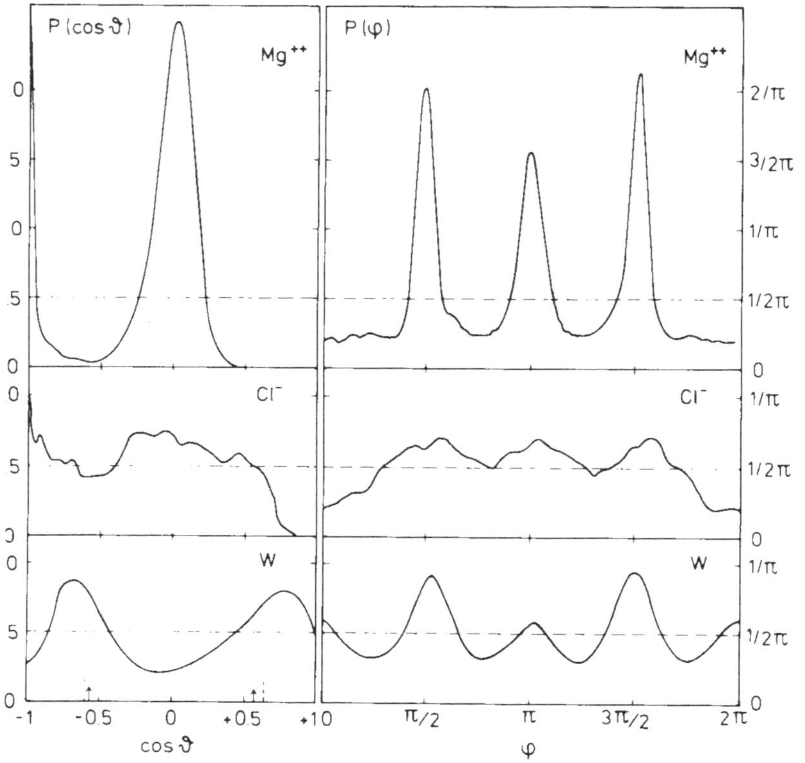

FIGURE 38. Probability density of oxygen atoms (from water molecules) around Mg^{2+}, Cl^- and water from the MD simulation of a 1.1-*m* solution of $MgCl_2$. (From Pálinkás, G., Radnai, T., Dietz, W., Szász, G. I., and Heinzinger, K., *Z. Naturforsch.*, 37a, 1049, 1982. With permission.)

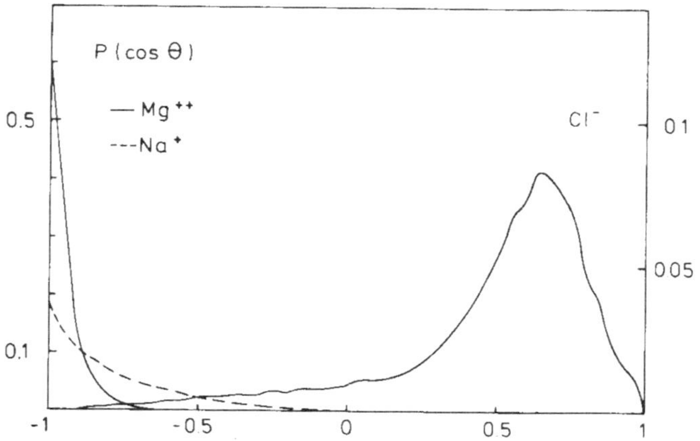

FIGURE 39. Orientation of the water molecules around ions: distribution of cosϑ for the water molecules in the first hydration shell of Mg^{2+}, Cl^-, and Na^+, as obtained from MD simulations. (From Dietz, W., Riede, W. O., and Heinzinger, K., *Z. Naturforsch.*, 37a, 1038, 1982. With permission.)

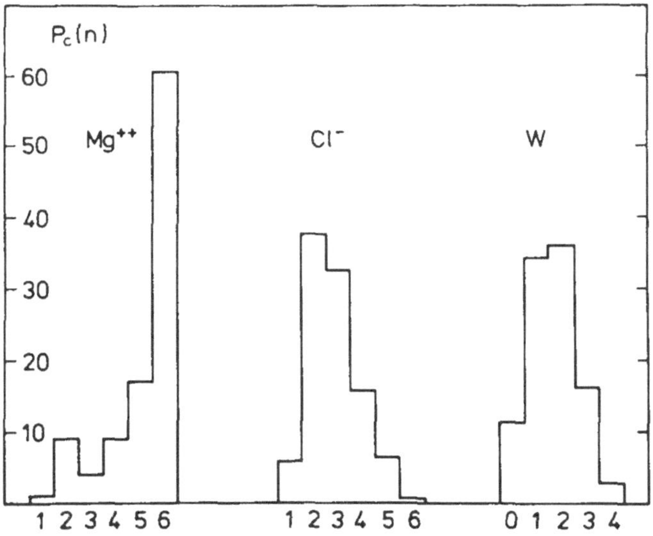

FIGURE 40. Probability of finding a given number of water molecules simultaneously occupying octahedral directions around Mg^{2+} and Cl^-, and tetrahedral directions around a water molecule. (From Pálinkás, G., Radnai, T., Dietz, W., Szász, G. I., and Heinzinger, K., Z. *Naturforsch.*, 37a, 1049, 1982. With permission.)

E. Monte Carlo Simulations

All MC studies on solutions so far published have explored the situation of one or two ions in water. In some cases the investigations dealt with small clusters made up by one ion and a few water molecules, and were aimed at gaining understanding about preferential configurations of the water molecules in the neighborhood of the ion. Though these studies have undoubtedly provided valuable information about ion-water interactions (tests of theoretical potentials, suggestions about importance of molecular configurations, and so on) they describe situations too far from real solutions. So, only the few papers that consider at least some tens of water molecules per system will be mentioned here, with the assumption that only in these cases some sort of comparison may be accomplished.

Two papers report simulations of aqueous solutions of LiF. Watts et al.[158] simulated a system consisting of 50 water molecules and two ions: Li^+ and F^- are set at fixed positions with distances of 3.0, 3.5, and 4.0 Å. Room temperature is assumed for all cases (one case is also simulated at higher temperature). The water molecules are free to move and to orient themselves around the ions. Their mutual positions are described using analytical formulas fitted to quantum mechanical potentials.[118-120] Also ion-water interactions are represented by theoretical potential functions, which take the triplet term water-$Li^+ - F^-$ into account beside the dominant pair terms.[125] The tridimensional picture of the environment of the ions provided by the simulation shows a well-defined shell of water molecules around each ion, with the oxygen atoms pointing towards the cation, and one of the hydrogens pointing towards the anion. The number of atoms nearest neighbors to the ions is four in every case, oxygen atoms at a distance of 2.0 Å in the case of Li^+ and hydrogen atoms at a distance of about 1.7 Å in the case of F^-. A second atom shell is also evident around the ions, which is differently populated in the two cases: eight hydrogen atoms around lithium, and four oxygen atoms around fluorine.

Another simulation of the same system has been carried out by the same group.[159] Several improvements have been introduced here with respect to the previous case. The system

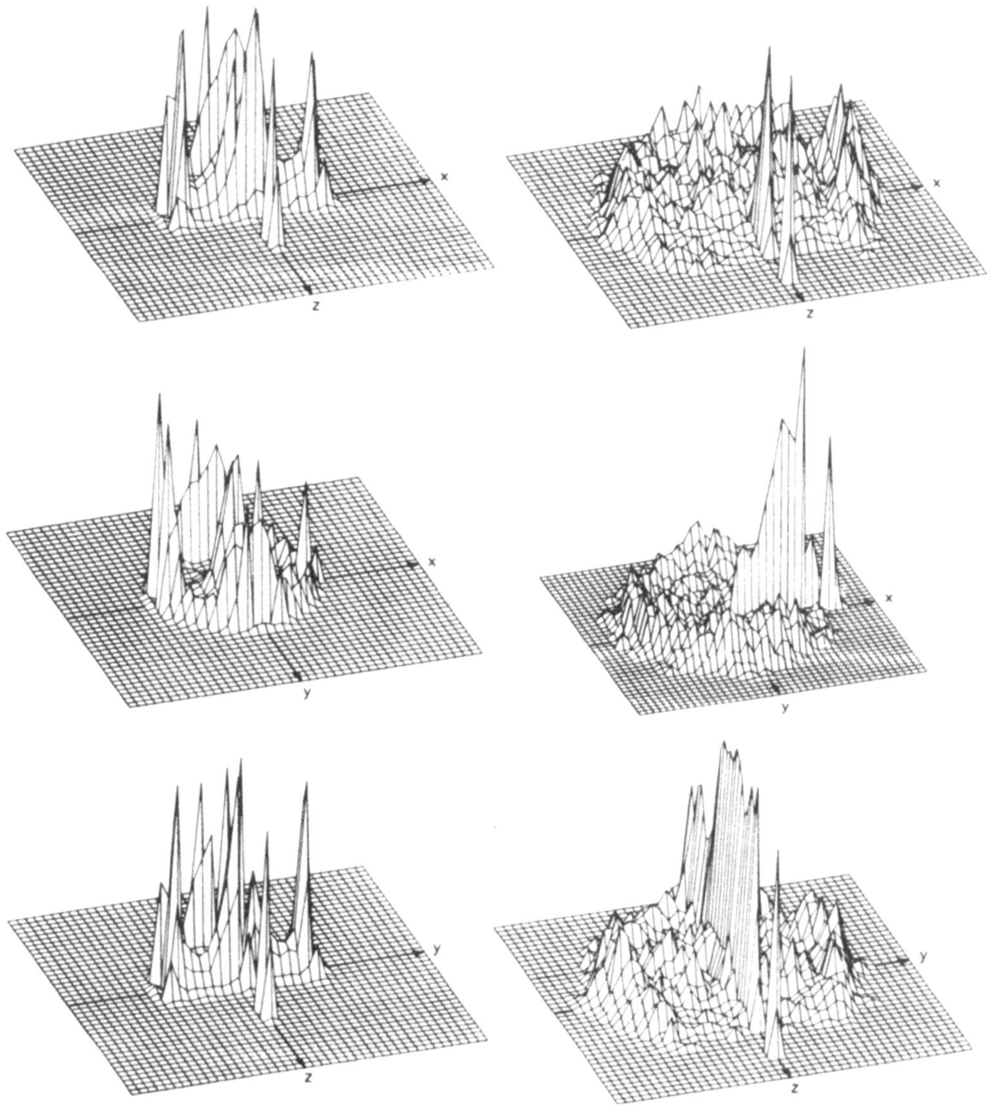

FIGURE 41. Three-dimensional drawings of the projections of distributions of oxygen atom positions around Mg^{2+} (left) and Cl^- (right) from the MD simulation of a 1.1-*m* solution of $MgCl_2$. (From Pálinkás, G., Radnai, T., Dietz, W., Szász, G. I., and Heinzinger, K., *Z. Naturforsch.*, 37a, 1049, 1982. With permission.)

comprises 200 water molecules and two ions. The separations between the ions are held fixed to 2.0, 6.0, and 10.0 Å; these values are somewhat representative of three realistic situations, namely the case of diatomic molecules, of separate atoms, and of dissociation limits. However, the major change regards the interaction potential of the water molecules, to which a term accounting for the dispersion effect has been added. This procedure should therefore lead to a more realistic description of the water. The changes in the parameters describing the near environment of the ions is dramatic: the $F^- - H$ distance decreases to 1.3 to 1.5 Å (coordination number 3 to 4) and the $Li^+ - O$ one decreases to 1.75 Å (coordination number equal to 4 to 6).

Similar systems, $Li^+ - Cl^-$ and $K^+ - F^-$ together with 50 water molecules, were also examined by Watts.[160] In this case, however, the investigation was directed towards the water

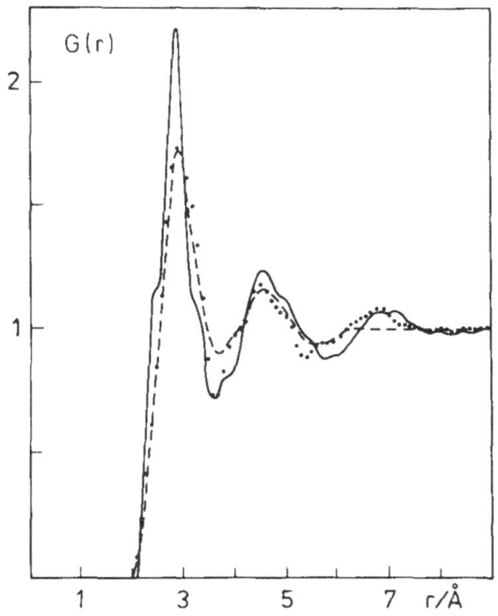

FIGURE 42. Total correlation functions for a 1.1-*m* solution of $CaCl_2$ obtained from XRD data (dots), model fit to XRD data (dashed line) and MD simulation (full line). (From Probst, M. M., Radnai, T., Heinzinger, K., Bopp, P., and Rode, B. M., *J. Chem. Phys.*, 89, 753, 1985. With permission.)

itself and the variations induced in its structure by the presence of ions of various sizes. Differences in the neighboring of the ions were indeed evidenced, suggesting that the hydration around the smaller ions was commanded by the larger ones.

A more systematic work has been made by Mezei and Beveridge.[161] They examined systems consisting of one ion and 215 water molecules, that is, situations similar to dilute aqueous solutions of Li^+, Na^+, K^+, F^-, and Cl^- ions. The ion-water and water-water potentials were derived by theoretical calculations. Some significant molecular structures were evidenced in these simulations, which, with some caution, may be viewed as representative of the neighborhood of an ion in a dilute solution. The interactions ion-center of mass of the water molecules are the ones that can be with some approximation compared to XRD results. Of course, no single ion-water radial distribution function can be obtained directly from XRD experiments, while only some of such curves have been obtained by neutron diffraction. In Figure 44, the results obtained for the chloride ion are compared with experimental data obtained by neutron diffraction. The agreement is good up to 4 Å, proving that the potential functions used are approaching realistic behaviors of ions in water. A comparative table (Table 13) may also be of interest, in which the values of ion-water peak maxima and of the coordination numbers calculated by MC simulation are reported. The reader may refer to results from XRD and ND for proper experimental references.

F. Final Comments

Some comments are undoubtedly in order; however, the entire matter is so complex and in such evolution that they seem of limited, at least temporal, validity. Indeed, the number of results already available is far from negligible, and some sort of classification might be useful. To this end, the type of potential functions used in the simulation studies could be

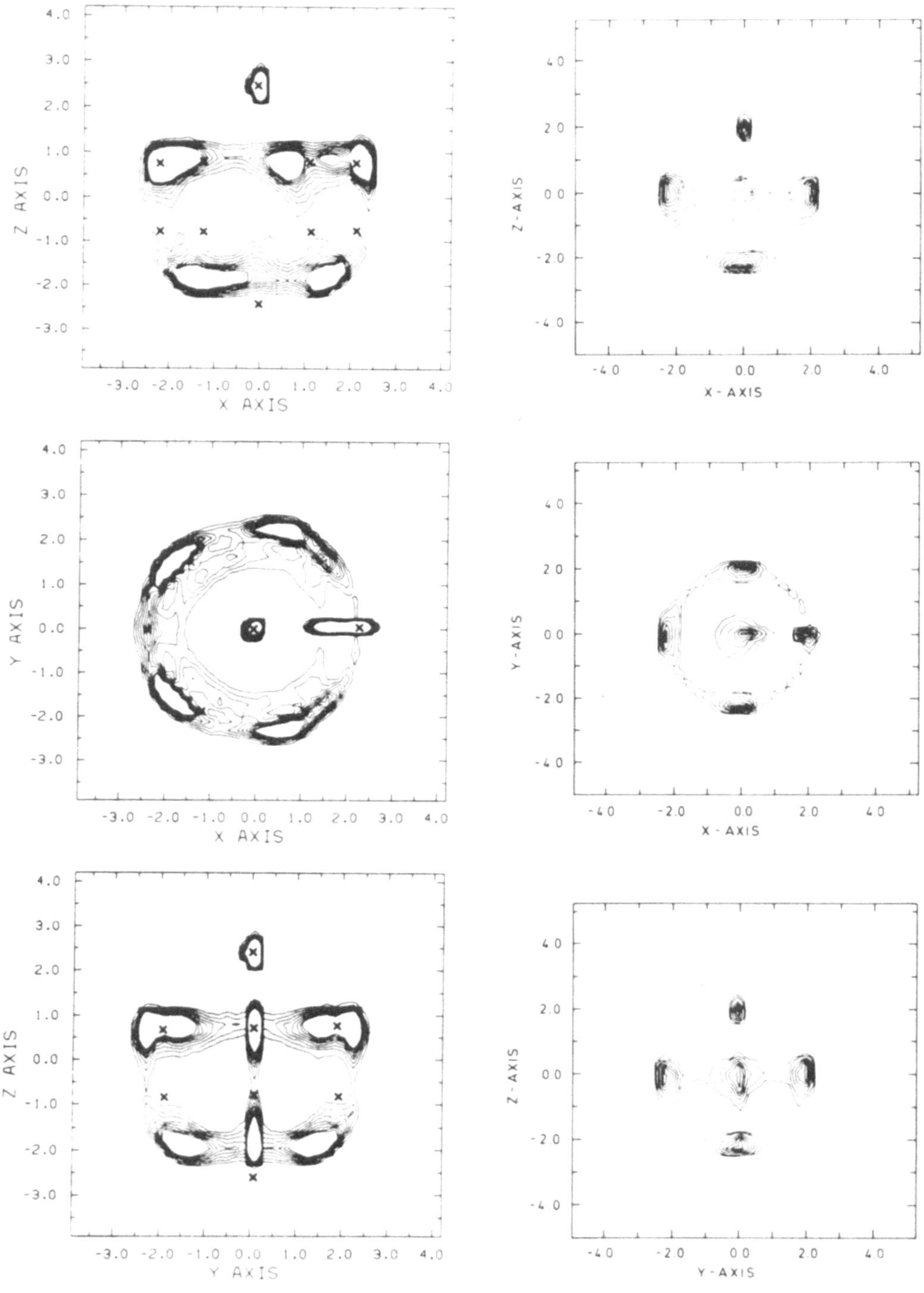

A B

FIGURE 43. Density maps of the projections of the oxygen atom positions of the nearest neighbor water molecules around two cations as obtained by MD simulations. A: Ca^{2+}; the crosses indicate positions expected for hexahedral coordination of the ion. B: Mg^{2+}. (From (A) Probst, M. M., Radnai, T., Heinzinger, K., Bopp, P., and Rode, B. M., *J. Chem. Phys.*, 89, 753, 1985 and (B) Pálinkás, G., Radnai, T., Dietz, W., Szász, G. I., and Heinzinger, K., *Z. Naturforsch.*, 37a, 1049, 1982. With permission.)

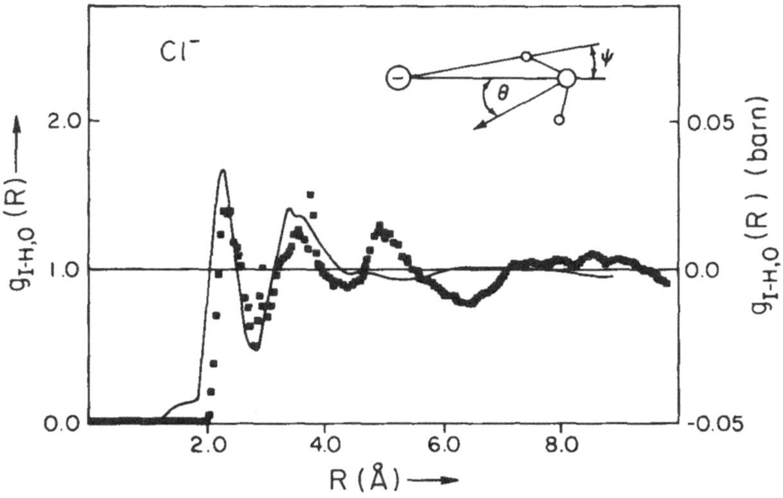

FIGURE 44. Comparison of the radial function obtained by MC simulation of the system $(Cl^-)_{aq}$ with experimental results obtained by neutron diffraction. (From Mezei, M. and Beveridge, D. L., *J. Chem. Phys.*, 74, 6902, 1981. With permission.)

Table 13
ION-WATER DISTANCES r AND COORDINATION NUMBERS N OBTAINED BY MC SIMULATIONS OF ONE-ION IN WATER SYSTEMS[161]

	r_{ion-w} (Å)	N_{ion-O}
Li^+	2.10	5.97 ± .02
Na^+	2.35	5.96 ± .02
K^+	2.71	6.27 ± .25
F^-	2.60	4.99 ± .10
Cl^-	3.25	8.36 ± .17

used as a discriminating factor. This choice sounds reasonable, but a survey of the matter shows clearly that few simulations can be defined as completely "theoretical" or completely "empirical", given the many different types of interactions involved in a solution. The only possible comment is that, predictably, the use of empirical potential functions seems to have given rise to more realistic simulations. Yet, the route of purely theoretical simulations able to display the properties of real systems is a fascinating goal, which explains the efforts of so many people involved in this research field. The number of attempts to simulate the liquid water illustrates the interest for these studies, but also the difficulties involved. The work is progressing, nevertheless; situations of simple ions in water have been described using several differently derived interaction potentials, and encouraging results have also been obtained with more complex ions like Ca^{2+}, and Ni^{2+}.[130] This research line is bound to be explored further in the future.

From the point of view of the "user" of the information provided by simulation studies, the discriminating point between investigations can only be one of **realism**. In this view, the long series of papers on solutions by Heinzinger et al.[139-157] represents the well from which to draw the information. By examining Heinzinger's papers, it is worth noting the progress of his simulations towards reality, particularly evident in the more recent works in which constant reference is made to experimental data. So, in the cases in which this comparison is judged satisfactory, the spatial description of the environment of the ions in solution is provided by these studies with details which no other investigative technique can give. The case of the $MgCl_2$ solution presented above is an example of the details that a good simulation can provide: coordination geometries, orientation of the water molecules around the ions, particulars about every type of displacements of particles around mean positions, and so on. Also the comparative examination of the behavior of Mg^{2+} and Ca^{2+} towards hydration is interesting, and though the simulation cannot explain the reason of so dissimilar environments, it partially clarifies the difficulties found in interpreting XRD data on $CaCl_2$ solutions.

In general, the relation between the simulation and the diffraction techniques seems to be of reciprocal need and advantage in that the simulation needs reliable experimental results to test its accuracy; in turn, the simulation explains experimental data in terms of detailed tridimensional descriptions of positions and orientations of the particles in the sample. From this point of view, it cannot be ignored that the development of adequate potential functions has been sometimes hampered by the lack of reliable experimental references. The easiest species to describe (alkali and halide ions) are in fact poorly seen by XRD, and only in some cases have recent ND studies provided reliable results. The situation is now improving, in that the simulation studies are expanding their field of application, and experimental data are increasingly available for their testing. It seems that we are on the right track and the future perspectives look good. A further push in the right direction may come by the technological progresses in the field of computers, which may remove many of the technical limitations connected with the simulation studies: limited number of particles, limited volume of the sample, severe cut-offs of long range interactions, short simulation time, and so on.

Other possible developments are expected in the long range. MD studies are in fact able to provide dynamic information on the systems; though the necessary reliability level of the results is here even greater than that needed for a purely static description of the system, the expectations are here that the simulation studies will contribute to fill the gap between the static description of a system and its dynamic behavior.

REFERENCES

1. **Enderby, J. E. and Neilson, G. W.,** X-ray and neutron scattering by aqueous solutions of electrolytes, in *Water: A Comprehensive Treatise*, Vol. 6, Franks, F., Ed., Plenum Press, New York, 1979, chap. 1.
2. **Ohtomo, N. and Arakawa, K.,** Neutron diffraction study of aqueous ionic solutions. I. Aqueous solutions of lithium chloride and cesium chloride, *Bull. Chem. Soc. Jpn.*, 52, 2755, 1979.
3. **Ohtomo, N. and Arakawa, K.,** Neutron diffraction study of aqueous ionic solutions. II. Aqueous solutions of sodium chloride and potassium chloride, *Bull. Chem. Soc. Jpn.*, 53, 1789, 1980.
4. **Narten, A. H., Vaslow, F., and Levy, H. A.,** Diffraction pattern and structure of aqueous lithium chloride solutions, *J. Chem. Phys.*, 58, 5017, 1973.
5. **Triolo, R. and Narten, A. H.,** Diffraction pattern and structure of aqueous hydrochloric acid solutions, *J. Chem. Phys.*, 63, 3624, 1975.
6. **Narten, A. H. and Hahn, R. L.,** Hydration of the Nd^{3+} ion in neodymium chloride solutions determined by neutron diffraction, *J. Phys. Chem.*, 87, 3193, 1983.
7. **Enderby, J. E., Howells, W. S., and Howe, R. A.,** The structure of aqueous solutions, *Chem. Phys. Lett.*, 21, 109, 1973.
8. **Soper, A. K., Neilson, G. W., Enderby, J. E., and Howe, R. A.,** A neutron diffraction study of hydration effects in aqueous solutions, *J. Phys. C*, 10, 1793, 1977.
9. **Newsome, J. R., Neilson, G. W., and Enderby, J. E.,** Lithium ions in aqueous solution, *J. Phys. C*, 13, L923, 1980.

10. **Ichikawa, K., Kameda, Y., Matsumoto, T., and Misawa, M.,** Indirect and direct correlations between unlike ions in incompletely hydrated solution, *J. Phys. C*, 17, L725, 1984.
11. **Copestake, A. P., Neilson, G. W., and Enderby, J. E.,** The structure of a highly concentrated solution of lithium chloride, *J. Phys. C*, 18, 4211, 1985.
12. **Neilson, G. W. and Skipper, N.,** K^+ coordination in aqueous solution, *Chem. Phys. Lett.*, 114, 35, 1985.
13. **Cummings, S., Enderby, J. E., and Howe, R. A.,** Ion hydration in aqueous $CaCl_2$ solutions, *J. Phys. C*, 13, 1, 1980.
14. **Hewish, N. A., Neilson, G. W., and Enderby, J. E.,** Environment of Ca^{2+} ions in aqueous solvent, *Nature*, 297, 138, 1982.
15. **Narten, A. H. and Hahn, R. L.,** Direct determination of ionic solvation from neutron diffraction, *Science*, 217, 1249, 1982.
16. **Annis, B. K., Hahn, R. L., and Narten, A. H.,** Hydration of the Dy^{3+} ion in dysprosium chloride solutions determined by neutron diffraction, *J. Chem. Phys.*, 82, 2086, 1985.
17. **Habenschuss, A. and Spedding, F. H.,** The coordination (hydration) of rare earth ions in aqueous chloride solutions from X-ray diffraction. III. $SmCl_3$, $EuCl_3$ and series behavior, *J. Chem. Phys.*, 73, 442, 1980.
18. **Neilson, G. W. and Enderby, J. E.,** The hydration of Ni^{2+} in aqueous solutions, *J. Phys. C*, 11, L625, 1978.
19. **Newsome, J. R., Neilson, G. W., Enderby, J. E., and Sandström, M.,** Ni^{2+} hydration in perchlorate and chloride solutions, *Chem. Phys. Lett.*, 82, 399, 1981.
20. **Neilson, G. W. and Enderby, J. E.,** The structure of an aqueous solution of nickel chloride, *Proc. R. Soc. London*, A390, 353, 1983.
21. **Neilson, G. W.,** Cu^{2+} coordination in aqueous solution, *J. Phys. C*, 15, L233, 1982.
22. **Magini, M.,** Hydration and complex formation study on concentrated MCl_2 solutions [M = Co(II), Ni(II), Cu(II)] by X-ray diffraction technique, *J. Chem. Phys.*, 74, 2523, 1981.
23. **Bol, W., Gerrits, G. J. A., and van Panthaleon van Eck, C. L.,** The hydration of divalent cations in aqueous solution. An X-ray investigation with isomorphous replacement, *J. Appl. Cryst.*, 3, 486, 1970.
24. **Neilson, G. W., Newsome, J. R., and Sandström, M.,** Neutron diffraction study of aqueous transition metal salt solutions by isomorphic substitution, *J. Chem. Soc. Faraday Trans. 2*, 77, 1245, 1981.
25. **Sandström, M., Neilson, G. W., Johansson, G., and Yamaguchi, T.,** Ag^+ hydration in perchlorate solution, *J. Phys. C*, 18, L1115, 1985.
26. **Cummings, S., Enderby, J. E., Neilson, G. W., Newsome, J. R., Howe, R. A., Howells, W. S., and Soper, A. K.,** Chloride ions in aqueous solutions, *Nature*, 287, 714, 1980.
27. **Biggin, S., Enderby, J. E., Hahn, R. L., and Narten, A. H.,** Hydration of the chloride ion in neodimium chloride solutions determined by neutron diffraction, *J. Phys. Chem.*, 88, 3634, 1984.
28. **Neilson, G. W. and Enderby, J. E.,** The structure around nitrate ions in concentrated aqueous solutions, *J. Phys. C*, 15, 2347, 1982.
29. **Brown, G. S. and Doniach, S.,** The principles of X-ray absorption spectroscopy, in *Synchrotron Radiation Research*, Winick, H. and Doniach, S., Eds., Plenum Press, New York, 1980, chap. 10.
30. **Koningsberger, D. C. and Prins, R.,** *X-ray Absorption: Principles, Applications, Techniques of EXAFS, SEXAFS and XANES*, John Wiley & Sons, New York, to be published.
31. **Kunz, C.,** Properties of synchrotron radiation, in *Synchrotron Radiation: Techniques and Applications*, Kunz, C., Ed., Springer-Verlag, Berlin, 1979.
32. **Koch, E. E., Eastman, D. E., and Farge, Y.,** Synchrotron radiation — a powerful tool in Science, in *Handbook on Synchrotron Radiation*, Koch, E. E., Ed., North-Holland, Amsterdam, 1983.
33. For a complete picture of recent progresses, see: *EXAFS and Near Edge Structure*, Proc. Int. Conf., Frascati, Italy, Bianconi, A., Incoccia, L., and Stipcich, S., Eds., Springer-Verlag, Berlin, 1983; *EXAFS and Near Edge Structure III*, Proc. Int. Conf., Stanford, CA, USA, Hodgson, K. O., Hedman, B., and Penner-Hahn, J. E., Eds., Springer-Verlag, Berlin, 1984; *EXAFS and Near Edge Structure IV*, Proc. Int. Conf., Fontevraud, France, Lagarde, P., Raoux, D., and Petiau, J., Eds., Les éditions de Physique, Les Ulis, 1986.
34. **Kutzler, F. W., Natoli, C. R., Misemer, D. K., Doniach, S., and Hodgson, K. O. I.,** Use of one-electron theory for the interpretation of near edge structure in K-shell X-ray absorption spectra of transition metal complexes, *J. Chem. Phys.*, 73, 3274, 1980.
35. **Belli, M., Scafati, A., Bianconi, A., Mobilio, S., Palladino, L., Reale, A., and Burattini, E.,** X-ray absorption near edge structures (XANES) in simple and complex Mn compounds, *Solid State Commun.*, 35, 355, 1980.
36. **Bianconi, A.,** X-ray absorption near edge structures (XANES) and their applications to local structure determination, in *EXAFS for Inorganic Systems*, Garner, C. O. and Hasnain, S. S., Eds., Daresbury report DL/SCI/R17, 1981, 13.
37. **Cramer, S. P., Eidem, P. K., Paffet, M. T., Winkler, J. R., Dori, Z., and Gray, H. B.,** X-ray absorption edge and EXAFS spectroscopic studies of molybdenum ions in aqueous solution, *J. Am. Chem. Soc.*, 105, 799, 1983.

38. **Durham, P. J., Pendry, J. B., and Hodges, C. H.,** XANES: determination of bond angles and multi-atom correlations in ordered and disordered systems, *Solid State Commun.*, 38, 159, 1981.
39. **Bianconi, A., Dell'Ariccia, M., Durham, P. J., and Pendry, J. B.,** Multiple-scattering resonances and structural effects in the X-ray absorption near-edge spectra of iron (II) and iron (III) hexacyanide complexes, *Phys. Rev. B*, 26, 6502, 1982.
40. **Natoli, C. R.,** Near edge absorption structure in the framework of the multiple scattering model. Potential resonance or barrier effects?, in *EXAFS and Near Edge Structure*, Bianconi, A., Incoccia, L., and Stipcich, S., Eds., Springer-Verlag, Berlin, 1983, 43.
41. **Pendry, J. B.,** The transition region between XANES and EXAFS, in *EXAFS and Near Edge Structure*, Bianconi, A., Incoccia, L., and Stipcich, S., Eds., Springer-Verlag, Berlin, 1983, 4.
42. **Bianconi, A., Dell'Ariccia, M., Gargano, A., and Natoli, C. R.,** Bond length determination using XANES, in *EXAFS and Near Edge Structure*, Bianconi, A., Incoccia, L., and Stipcich, S., Eds., Springer-Verlag, Berlin, 1983, 57.
43. **Garcia, J., Benfatto, M., Natoli, C. R., Bianconi, A., Davoli, I., and Marcelli, A.,** Three particle correlation function of metal ions in tetrahedral coordination determined by XANES, *Solid State Commun.*, 58, 595, 1986.
44. **Benfatto, M., Natoli, C. R., Bianconi, A., Garcia, J., Marcelli, A., Fanfoni, M., and Davoli, I.,** Multiple scattering regime and higher order correlation in X-ray absorption spectra of liquid solutions, to be published.
45. **Sayers, D. E., Stern, E. A., and Lytle, F. W.,** New technique for investigating noncrystalline structures: Fourier analysis of the extended X-ray absorption fine structure, *Phys. Rev. Lett.*, 27, 1204, 1971.
46. **Stern, E. A.,** Theory of the extended X-ray absorption fine structure, *Phys. Rev. B*, 10, 3027, 1974.
47. **Stern, E. A., Sayers, D. E., and Lytle, F. W.,** Extended X-ray absorption fine structure technique. III. Determination of physical parameters, *Phys. Rev. B*, 11, 4836, 1975.
48. **Ashley, C. A., and Doniach, S.,** Theory of extended X-ray absorption edge fine structure (EXAFS) in crystalline solids, *Phys. Rev. B*, 11, 1279, 1975.
49. **Lee, P. A. and Pendry, J. B.,** Theory of extended X-ray absorption fine structure, *Phys. Rev. B*, 11, 2795, 1975.
50. **Lytle, F. W., Sayers, D. E., and Stern, E. A.,** Extended X-ray absorption fine structure technique. II. Experimental practice and selected results, *Phys. Rev. B*, 11, 4836, 1975.
51. **Teo, B. K.,** Extended X-ray absorption fine structure: techniques and applications, in *EXAFS Spectroscopy*, Teo, B. K. and Joy, D. C., Eds., Plenum Press, New York, 1981, chap. 3.
52. **Lee, P. A., Citrin, P. H., Eisenberger, P., and Kincaid, B. M.,** Extended X-ray absorption fine structure — its strengths and limitations as a structural tool, *Rev. Mod. Phys.*, 53, 769, 1981.
53. **Hayes, T. M. and Boyce, J. B.,** Extended X-ray absorption fine structure, in *Solid State Physics*, Vol. 37, Ehrenreich, H., Seitz, F., and Turnbull, D., Eds., Academic Press, New York, 1982, 173.
54. **Stern, E. A. and Heald, S. M.,** Basic principles and applications of EXAFS, in *Handbook on Synchrotron Radiation*, Koch, E. E., Ed., North-Holland, Amsterdam, 1983, chap. 10.
55. **Teo, B. K.,** *EXAFS: Basic Principles and Data Analysis*, Springer-Verlag, Berlin, 1986.
56. **Babanov, Y. A., Vasin, V. V., Ageev, A. L., and Ershov, N. V.,** A new interpretation of EXAFS in real space. I. General formalism, *Phys. Status Solidi*, 105, 747, 1981.
57. **Ershov, N. V., Ageev, A. L., Vasin, V. V., and Babanov, Y. A.,** A new interpretation of EXAFS in real space. II. A comparison of the regularization technique with the Fourier transformation method, *Phys. Status Solidi*, 108, 103, 1981.
58. **Ageev, A. L., Babanov, Y. A., Vasin, V. V., Ershov, N. V., and Serikov, A. V.,** Amorphous problem in EXAFS data analysis, *Phys. Status Solidi*, 117, 345, 1983.
59. **Licheri, G. and Pinna, G.,** unpublished data.
60. **Lee, P. A. and Beni, G.,** New method for the calculation of atomic phase shifts: application to extended X-ray absorption fine structure (EXAFS) in molecules and crystals, *Phys. Rev. B*, 15, 2862, 1977.
61. **Lee, P. A., Teo, B. K., and Simons, A. L.,** EXAFS: a new parameterization of phase shifts, *J. Am. Chem. Soc.*, 99, 3856, 1977.
62. **Teo, B. K. and Lee, P. A.,** Ab initio calculations of amplitude and phase shift functions for extended X-ray absorption fine structure spectroscopy, *J. Am. Chem. Soc.*, 101, 2815, 1979.
63. **Citrin, P. H., Eisenberger, P., and Kincaid, B. M.,** Transferability of phase shifts in extended X-ray absorption fine structure, *Phys. Rev. Lett.*, 36, 1346, 1976.
64. **Martens, G., Rabe, P., Schwentner, N., and Werner, A.,** Improved EXAFS studies applied to the investigation of Cu-O, Cu-N, and Cu-Br bond lengths, *Phys. Rev. B*, 17, 1481, 1978.
65. **Rabe, P.,** Determination of bond lengths from EXAFS with high resolution, *Jpn. J. Appl. Phys.*, 17, 22, 1978.
66. **Cramer, P.,** Interpretation of EXAFS using model compounds, in *EXAFS for Inorganic Systems*, Garner, C. D. and Hasnain, S. S., Eds., Daresbury report DL/SCI/R17, 1981, 47.

67. **Teo, B. K., Lee, P. A., Simons, A. I., Eisenberger, P., and Kincaid, B. M.,** EXAFS: approximations, parameterizations, and chemical transferability of amplitude functions, *J. Am. Chem. Soc.*, 99, 3854, 1977.
68. **Eisenberger, P. and Lengeler, B.,** Extended X-ray absorption fine structure determination of coordination numbers: limitations, *Phys. Rev. B*, 22, 3551, 1980.
69. **Stern, E. A.,** Limitations of EXAFS: real and imagined, in *EXAFS for Inorganic Systems*, Garner, C. D., and Hasnain, S. S., Eds., Daresbury report DL/SCI/R17, 1981, 40.
70. **Stern, E. A., Bunker, B., and Heald, S. M.,** Understanding the causes of non-transferability of EXAFS amplitudes, in *EXAFS Spectroscopy*, Teo, B. K. and Joy, D. C., Eds., Plenum Press, New York, 1981, chap. 4.
71. **Stern, E. A. and Kim, K.,** Thickness effects on the extended absorption fine structure amplitude, *Phys. Rev. B*, 23, 3781, 1983.
72. **Goulon, J., Goulon-Ginet, C., Cortes, R., and Dubois, J. M.,** Thickness effects on the determination of EXAFS amplitudes, in *EXAFS and Near Edge Structure*, Bianconi, A., Incoccia, L., and Stipcich, S., Eds., Springer-Verlag, Berlin, 1983, 96.
73. **Heald, S. M.,** Thickness effects in X-ray absorption measurements, in *EXAFS and Near Edge Structure*, Bianconi, A., Incoccia, L., and Stipcich, S., Eds., Springer-Verlag, Berlin, 1983, 98.
74. **Lu, K. and Stern, E. A.,** Size effects of powdered samples on EXAFS amplitude, *Nucl. Instrum. Methods*, 212, 475, 1983.
75. **Hayes, T. M., Sen, P. N., and Hunter, S. H.,** Structure determination using EXAFS in real space: Ge, *J. Phys. C*, 9, 4357, 1976.
76. **Hayes, T. M.,** EXAFS studies of disordered solids, *J. Non-Cryst. Solids*, 31, 57, 1978.
77. **Pettifer, R. F.,** EXAFS studies of glasses, in *EXAFS for Inorganic Systems*, Garner, C. D. and Hasnain, S. S., Eds., Daresbury report DL/SCI/R17, 1981, 57.
78. **Mobilio, S. and Incoccia, L.,** Radial distribution function by EXAFS: asimmetry in metallic glasses, *Nuovo Cimento*, 3D, 846, 1984.
79. **Licheri, G. and Pinna, G.,** EXAFS and X-ray diffraction in solutions, in *EXAFS and Near Edge Structure*, Bianconi, A., Incoccia, L., and Stipcich, S., Eds., Springer-Verlag, Berlin, 1983, 240.
80. **Frahm, R., Haensel, R., and Rabe, P.,** A model comparison between EXAFS and X-ray scattering, in *EXAFS and Near Edge Structure*, Bianconi, A., Incoccia, L., and Stipcich, S., Eds., Springer-Verlag, Berlin, 1983, 107.
81. **Eisenberger, P. and Kincaid, B. M.,** Synchrotron radiation studies of X-ray absorption spectra of ions in aqueous solutions, *Chem. Phys. Lett.*, 36, 134, 1975.
82. **Sandstrom, D. R., Stults, B. R., and Greegor, R. B.,** Structural evidence for solutions from EXAFS measurements, in *EXAFS Spectroscopy*, Teo, B. K. and Joy, D. C., Eds., Plenum Press, New York, 1981, chap. 9.
83. **Sham, T. K., Hastings, J. B., and Perlman, M. L.,** Structure and dynamic behavior of transition-metal ions in aqueous solutions: an EXAFS study of electron-exchange reactions, *J. Am. Chem. Soc.*, 102, 5904, 1980.
84. **Sham, T. K., Hastings, J. B., and Perlman, M. L.,** Application of the EXAFS method to Jahn-Teller ions: static and dynamic behavior of $Cu(H_2O)_6{}^{2+}$ and $Cr(H_2O)_6{}^{2+}$ in aqueous solution, *Chem. Phys. Lett.*, 83, 391, 1981.
85. **Sandstrom, D. R., Dodgen, H. W., and Lytle, F. W.,** Study of Ni(II) coordination in aqueous solutions by EXAFS analysis, *J. Chem. Phys.*, 67, 473, 1977.
86. **Licheri, G., Pinna, G., Navarra, G., and Vlaic, G.,** EXAFS study of Ni^{2+} coordination in nitrate solutions, *Z. Naturforsch.*, 38a, 559, 1983.
87. **Licheri, G., Paschina, G., Piccaluga, G., Pinna, G., and Vlaic, G.,** EXAFS study of Ni^{2+} coordination in concentrated aqueous solutions, *Chem. Phys. Lett.*, 83, 384, 1981.
88. **Sandstrom, D. R.,** Ni^{2+} coordination in aqueous $NiCl_2$ solutions: study of the extended X-ray absorption fine structure, *J. Chem. Phys.*, 71, 2381, 1979.
89. **Licheri, G., Paschina, G., Piccaluga, G., and Pinna, G.,** EXAFS study of Ni-Cl bonding in Ni(II) aqueous solutions at increasing Cl^-/Ni^{2+} ratios, *J. Chem. Phys.*, 79, 2168, 1983.
90. **Sano, M., Taniguchi, K., and Yamatera, H.,** The EXAFS study of Cu(II) aqueous solution using a position sensitive detector, *Chem. Lett.*, 1285, 1980.
91. **Yamaguchi, T., Lindqvist, O., Boyce, J. B., and Claeson, T.,** Determination of the hydration structure of silver ions in aqueous silver perchlorate and nitrate solutions from EXAFS using synchrotron radiation, *Acta Chem. Scand.*, A38, 423, 1984.
92. **Yamaguchi, T., Lindqvist, O., Claeson, T., and Boyce, J. B.,** EXAFS and X-ray diffraction studies of the hydration structure of stereochemically active Sn(II) ions in aqueous solution, *Chem. Phys. Lett.*, 93, 528, 1982.
93. **Yamaguchi, T., Lindqvist, O., Boyce, J. B., and Claeson, T.,** EXAFS, X-ray, and neutron diffraction of electrolyte solutions, in *EXAFS and Near Edge Structure III*, Hodgson, K. O., Hedman, B., and Penner-Hahn, J. E., Eds., Springer-Verlag, Berlin, 1985, 417.

94. **Lagarde, P., Fontaine, A., Raoux, D., Sadoc, A., and Migliardo, P.,** EXAFS studies of strong electrolytic solutions, *J. Chem. Phys.*, 72, 3061, 1980.
95. **Fontana, M. P., Maisano, G., Migliardo, P., and Wanderlingh, F.,** Raman spectroscopy and local order in aqueous solutions of strong II-I electrolytes, *J. Chem. Phys.*, 69, 676, 1978.
96. **Joyner, R. W.,** An extended X-ray absorption fine structure (EXAFS) study of copper(II) sulphate pentahydrate, *Chem. Phys. Lett.*, 72, 162, 1980.
97. **Maeda, M., Maegawa, Y., Yamaguchi, T., and Ohtaki, H.,** X-ray diffraction studies on the structures of diammine- and aquasilver(I) complexes in aqueous solutions, *Bull. Chem. Soc. Jpn.*, 52, 2545, 1979.
98. **Yamaguchi, T., Johansson, G., Holmberg, B., Maeda, M., and Ohtaki, H.,** The coordination of complex formation of silver(I) in aqueous perchlorate, nitrate and iodide solutions, *Acta Chem. Scand.*, A38, 437, 1984.
99. **Sandstrom, D. R.,** EXAFS studies of electrolyte solutions, in *EXAFS and Near Edge Structure III*, Hodgson, K. O., Hedman, B., and Penner-Hahn, J. E., Eds., Springer-Verlag, Berlin, 1985, 409.
100. **Asakura, K., Nomura, M., and Kuroda, H.,** Fe K-edge XANES and EXAFS of the X-ray absorption spectra of $FeCl_3$ aqueous solutions. A structural study of the solute, iron(III) chloro complexes, *Bull. Chem. Soc. Jpn.*, 58, 1543, 1985.
101. **Magini, M., Paschina, G., and Piccaluga, G.,** Ni-Cl bonding in concentrated Ni(II) aqueous solutions at high Cl^-/Ni^{2+} ratios. An X-ray diffraction investigation, *J. Chem. Phys.*, 76, 1116, 1982.
102. **Caminiti, R., Licheri, G., Paschina, G., Piccaluga, G., and Pinna, G.,** X-ray diffraction and structural properties of aqueous solutions of divalent metal chlorides, *Z. Naturforsch.*, 35a, 1361, 1980.
103. **Corrias, A., Musinu, A., and Pinna, G.,** EXAFS and X-ray diffraction study of a mixed $NiCl_2$-$CoCl_2$ solution, *Chem. Phys. Lett.*, 120, 295, 1985.
104. **Fontaine, A., Lagarde, P., Raoux, D., Fontana, M. P., Maisano, G., Migliardo, P., and Wanderlingh, F.,** Extended X-ray-Absorption Fine Structure studies of local ordering in highly concentrated aqueous solutions of $CuBr_2$, *Phys. Rev. Lett.*, 41, 504, 1978.
105. **Galli, G., Maisano, G., Migliardo, P., Vasi, C., and Wanderlingh, F.,** Temperature effects on the local structure of $CuBr_2$ aqueous solutions by EXAFS, *Solid State Commun.*, 42, 213, 1982.
106. **Alliota, F., Galli, G., Maisano, G., Migliardo, P., Vasi, C., and Wanderlingh, F.,** Local order and vibrational dynamics in strong II-I mixed electrolytic aqueous solutions, *Nuovo Cimento*, 2D, 103, 1983.
107. **Sadoc, A., Lagarde, P., and Vlaic, G.,** EXAFS evidence for local order in aqueous solutions of $CdBr_2$, *J. Phys. C*, 18, 23, 1985.
108. **Ichihashi, M., Wakita, H., Mibuchi, T., and Masuda, I.,** Coordination structure of Cu^{2+} ion in highly concentrated aqueous $CuBr_2$ solutions determined by X-ray diffraction analysis, *Bull. Chem. Soc. Jpn.*, 55, 3160, 1982.
109. **Goggin, P. L., Johansson, G., Maeda, M., and Wakita, H.,** The structures of zinc bromide complexes in aqueous solution, *Acta Chem. Scand.*, A38, 625, 1984.
110. **Lykos, P., Ed.,** *Computer Modeling of Matter*, Am. Chem. Soc. Symp. Series, Washington, D.C., 1978.
111. **Catlow, C. R. A. and Mackrodt, W. C., Eds.,** Computer simulation of condensed matter, Proc. NATO Adv. Res. Workshop on computer simulation of condensed matter, East Anglia, UK, 1984, *Physica*, 131B, 1985.
112. **Alder, B. J. and Wainwright, T. E.,** Studies in molecular dynamics. I. General method, *J. Chem. Phys.*, 31, 459, 1959.
113. **Metropolis, N., Rosenbluth, A. W., Rosenbluth, M. N., Teller, A. H., and Teller, E.,** Equation of state calculations by fast computing machines, *J. Chem. Phys.*, 21, 1087, 1953.
114. **Rushbrooke, G. S.,** Equilibrium theories of the liquid state, in *Physics of Simple Liquids*, Temperley, H. N. V., Rowlinson, J. S., and Rushbrooke, G. S., Eds., North-Holland, Amsterdam, 1968, chap. 2.
112. **Alder, B. J. and Wainwright, T. E.,** Studies in molecular dynamics. I. General method, *J. Chem. Phys.*, 31, 459, 1959.
113. **Metropolis, N., Rosenbluth, A. W., Rosenbluth, M. N., Teller, A. H., and Teller, E.,** Equation of state calculations by fast computing machines, *J. Chem. Phys.*, 21, 1087, 1953.
114. **Rushbrooke, G. S.,** Equilibrium theories of the liquid state, in *Physics of Simple Liquids*, Temperley, H. N. V., Rowlinson, J. S., and Rushbrooke, G. S., Eds., North-Holland, Amsterdam, 1968, chap. 2.
115. **Croxton, C.,** *Introduction to Liquid State Physics*, John Wiley & Sons, London, 1975.
116. **Barker, J. A. and Henderson, D.,** What is "liquid"? Understanding the states of matter, *Rev. Mod. Phys.*, 48, 587, 1976.
117. **Wood, D. W.,** Computer simulation of water and aqueous solutions, in *Water: A Comprehensive Treatise*, Vol. 6, Franks, F., Ed., Plenum Press, New York, 1979, chap. 6.
118. **Popkie, H., Kistenmacher, H., Clementi, E.,** Study of the structure of molecular complexes. IV. The Hartree-Fock potential for the water dimer and its application to the liquid state, *J. Chem. Phys.*, 59, 1325, 1973.

119. **Kistenmacher, H., Lie, G. C., Popkie, H., and Clementi, E.,** Study of the structure of molecular complexes. VI. Dimers and small clusters of water molecules in the Hartree-Fock approximation, *J. Chem. Phys.*, 61, 546, 1974.
120. **Kistenmacher, H., Popkie, H., Clementi, E., and Watts, R. O.,** Study of the structure of molecular complexes. VII. Effect of correlation energy corrections to Hartree-Fock water-water potential on Monte Carlo simulations of liquid water, *J. Chem. Phys.*, 60, 4455, 1974.
121. **Lie, G. C. and Clementi, E.,** Study of the structure of molecular complexes. XII. Structure of liquid water obtained by Monte Carlo simulation with Hartree-Fock potential corrected by inclusion of dispersion forces, *J. Chem. Phys.*, 62, 2195, 1975.
122. **Matsuoka, O., Clementi, E., and Yoshimine, M.,** CI study of the water dimer potential surface, *J. Chem. Phys.*, 64, 1351, 1976.
123. **Lie, G. C., Clementi, E., and Yoshimine, M.,** Study of the structure of molecular complexes. XIII. Monte Carlo simulation of liquid water with a configuration interaction pair potential, *J. Chem. Phys.*, 64, 2314, 1976.
124. **Kistenmacher, H., Popkie, H., and Clementi, E.,** Study of the structure of molecular complexes. V. Heat of formation for the Li^+, Na^+, K^+, F^-, and Cl^- ion complexes with a single water molecule, *J. Chem. Phys.*, 59, 5842, 1973.
125. **Kress, J. W., Clementi, E., Kozak, J. J., and Schwartz, M. E.,** Study of the structure of molecular complexes. IX. The Hartree-Fock energy surface for H_2O-Li-F complex, *J. Chem. Phys.*, 63, 3907, 1975.
126. **Ben-Naim, A. and Stillinger, F. H.,** in *Water and Water Solutions*, Horne, R. A., Ed., Wiley-Interscience, New York, 1972, chap. 8.
127. **Stillinger, F. H. and Rahman, A.,** Improved simulation of liquid water by molecular dynamics, *J. Chem. Phys.*, 60, 1545, 1974.
128. **Lemberg, H. L. and Stillinger, F. H.,** Central-force model for liquid water, *J. Chem. Phys.*, 62, 1677, 1975.
129. **Rahman, A., Stillinger, F. H., and Lemberg, H. L.,** Study of a central force model for liquid water by molecular dynamics, *J. Chem. Phys.*, 63, 5223, 1975.
130. **Stillinger, F. H. and Rahman, A.,** Revised central force potentials for water, *J. Chem. Phys.*, 68, 666, 1978.
131. **Jorgensen, W. L.,** Transferable intermolecular potential functions for water, alcohols, and ethers. Application to liquid water, *J. Am. Chem. Soc.*, 103, 335, 1981.
132. **Jorgensen, W. L.,** Revised TIPS for simulations of liquid water and aqueous solutions, *J. Chem. Phys.*, 77, 4156, 1982.
133. **Jorgensen, W. L., Chandrasekhar, J., Madura, J. D., Impey, R. W., and Klein, M. L.,** Comparison of simple potential functions for simulating liquid water, *J. Chem. Phys.*, 79, 926, 1983.
134. **Briant, C. L. and Burton, J. J.,** Molecular dynamics study of the effects of ions on water microclusters, *J. Chem. Phys.*, 64, 2888, 1976.
135. **Impey, R. W., Madden, P. A., and McDonald, I. R.,** Hydration and mobility of ions in solution, *J. Phys. Chem.*, 87, 5071, 1983.
136. **Nguyen, H. L. and Adelman, S. A.,** Studies of solvated ion motion: molecular dynamics results for dilute aqueous solutions of alkali and halide ions, *J. Chem. Phys.*, 81, 4564, 1984.
137. **Bounds, D. G.,** Molecular dynamics studies of aqueous solutions, *J. Phys. (Paris) C7*, 45 (Supp. 9), 113, 1984.
138. **Bounds, D. G.,** A molecular dynamics study of the structure of water around the ions Li^+, Na^+, K^+, Ca^{2+}, Ni^{2+} and Cl^-, *Mol. Phys.*, 54, 1335, 1985.
139. **Heinzinger, K. and Vogel, P. C.,** A molecular dynamics study of aqueous solutions. I. First results for LiCl in H_2O, *Z. Naturforsch.*, 29a, 1164, 1974.
140. **Vogel, P. C. and Heinzinger, K.,** A molecular dynamics study of aqueous solutions. II. Cesium chloride in H_2O, *Z. Naturforsch.*, 30a, 789, 1975.
141. **Heinzinger, K. and Vogel, P. C.,** A molecular dynamics study of aqueous solutions. III. A comparison of selected alkali halides, *Z. Naturforsch.*, 31a, 463, 1976.
142. **Vogel, P. C. and Heinzinger, K.,** A molecular dynamics study of aqueous solutions. IV. Sodium chloride at two different concentrations, *Z. Naturforsch.*, 31a, 476, 1976.
143. **Heinzinger, K.,** A molecular dynamics study of aqueous solutions. V. Angular distribution of the water dipoles in the hydration shells of various alkali- and halide ions, *Z. Naturforsch.*, 31a, 1073, 1976.
144. **Bopp, P., Heinzinger, K., and Jancsò, G.,** A molecular dynamics study of aqueous solutions. VI. Remarks on the hydration numbers of alkali and halide ions, *Z. Naturforsch.*, 32a, 620, 1977.
145. **Pálinkás, G., Riede, W. O., and Heinzinger, K.,** A molecular dynamics study of aqueous solutions. VII. Improved simulation and comparison with X-ray investigation of a NaCl solution, *Z. Naturforsch.*, 32a, 1137, 1977.
146. **Pálinkás, G., Radnai, T., and Hajdu, F.,** Ion-solvent and solvent-solvent interactions. X-ray study of aqueous alkali chloride solutions, *Z. Naturforsch.*, 35a, 107, 1980.

147. **Heinzinger, K.**, Molecular dynamics studies of liquid structures, in *Diffraction Studies on Non-Crystalline Substances*, Hargittai, I. and Orville-Thomas, W. J., Eds., Elsevier, Amsterdam, 1981, 401.
148. **Bopp, P., Dietz, W., and Heinzinger, K.**, A molecular dynamics study of aqueous solutions. X. First results for a NaCl solution with a central force model for water, *Z. Naturforsch.*, 34a, 1424, 1979.
149. **Szász, G. I., Heinzinger, K., and Pálinkás, G.**, The structure of the hydration shell of the lithium ion, *Chem. Phys. Lett.*, 78, 194, 1981.
150. **Szász, G. I., Heinzinger, K., and Riede, W. O.**, Structural properties of an aqueous LiI solution derived from a molecular dynamics simulation, *Z. Naturforsch.*, 36a, 1067, 1981.
151. **Radnai, T., Pálinkás, G., Szász, G. I., and Heinzinger, K.**, The second hydration shell of Li^+ in aqueous LiI from X-ray and MD studies, *Z. Naturforsch.*, 36a, 1076, 1981.
152. **Bopp, P., Jancsò, J., and Heinzinger, K.**, An improved potential for non-rigid water molecules in the liquid phase, *Chem. Phys. Lett.*, 98, 129, 1983.
153. **Bopp, P., Okada, I., Ohtaki, H., and Heinzinger, K.**, A molecular dynamics study of the structure of an $LiCl \cdot 4H_2O$ solution, *Z. Naturforsch.*, 40a, 116, 1985.
154. **Dietz, W., Riede, W. O., and Heinzinger, K.**, Molecular Dynamics simulation of an aqueous $MgCl_2$ solution. Structural results, *Z. Naturforsch.*, 37a, 1038, 1982.
155. **Pálinkás, G., Radnai, T., Dietz, W., Szász, G. I., and Heinzinger, K.**, Hydration shell structures in an $MgCl_2$ solution from X-ray and MD studies, *Z. Naturforsch.*, 37a, 1049, 1982.
156. **Szász, G. I., Dietz, W., Heinzinger, K., Pálinkás, G., and Radnai, T.**, On the orientation of water molecules in the hydration shell of the ions in a $MgCl_2$ solution, *Chem. Phys. Lett.*, 92, 388, 1982.
157. **Probst, M. M., Radnai, T., Heinzinger, K., Bopp, P., and Rode, B. M.**, Molecular dynamics and X-ray investigation of an aqueous $CaCl_2$ solution, *J. Phys. Chem.*, 89, 753, 1985.
158. **Watts, R. O., Clementi, E., and Fromm, J.**, A theoretical study of the lithium fluoride molecule in water, *J. Chem. Phys.*, 61, 2550, 1974.
159. **Fromm, J., Clementi, E., and Watts, R. O.**, Coordination number for the Li^+-F^- ion pair in water, *J. Chem. Phys.*, 62, 1388, 1975.
160. **Watts, R. O.**, The effect of ion pairs on water structure, *Mol. Phys.*, 32, 659, 1976.
161. **Mezei, M. and Beveridge, D. L.**, Monte Carlo studies of the structure of dilute aqueous solutions of Li^+, Na^+, K^+, F^-, and Cl^-, *J. Chem. Phys.*, 74, 6902, 1981.

INDEX

A

Absolute intensities, 189
Absorption
  atomic, 215
  preedge, 215
  in reflection geometry, 21—22
  in small sample, 6
  in transmission geometry, 22—23
Absorption coefficient, 210, 212
Absorption correction, 20—22
Absorption edge, 8, 19, 210—211
Ag, see Silver
Air scattering corrections, 20
Al, see Aluminium
Alkali halide solutions, 191—192
Alkali ions, 60—66
  cesium, 65
  lithium, 60—62
  potassium, 65
  sodium, 62, 64—65
Alkaline earth ions, 66—76
  calcium, 70—73
  magnesium, 66—70
  strontium, 73—76
Alkaline hydrolysis, 83
Aluminium (Al), 77—79, 173, 177
Aluminium chloride solutions
  coordination of, 78, 162
  correlation functions for, 77
  partial distribution function for, 79, 80
Aluminium nitrate solutions, 77—79
Aluminium(III) nitrate solution, 79, 80
Ammonium solutions, 166, 168
Anomalous dispersion, 8, 42
Aqueous solutions
  molecular dynamics simulations of, 236—246
    one-ion studies in, 237
    solution studies in, 237—246
      1:1 salt, 237—242
      2:1 salt, 242—246
  structural information on, see Structural information
Atomic absorption, 215
Atomic distribution functions, 1
Atomic position, 1
Atomic scattering factor, 15, 16
Au, see Gold
Auger electrons, 210
Average number density, 2, 5

B

Background corrections, 18—20
Backscattering amplitudes, 218
$BaCl_2$, 207
BNS model of water molecule, 233
Boltzman factor, 231
Bond lengths, 170
"Bond valence" point of view, 172—174
Bond valences for ion-water pair distances, 171—173
Br, see Bromine
Bragg angle, 23
Bragg-Brentano reflection geometry, 17
Breit-Dirac recoil factor, 25
Bromine (Br), 164—165, 237

C

Ca, see Calcium
Cadmium (Cd), 148—155
  bond valences for, 177
  complex formation of, 148—152
    with halide ions, 148—150
    with nitrate ions, 151
    with phosphate ions, 151
    with sulfate ions, 150—151
  hydration of, 148
  XRD data on, 173
Cadmium bromide solutions, 227
Cadmium chloride solution, 148—149
Cadmium iodide solution, 150
Cadmium nitrate solutions, 148, 151, 154—155
Cadmium perchlorate solutions, 148, 150—151
Cadmium phosphate solution, 151, 154
Cadmium sulfate solutions, 150—154
Calcium (Ca), 70—73, 196
  coordination of, 200
  coordination number for, 238
  Monte Carlo simulations with, 249
  one-ion study with, 237
  oxygen atoms around, 244
  XRD data on, 173
Calcium bromide solutions, 71, 164
Calcium chloride solutions
  concentrated, 71
  coordination number in, 209
  coordination of, 73, 162, 207
  correlations functions for, 70, 72—74, 248
  distribution curves for, 196, 200, 206—207
  with magnesium, 70
  scattering function for, 198
  simulated, 243
  structure functions for, 36, 38
  weighting factors for, 43
Calibration parameters, 24, 189
Cations, 178
Cd, see Cadmium
Ce, 87
$Ce(NO_3)_3$ solution, 87—88
Central force model, 235, 238—239
Central force pair potential, 240, 242
Cesium (Cs), 65

coordination parameters for, 193
one-ion study with, 237
scattering factors for, 8
Cesium chloride solution, 162, 237
Cesium fluoride solution, 237
Cesium nitrate solution, 169
Chemical isomorphous substitutions, 12
Chloride solutions, 208
Chlorine (Cl), 160—164
coordination number for, 202, 238, 250
coordination of, 207
coordination parameters for, 193
first hydration shell of, 245
hydration of, 161—162
Monte Carlo simulations with, 248
one-ion study with, 237
oxygen atoms around, 247
radial function of, 250
water molecules around, 246
Chloro-complexation, 224—225
Chromium (Cr), 89—97
bond valences for, 177
chloride complex formation of, 91—97, 180—181
hydration of, 89—91, 94
XRD data on, 173
Chromium chloride solution, 89—97
coordination of, 96, 162—163
correlation functions for, 90, 92
radial distribution functions for, 46, 48, 92, 95
Chromium nitrate solutions
coordination of, 94
correlation functions for, 90, 93, 165
Cr—O distance in, 220
Fourier transform in, 216—217
partial distribution function for, 80
structural parameters for, 222
Chromium perchlorate solutions, 222
Chromium sulfate solution, 91
Cl, see Chlorine
Close hydration, 53
Co, see Cobalt
Cobalt (Co), 112—120
bond valences for, 177
complex formation of, 113—119
with bromide ions, 117—119
with chloride ions, 113—117
hydration of, 112, 114
XRD data on, 173
Cobalt bromide solutions, 117—120
Cobalt chloride-lithium chloride solutions, 116—118
Cobalt chloride solutions, 112—117
coordination of, 118, 163
mixed with $NiCl_2$ solutions, 229
radial distribution curve for, 134
Cobalt nitrate solution, 222
Cobalt perchlorate solution, 112, 115
Coherent independent scattering, 10
Coherent scattering length, 16
Complex formation, 58, 179—181
Composition coefficient, 4—5, 12, 46—47
Composition unit (CU), 4, 9—10
Compton scattering, 24—26
Computer simulation methods, 189, 230—251
interaction potentials in, 232—236
Monte Carlo simulations in, 246—248
molecular dynamics and Monte Carlo techniques in, 231—232
molecular dynamics simulations in, 236—246
one-ion studies in, 237
solution studies in, 237—246
Conditional probability density, 2
Configuration-interaction, 233
Continuous distribution of distances, 181
Convergence factor, 13, 30
Coordination distance, 45
Coordination numbers, 170—173, 214
mean, 46
obtaining, 218, 238
running, 239, 244
second, 174
simulated, 250
values of, 230
Coordination shells, 45, 49, 214
first, 170—174
second, 174—179
Copper (Cu), 127, 129—142, 199, 201
complex formation of, 132, 134—139
hydration of, 127, 129—133
Copper bromide solutions
coordination of, 139
Cu—O distance in, 221
distribution curves for, 135, 137, 138
hydration number in, 226—227
structural parameters for, 222
Copper chloride solutions
coordination of, 205
distribution curve for, 134—136
distribution function for, 199, 205
interatomic distances in, 209
Copper nitrate solutions
coordination of, 141
correlation functions for, 136, 140
structure function for, 36
Copper perchlorate solution
coordination number in, 221
coordination of, 133
radial distribution curve for, 129—130
radial distribution functions for, 131
structural parameters for, 222
Copper sulfate solution
coordination of, 133
Cu—O distances in, 221
Jahn-Teller effect in, 47, 129
radial distribution functions for, 49, 132
Correlation functions
for $AlCl_3$ solutions, 77
for ammonium solutions, 168
behavior of, 2—3
for $CaCl_2$ solutions, 70, 72—74, 248
for $CdCl_2$ solution, 149
for $Cd(NO_3)_2$ solution, 155
for $CoCl_2$-LiCl solutions, 116—117

for $CoCl_2$ solution, 134
in computer simulation studies, 231
for $CrCl_3$ solutions, 90, 92
for $Cr(NO_3)_3$ solution, 93
for $Cu(NO_3)_2$ solution, 93
for gaseous, noncrystalline, and crystalline substances, 3, 4, 41
for HCl solutions, 193
for $In_2(SO_4])_3$ solution, 79, 81
for LiCl solutions, 47, 61—62, 241
for LiI solutioin, 241
for $MgCl_2$-$CaCl_2$ solution, 70, 71
for $MgCl_2$ solutions, 67, 243
for $MnSO_4$ solutions, 100
for $NiCl_2$ solutions, 120, 122
pair, 2, 220, 239
partial, 5, 11
radial, 214
for rare earth chloride solutions, 86—89
for $SrCl_2$ solutions, 73—75
total, 14
for $ZnSO_4$ solutions, 142, 144, 167
Cosmic radiation, 19
Cr, see Chromium
Crystal axes, 1
Crystalline long range order, 1
Crystalline substances, 3, 4, 218—219
Cs, see Cesium
Cu, see Copper
CU, see Composition unit
Curve fitting, 47—54
analysis of dominant contribution in, 50—52
interpretative models of, 52—54
mathematical modeling in, 47—50

D

Debye scattering equation, 9
Differences curves
for $CdSO_4$ solutions, 150
for $FeCl_3$ solutions, 108
for $NiBr_2$ solutions, 175—176
Diffracted beam, 25—26
Diffracted intensity, 8—9
Diffraction geometry, 6—7
Diffraction techniques, 251
Dipole approximation, 212
Discrete shells, 50
Discrete structure, 50
Dispersion, anomalous, 8, 42
Distribution of distances, 181
Distribution functions, 1—5
atomic, 1
pair, 5, 11
partial, 79, 80
radial, see Radial distribution function
total, 190
Dominant contributions, 50—52
Dy, 173
$DyCl_3$ solution, 88, 196

E

Electrolyte solutions, 58
Electron density, 6—7
Electron density difference functions, 66, 68
Electron units, 6
Electrons, effective number of, 46
Electrostatic contribution, 234
Energy transfer, 6, 191
Equilibrium constants, 97, 108, 179
Er, 173
$ErCl_3$ solution, 88
Ergodic hypothesis, 9
Er perchlorate solution, 88
Eu, 87, 173
$EuCl_3$ solution, 88
EXAFS, see Extended X-ray absorption fine structure
EXAFS interference function, 219
EXAFS spectra, 225, 227, 228
EXAFS spectroscopy, 210—230
basic concepts and relations in, 212—215
compared with XRD, 218—221
data analysis on, 215—218
limitations of, 228—230
review of results with, 221—228
X-ray absorption spectroscopy in, 210—212
Experimental data interpretation, 41—55
analysis of radial distribution in, 44—47
curve fitting in, 47—54
analysis of dominant contributions in, 50—52
mathematical modeling of, 47—50
interpretative models of, 52—54
detectability of partial contributions in, 42—44
Extended X-ray absorption fine structure (EXAFS), 83, 189, 213

F

F, see Fluorine
False peak, 31
Fe(II), see Iron(II)
Fe(III), see Iron(III)
Fermi Golden Rule, 212
Fermi pseudopotentials, 16
First coordination shell, see Coordination shells
First neighbor model 1 (FNM1), 53
First neighbor model, modified (FNM2), 53—54, 238
First order difference method, 44
First order difference scattering function, 193, 197, 198
First shell filtered spectra of Ni solutions, 223
First shell peak, 46
Fluorescence correction, 19—20
Fluorescence radiation, 19, 26
Fluorine (F), 160
Monte Carlo simulations with, 248
one-ion study with, 237
simulated coordination number for, 238, 250
FNM1, see First neighbor model 1
FNM2, see First neighbor model, modified
Focus slit, 17
Forward scattering peak, 10
Fourier filtered spectrum, 216
Fourier transform, 7
in $Cr(NO_3)_3$ solution, 217

of distribution functions, 12—13
in $NiCl_2$ solutions, 228
peaks in, 218
in radial distribution function, 29—32, 34
sine, 11
structure function of, 16
termination errors in, 165
Free water, 53
Full multiple scattering, 210

G

Gaseous substances, 3, 4
Geometry
diffraction, 6—7
of hydration polyhedra, 172
reflection, see Reflection geometry
transmission, 20—23
Gold (Au), 138, 140
Group IB ions, 127, 129—142
copper, 127, 129—141, see also Copper
gold, 138, 139
silver, 137—138
Group IIB ions, 142—160
cadmium, 148—155, see also Cadmium
mercury(I), 152—153, 155
mercury(II), 155—156
zinc, 142—148
Group IIIA ions, 77—83
aluminium, 77—79
indium, 79—82
thallium, 81, 83
Group IVA ions, 83, 85
Group VIB ions, 88—97, see also Chromium
Group VIIB ions, 97—101, see also Manganese
Group VIII ions, 100—128
cobalt, 112—120, see also Cobalt
iron(II), 100, 102, 103
iron(III), 102—112, see also Iron(III)
nickel, 119—128, see also Nickel
platinum, 127
rhodium, 127, 128

H

HAA, see High-angle analysis
Habenschuss-Spedding scaling method, 27—29
Halide ions, 160—165
bromine, 164—165
chlorine, 160—164
complex formation with, 179
fluorine, 160
parameter values for, 238
Halo-complexes, 170
H-bonds
long, 178
short and linear, 90
strong, 70, 80, 177
very short, 59
HCl, see Hydrochloric acid
Hg, see Mercury
High-angle analysis (HAA), 50, 102, 149, 155
High-angle scaling method, 26—27
$H_2O_IH_2O_{II}$ problem, 177—178
Hot crystals, 54
Hydrated ions, independence of, 174
Hydration and complex formation, 57—187
of alkali ions, 60—66
cesium, 65
lithium, 60—62
potassium, 65
sodium, 62, 64—65
of alkaline earth ions, 66—76
calcium, 70—73
magnesium, 66—70
strontium, 73—76
general discussion, 169—182
of complex formation, 179—181
of continuous distribution of distances, 181
of first coordination shell, 170—174
of second coordination shell, 174—179
of Group IB ions, 127, 129—142
copper, 127, 129—141, see also Copper
gold, 138, 139
silver, 137—138
of Group IIB ions, 142—160
cadmium, 148—155, see also Cadmium
mercury(I), 152—153, 155
mercury(II), 155—156
zinc, 142—148
of Group IIIA ions, 77—83
aluminium, 77—79
indium, 79—82
thallium, 81, 83
of Group IVA ions, 83, 85
of Group VIB ions, 88—97, see also Chromium
of Group VIIB ions, 97—101, see also Manganese
of Group VIII ions, 100—128
cobalt, 112—120, see also Cobalt
iron(II), 100, 102, 103
iron(III), 102—112, see also Iron(III)
nickel, 119—128, see also Nickel
platinum, 127
rhodium, 127, 128
of halide ions, 160—165
bromine, 164—165
chlorine, 160—164
fluorine, 160
iodine, 165
of H_3O^+ ion, 59—60
of oxyanions, 165—169
of rare earth ions, 85—88
Hydration numbers, 57, 170—171
Hydration shell, second, 66
Hydrochloric acid solutions, 59—60, 192—193

I

I, see Iodine
Ice-I-like model, 164
In, see Indium
Incoherent contribution, see Incoherent scattering
Incoherent scattering, 24—26, 189, 191
Independent scattering, 24, 28
Indium (In), 79—82, 173, 177
Indium bromide solution, 79

Indium chloride solution, 79
Indium percholate solution, 79
Indium sulfate solution, 79, 81—82
Inelastic scattering, 191, 214
Inner sphere complex formation, 42, 46, 53, 79
Inner sphere complexes, 174
Instrumental apparatus and data collection, 16—18
Integral scaling method, 26—27
Interaction potentials, 232—236
Interatomic local order, 44
Interference functions, 11, 12, 219
Intermediate multiple scattering, 210, 212
Intermolecular potential energy, 232
Interpretative models, 52—54
Iodine (I), 165, 237
Ion-halide interactions, 224—228
Ion-hydrogen correlation functions, 244
Ion-ion interactions, 235
Ion-ion positional correlations, 73
Ion-ligand distances, 170
Ion-oxygen correlation functions, 244
Ion-oxygen distances, 238
Ion-oxygen interactions, 221—223
Ion-water distances, 171, 173, 250
Ion-water interactions, 235, 237
Ion-water potential model, 234
Iron(II), 100, 102—103, 173
Iron(II) bromide solution, 100, 103
Iron(II) perchlorate solution, 100
Iron(III), 102—112
  bond valences for, 177
  complex formation of, 102—112
    with chloride ions, 102—109
    with sulfate ions, 109, 111—112
  hydration of, 102, 106
  XRD data on, 173
Iron(III) chloride solutions, 189, 224
  coordination of, 110
  differences curves for, 108
  dimeric form, 109
  EXAFS spectra for, 225—226
  EXANS spectra for, 226
  radial distribution curves for, 190
  radial distribution functions of, 107
Iron(III) nitrate solutions, 102, 104—106
Iron(III) perchlorate solutions
  Fe—H_2O distances in, 102
  radial distribution functions for, 105, 112
  structural parameters for, 222
Iron(III) sulfate solutions
  coordination of, 114
  model for, 111, 113
  radial distribution functions for, 109, 111, 112
IR spectroscopies, 143
Isomorphous replacement, 66, 71
Isomorphous substitution, 44, 148, 199
Isostructurality, 167
Isotopic substitutions, 12, 44, 192—194, 231

J

Jahn-Teller distortion
  in $CuCl_2$ solution, 132, 199
  in Cu perchlorate solution, 174, 209
  in Cu(II) salts, 169
  in $CuSO_4$ solution, 47, 129, 174

K

K, see Potassium

L

La, 85, 87, 173
$LaCl_3$ solution, 88
Lanthanides, 85—88, 169, 173
La perchlorate solution, 88
Lattice, 9, 120
Laue equation, 9
Lennard-Jones interaction term, 233—234
Lithium (Li), 60—62, 194—195, 237
  coordination parameters for, 193, 238
  hydration of, 198
  simulated coordination numbers for, 248, 250
  XRD data on, 173
Lithium bromide solution coordination, 164
Lithium chloride solutions
  coordination of, 63, 162
  correlation functions for, 46, 47, 60—62, 241
  distribution functions for, 195, 198, 202
  simulated, 237, 240
  structure functions for, 61
Lithium chloride-water coordination, 207
Lithium fluoride solutions, 246
Lithium iodide solutions, 237, 239—241
Lu, 85, 173
$LuCl_3$ solution, 88

M

Magnesium (Mg), 66—70
  bond valences for, 177
  simulated aqueous solutions of, 244, 249
  XRD data on, 173
Magnesium acetate solution, 70
Magnesium chloride solutions
  with calcium chloride, 70
  coordination of, 69, 162
  correlation function for, 66, 67
  distribution of O_2 atoms in, 247
  simulated, 242—245
  structure function for, 36—37
Magnesium nitrate solution, 66, 69
Magnesium sulfate solution, 66, 69, 169
Manganese (Mn), 97—100
  bond valences for, 177
  complex formation of, 97—99
  hydration of, 97
  XRD data on, 173
Manganese chloride solutions, 97—99
Manganese nitrate solutions, 98—99, 101
Manganese perchlorate solution, 222
Manganese sulfate solutions, 97, 100, 101
MC, see Monte Carlo
MD, see Molecular dynamics
$MeBr_2$ solutions, 226

Melted salts, 58, 169
Mercury(I) [Hg(I)], 152—153, 155
Mercury(I) perchlorate solutions, 152, 156
Mercury(II) [Hg(II)], 155—160
complex formation of, 156—160
hydration of, 155—156
Mercury(II) bromide solution, 157—159
Mercury(II) chloride solution, 157—159
Mercury(II) iodide solution, 157—159
Mg, see Magnesium
Mn, see Manganese
Modification function, 13, 30, 41
Modified partial functions, 15
Molecular dynamics (MD), 60, 231—232
Molecular dynamics simulations, 236—246
one-ion studies of, 237
solution studies of, 237—246
1:1 salt, 237—242
2:1 salt, 242—246
Molybdenum, 26, 210, 211
Momentum transfer, 6
Monochromator
to avoid fluorescence, 19
incoherent scattering with, 25—26
polarization with, 23
Monochromator crystal, 17—18
Monochromator discrimination function, 25, 27, 34
Monochromator function, 29
Monochromator pass-band, 25
Monte Carlo (MC) simulation, 246—248
Monte Carlo techniques, 231—232
Multiple scattering, 24, 189, 191
Multiple scattering correction, 29

N

Na, see Sodium
Nd, 85, 196
Nd(III), 196
ND, see Neutron diffraction
$NdCl_3$ solution
coordination of, 88, 201
neutron scattering cross sections of, 194—195
radial distribution curve for, 201
NDIS, see Neutron diffraction with isotopic substitution
Neutron diffraction (ND), 16, 189—209
compared with XRD studies, 194—203, 205, 208—210
on calcium, 196
on copper, 199, 201
on chlorine, 202
on lithium, 194—195
on nickel, 197, 199
on nitrate, 203
on potassium, 195—196
on rare earth ions, 196—197
on silver, 201—202
coordination parameters from, 193
final comments on, 203, 205, 208—210
with isotopic substitution (NDIS), 192—194
single experiments with, 189—192
Neutron scattering amplitudes, 190
Neutron scattering cross sections, 195
Neutron scattering factors, 190—191
Ni, see Nickel
Nickel (Ni), 119—127, 197, 199, 237
bond valences for, 177
complex formation of, 120—128
with bromide ions, 123, 125, 126
with chloride ions, 120, 122—124
with sulfate and phosphate ions, 125, 127, 128
coordination numbers of, 238
hydration of, 119—121
XRD data on, 173
Nickel bromide solutions
coordination of, 126, 164
experimental data from, 19, 28
radial distribution functions in, 30—32, 34, 123, 125
structure functions in, 28, 35
Nickel chloride solutions
$CoCl_2$ solution and, 229
coordination of, 124, 163, 204
correlation functions for, 120, 122
curve-fitting analysis of, 228
EXAFS spectra for, 227, 228
first shell filtered spectra of, 221, 223
ion-oxygen interaction in, 222
measurement at Cl K-edge in, 224—225, 227
neutron scattering cross sections of, 196
radial distribution curve for, 134, 208
radial distribution functions for, 122—123, 125, 202—203
scattering functions in, 194, 197
Nickel chloride-water coordination, 207
Nickel nitrate solution, 221—223
Nickel perchlorate solution, 120, 121, 197, 204
Nickel phosphate solutions, 127—128
Nickel sulfate solution, 125
Nitrate, 203, 208
Noncrystalline system
correlation function for, 3, 4
multicomponent, 3—5
one-component, 2—3
X-ray diffraction theory for, 6—16
for diffracted intensity, 8—9
for neutron diffraction, 16
for one-component system, 11
for partial structure functions, 11—12
for radial functions, 12—15
for scattering factors, 6—8
for scattering process, 6
for structure function, 9—10
for WKM approximation, 15—16
Nontransition metals, 58
Normal beam transmission, 17

O

Opening slit, 17—18
Orientation of water molecules, 70

Outer-complex formation, 181
Oxonium ions, 59—60
Oxyanions, 154, 165—169, 179

P

Pair correlation functions, 2, 220
Pair distribution function, 5, 11
Pair potential function, 235
Pairwise additive forces, 232
Parameters, 181—182, 189
Partial contributions, 42—44
Partial correlation functions, 5, 11
Partial distribution function, 79, 80
Partial functions, modified, 15
Partial interference functions, 11
Partial pair correlation functions, 2
Partial radial density, 5
Partial residual distribution curve for Cu, 129, 131
Partial structure factors, 11
Partial structure function, 42, 48, 192
Pb salt solutions, 83
Peak-shape function, 13, 46, 219
Peaks-shape analysis (PSA), 51, 111
Phase shift, 214, 217
Phonon inelastic scattering, 6
Photoelectron mean free path, 214
Photoelectron wave vector, 214
Placzek correction, 191, 194
Plane wave approximation, 213
Platinum (Pt), 127
Point charge model, 233
Polarization, 23
Polarization correction factor, 23, 24
Polynuclear complexes, 58
Potassium (K), 65, 195—196, 237
  coordination parameters for, 193, 238
  simulated coordination numbers for, 248, 250
Potassium chloride coordination, 162
Potassium chloride solutions, 195, 199
Pr, 87, 173
$PrCl_3$ solution, 88
Preedge absorption, 215
Probability density, conditional, 2
PSA, see Peaks-shape analysis
Pt, see Platinum

Q

Quasi atom, 42
Quasi-lattice model, 54
Quasi-lattice structure, 120
Quasi-molecules, 53

R

Radial density, partial, 5
Radial distribution curves
  for $CaCl_2$ solution, 206
  for $CoCl_2$ solution, 134
  for $CuBr_2$ solution, 137, 138
  for $CuCl_2$ solution, 134—136, 205
  for Cu perchlorate solutions, 129, 130
  for $FeCl_3$ solution, 190
  for NaCl solution, 206
  for $NaNO_3$ solution, 207
  for $NdCl_3$ solution, 201
  for $NiBr_2$ solution, 35
  for $NiCl_2$ solution, 134
Radial distribution function, 3, 12
  analysis of, 44—47
  for Cd percholate solution, 151
  for $CdSO_4$ solution, 152, 153
  for $CoBr_2$ solutions, 119
  for $CoCl_2$ solutions, 115
  for Co perchlorate solutions, 115
  for $CrCl_3$ solutions, 46, 48, 95
  for Cu perchlorate solutions, 129, 131
  for $CuSO_4$ solution, 49, 132
  errors in, 30—33
  for $FeCl_3$ solutions, 107
  for $Fe(NO_3)_3$ solutions, 104, 105
  for Fe(III) perchlorate solution, 105, 112
  for $Fe_2(SO_4)_3$ solutions, 111, 112
  from Fourier transform, 12, 29—30, 34
  for $Hg(II)Br_2$ solution, 159
  for $Hg(II)Cl_2$ solution, 159
  for $Hg(II)I_2$ solution, 159
  for $MnCl_2$ solutions, 97, 98
  modified, 216
  for $NiBr_2$ solution, 30—32, 34, 125
  for $NiCl_2$ solutions, 122—123
  of simulated aqueous Cl^- solutions, 248, 250
  spurious peaks in, 33—34
  total, 14—15, 41
  for $ZnBr_2$ solution, 147
  for Zn perchlorate solution, 147
  for $ZnSO_4$ solution, 49, 132
Radial functions, 12—15
Radial number density, 3
Radial pair correlation functions, 214
Raman spectroscopy, 142—143, 151
Range of order, 44, 50
Rare earth chloride solutions, 86—89
Rare earth ions, 85—88, 196—197, 201
Rb, 237
RbCl-water coordination, 207
Receiving slit, 17—19
Reciprocal lattice, 9
Reciprocal space, 6
Reduced intensity, 10
Reflection diffractometer, 36
Reflection geometry, 17, 20—22
Reinversion procedure, 34
Rh, see Rhodium
Rhodium (Rh), 127, 128, 177
Root mean square deviations, 171—172
"Running coordination number", 65

S

Sample thickness, 20—23
Scaling to absolute units, 26—29
  methods for, 26—29
  multiple scattering correction for, 29
Scattering

Compton, 24—25
elastic, 6, 214
incoherent, 24—25, 191
independent, 10, 23—26, 28
multiple, 24, 189, 191
correction for, 29
full, 210
intermediate, 210, 212
phonon inelastic, 6
total, 191
volume, 10
zero angle, 10
Scattering air volume, 20—21
Scattering amplitude, 6, 7, 190
Scattering angle, 6
Scattering cross-sections, 189
Scattering equation, Debye, 9
Scattering factors, 6—8
atomic, 15, 16
change in, 44
neutron, 190—191
real, 9
varying, 12
Scattering functions, 193—194
Scattering length, coherent, 16
Scattering peak, forward, 10
Scattering process, 6, 191
Scattering vector, 6, 16, 220
Second coordination shell, see Coordination shells
Second order difference scattering functions, 194
Sharpening factor, 13, 30
Silver (Ag), 137—138, 201—202
Silver nitrate solution, 222, 223
Silver perchlorate solution, 201, 222
Simulation, see Computer simulation
Slits, 17—19
Slit width, 20
Sm, 87
$SmCl_3$ solution, 88
Sm perchlorate solution, 88
$Sn(ClO_4)_2$ solution, 83, 222—223
Sodium (Na), 62, 64—65, 237
coordination numbers for, 248, 250
coordination parameters for, 193, 238
scattering factors for, 8
Sodium bromide solution, 222
Sodium chloride solution
coordination of, 162, 239
correlation functions for, 64, 239
NaH_2O distances in, 62
neutron scattering cross sections of, 196
radial distribution curve for, 206
simulated, 237, 240
Sodium chloride-water coordination, 207
Sodium halide solutions, 62
Sodium nitrate solutions, 64, 203, 207
Spurious details, 41
Spurious peaks, 33—34
Sr, see Strontium
Stability constants, 83, 93, 125
Static approximation, 191
Static structure, 6
Statistical errors, 32—33
Strontium (Sr), 73—76, 173, 177
Strontium bromide solutions, 222, 227
Strontium chloride solutions, 73—76, 162
Structural information on aqueous solutions, 189—257
from computer simulation studies, 230—251
final comments on, 248—251
interaction potentials, 232—236
molecular dynamics and Monte Carlo technique, 231—232
molecular dynamics simulations, 236—246
Monte Carlo simulation, 246—248
from EXAFS spectroscopy, 210—230
basic concepts and relations, 212—215
compared with XRD, 218—221
data analysis, 215—218
final comments on, 228—230
review of results with, 221—228
X-ray absorption spectroscopy, 210—212
from neutron diffraction, 189—209
compared with XRD studies, 194—203
final comments on, 203, 205, 208—210
with isotopic substitution, 192—194
single experiments, 189—192
Structural model, 52
Structural units, 174
Structure factors, partial, 11
Structure function, 9—10, 26—29, 41
errors in, 32—33
for $CaCl_2$ solution, 38
for $Cu(NO_3)_2$ solution, 36
for LiCl solution, 61
for $MgCl_2$ solution, 36—37
for $NaCl_2$ solution, simulated, 240
for $NiBr_2$ solution, 28, 35
partial, see Partial structure function
total, 44, 48, 189
for uranyl(VI) chloride solution, 20
for $Zn(ClO_4)_2$ solution, 50, 51
for $ZnSO_4$ solutions, 143, 166
Structure making power of cations, 178
Superconcentrated solutions, 58
Switching function, 235
Symmetrical reflection, 17—18
Symmetrical reflection goniometer, 18
Symmetrical transmission, 17
Synchroton radiation, 43, 210

T

Tb, 85, 173
$TbCl_3$ solution, 88
Tb perchlorate solution, 88
Termination errors, 11, 30—32, 41, 165
Tetrabromo-auric acid, 138
Tetrachloro-auric acid, 138
Tetrachlorocobaltate ions, 59
Thallium (Tl), 81, 83
Thallium(III) bromide complexes, 83—85

Thallium(III) chloride complexes, 83, 84
Thallium(III) percholate solutions, 81
Thermodynamic equilibrium, 180
Threshold energy, 215, 217
Tilt angle, 70, 79
Tl, see Thallium
Tm, 173
$TmCl_3$ solution, 88
Total correlation function, 14, 240—241
Total distribution function, 190
Total interference function, 12
Total intermolecular potential energy, 232
Total radial distribution function, 14—15, 41
Total scattering, 191
Total structure function, 44, 48, 189
Transition metals, 58
Transmission diffractometer, 37
Transmission geometry, 20—23
Truncation error, 31
TSM, see Two shell model
Two shell model (TSM), 54

U

Unit cell, 1
Uranyl(VI) chloride solution, 20

V

Volume scattering, 10

W

Water molecule
  BNS model of, 233—235
  conformation of, 174
  orientation of, 70, 243, 245, 246
  tetrahedral charge distribution of, 235
Water-water interactions, 237
Water-water pair potentials, 232—233
Wave vector, 6
Weighting factors, 12, 42, 43
Window function, 216
WKM approximation, 15—16

X

XANES, see X-ray absorption near edge structure
X-ray absorption near edge structure (XANES), 210
X-ray absorption spectroscopy (XAS), 210—212
X-ray diffraction (XRD), 57
  ND experiments with, 192
  of noncrystalline systems, 1—39
    data treatment in, 18—38
      evaluation of independent scattering in, 23—26
      preliminary corrections of, 18—23
      radial distribution function in, 29—34
      reproducibility of XRD results in, 34—38
      scaling to absolute units in, 26—29
    distribution functions in, 1—5
    instrumental apparatus and data collection for, 16—18
    theory of, 6—16
      for diffracted intensity, 8—9
      for neutron diffraction, 16
      for one-component system, 11
      for partial structure functions, 11—12
      for radial functions, 12—15
      for scattering factors, 6—8
      for scattering process, 6
      for structure function, 9—10
      for WKM approximation, 15—16
  potentialities of, 65—66, 169
  reproducibility of results of, 34—38
X-ray photons, 210
X-ray spectrum, 25, 26, 210, 211
XRD, see X-ray diffraction

Z

Zero angle scattering, 10
Zinc (Zn), 142—148
  bond valences for, 177
  halide complex formation of, 142—148
  hydration of, 142, 145
  XRD data on, 173
Zinc bromide solutions
  coordination of, 146
  EXAFS studies of, 227—228
  radial distribution functions for, 147
  XRD studies of, 142—144
Zinc perchlorate solution
  Cl/Zn ratio in, 143
  coordination of, 146
  peaks-shape analysis for, 51—52
  radial distribution functions for, 147
  structure function for, 50, 51
Zinc sulfate solutions
  coordination of, 145
  correlation functions for, 142, 144, 167
  radial distribution functions for, 49, 132
  structure functions for, 142—143, 165—166
  ZnH_2O_1 interactions in, 182
Zirconium filter, 25
Zirconium K-edge, 26
Zn, see Zinc